绿色家园的保障

——美国和加拿大50个天堂级的国家公园

主　编　田　耀
副主编　蒋　瑜　侯　婷　张　颖　田浩良
编　者　田　路　汪　静　卢茜茜　赵　柯
薛弟洲　田增润

HEUP 哈爾濱工程大學出版社

内容简介

本书采用中文导读和英语叙述相结合的方式，分别介绍了美国和加拿大各25个具有代表性的国家公园。读者通过阅读本书可以深深了解到国家公园在环境保护中所发挥出的巨大作用，同时也可以体会到美加这两个世界大国的历届政府对于环境保护的高度重视。本书采用简练、准确精美的语言，配有必要的关键词注解和难句的中文翻译，使读者能在轻松愉快的背景下，提高英语修养和文化素质，加深对美国文化方面的修养。

本书是了解美国环境保护历程的必备书籍，可以作为高等学校英语专业、非英语专业大学生英语选修课教材，也可以作为英语爱好者和环保人员参考教材。

图书在版编目(CIP)数据

绿色家园的保障：美国和加拿大50个天堂级的国家公园/田耀主编.—哈尔滨：哈尔滨工程大学出版社，2015.1
ISBN 978-7-5661-0971-2

Ⅰ.①绿… Ⅱ.①田… Ⅲ.①国家公园-介绍-美国②国家公园-介绍-加拿大 Ⅳ.①S759.997.12②S759.997.11

中国版本图书馆CIP数据核字(2015)第016537号

出版发行 哈尔滨工程大学出版社
社　　址 哈尔滨市南岗区东大直街124号
邮政编码 150001
发行电话 0451-82519328
传　　真 0451-82519699
经　　销 新华书店
印　　刷 黑龙江省地质测绘印制中心
开　　本 787mm×960mm 1/16
印　　张 22.75
字　　数 776千字
版　　次 2015年1月第1版
印　　次 2015年1月第1次印刷
定　　价 48.00元
http://www.hrbeupress.com
E-mail:heupress@hrbeu.edu.cn

前　言

PREFACE

美国和加拿大的国民经济处于世界前列。在经济发展过程中，广袤、美丽的自然也遭受过不同程度的污染。但通过几十年的不懈努力，目前北美已经成为生态环境质量最好的地区之一。欣赏北美地区的国家公园之美，深入了解各个国家公园的发展历程，既是一种美的享受，也是一种环保教育。此外，对于我国当前和今后一段时间内环境保护工作具有重要借鉴意义。

该书由两大篇组成，美国篇介绍了美国的25个国家公园，加拿大篇介绍了加拿大的25个国家公园。在美国篇中，编者选取了美国较有代表性的国家公园，如黄石国家公园、死谷国家公园、约塞米蒂国家公园、大峡谷国家公园、冰川国家公园、落矶山国家公园等。在加拿大篇中，编者选取了加拿大较有代表性的班夫国家公园、布鲁斯半岛国家公园、爱德华王子岛国家公园、哈纳斯国家公园和海达文化遗址等。

本书用英语介绍美国和加拿大的各个国家公园，并附有中文导读和文后注释，对一些较难理解或关键性的语句进行了必要的中文翻译，使形式更为简洁，意义更为明确，使读者在轻松愉快、节奏明快的中英文享受中扩展视野，提高英语语言的修养，欣赏自然之美，加深对北美国家公园的了解。

该书是天津市高等学校创新团队建设规划项目成果之一，项目编号：TD12－5045。

该书的读者范围广泛，适合于任何专业的在校大学生旅友和广大英语爱好者。

编　者

2015年1月

前言

目录
CONTENTS

PART I 美 国 篇

PART Ⅱ 加拿大篇

PART Ⅰ

美 国 篇

Chapter 1

Yellowstone National Park
黄石国家公园

【简介】

黄石国家公园成立于1872年,是世界第一座国家公园。它被美国人自豪地称为"地球上最独一无二的神奇乐园"。黄石公园位于美国中西部怀俄明州的西北角,并向西北方向延伸到爱达荷州和蒙大拿州,面积达7 988平方千米。1978年被列为世界自然遗产。黄石河、黄石湖纵贯其中,有峡谷、瀑布、温泉以及间歇喷泉等,景色秀丽,引人入胜。其中尤以每小时喷水一次的"老忠实泉"最著名。园内森林茂密,还牧养了一些残存的野生动物,如美洲野牛等,供人观赏。

Yellowstone National Park, established by the U. S. Congress and signed into law by President Ulysses S. Grant on March 1, 1872, is a national park located primarily in the U. S. state of Wyoming, though it also extends into Montana and Idaho. Yellowstone park was the first national park in the world, and is known for its wildlife and its many geothermal(地热的) features, especially Old Faithful Geyser, one of the most popular features in the park. It has many types of ecosystems, but the subalpine(亚高山带的)forest is dominant.

Yellowstone National Park spans an area of 3,468 square miles, comprising lakes, canyons, rivers and mountain ranges. Yellowstone Lake is one of the largest high-altitude lakes in North America and is centered over the Yellowstone Caldera(火山喷口), the largest super volcano on the continent. The Caldera is considered an active volcano; it has erupted(喷发)with tremendous force several times in the last two million years. Half of the world's geothermal features are in Yellowstone, fueled by this ongoing volcanism. Lava flows and rocks from volcanic eruptions cover most of the land area of Yellowstone. The park is the centerpiece(核心) of the Greater Yellowstone Ecosystem, the largest remaining, nearly-intact ecosystem in the Earth's northern temperate zone.

Old Faithful erupts(喷出) more frequently than any of the other big geysers, although it is not the largest or most regular geyser in the park. Its average interval between eruptions is about 91 minutes, varying from 65 – 92 minutes. An eruption lasts 5.5 to 5 minutes, expels 3,700 – 8,400 gallons (14,000 – 32,000 liters) of boiling water, and reaches heights of 106 – 184 feet (30 – 55 m). It was named for its consistent performance by members of the Washburn Expedition in 1870. Although its average interval has lengthened through the years (due to earthquakes and vandalism(毁坏)), Old Faithful is still as spectacular and predictable as it was a century ago.

Native Americans have lived in the Yellowstone region for at least 11,000 years. The region was bypassed(忽略) during the Lewis and Clark Expedition in the early 19th century. Aside from visits by mountain men during the early-to-mid-19th century, organized exploration did not begin until the late 1860s. The U. S. Army was commissioned(委任) to oversee(监督) the park just after its establishment. In 1917, administration(管理) of the park was transferred to the National Park Service, which had been created the previous year. Hundreds of structures have been built and are protected for their architectural and historical significance, and researchers have examined more than 1,000 archaeological sites.

The park is located at the headwaters(源头) of the Yellowstone River, from which it takes its historical name. Near the end of the 18th century, French trappers(猎人) named the river "Roche Jaune", which is probably a translation of the Minnetaree name "Mi tsi a-da-zi" (Rock Yellow River). Later, American trappers rendered the French name in English as "Yellow Stone". Although it is commonly believed that the river was named for the yellow rocks seen in the Grand Canyon of the Yellowstone, the Native American name source is not clear.

Hundreds of species of mammals, birds, fish and reptiles (爬行动物) have been documented, including several that are either endangered or threatened. The vast forests and grasslands also include unique species of plants. Grizzlies(灰熊), wolves, and free-ranging herds(兽群) of bison(北美野牛) and elk(麋鹿) live in the park.

The Grizzly Bear is a powerful predator(食肉动物), capable of out sprinting(疾跑) a horse, and weighing as much as 350 – 600 pounds. Grizzlies are omnivorous(杂事的), meaning they eat both meat and plants. In Yellowstone, grizzlies feed on elk, trout, bison carrion, pine nuts, grasses, roots, and berries.

Certain characteristics distinguish grizzlies from black bears. The grizzly is larger, both in girth and weight. The grizzly's coat ranges from tawny(黄褐色) cinnamon(肉桂色) to light brown or even black. Some grizzlies are flecked(有斑点的) with recognizable "silver tips", creating the "grizzled" look. A feature also associated with grizzlies is the shoulder hump(突起物), and area of well-defined muscle. Another is his dish-shaped face.

Grizzlies hibernate(冬眠) in dens(洞穴) during the winter. This is when the cubs(幼崽) are born. Emerging from their dens in the spring with one to three cubs, the mothers are very protective. They will not hesitate to charge intruders which they perceive as threatening their cubs. Grizzly Bears range throughout Yellowstone but are most often seen in and around the Dunraven Pass area and just past the turn off to Mount Washburn. Another good area is across the Yellowstone River in Hayden Valley and in the Fishing Bridge area. Also in Lamar Valley, a good pair of binoculars(双筒望远镜) or a spotting scope(观测镜) is a must.

Undoubtedly, one of the more popular species now found in Yellowstone Park is thegray wolf. The adult male gray wolf stands 26 to 38 inches high at the shoulder and is usually 40 to 58 inches in length (excluding the tail). Males are 15 to 20 percent larger than females. The weight of gray wolves varies from 40 to 175 pounds. Coyotes(土狼) are often mistaken for wolves; however, the coyote is actually much smaller at only 20 to 30 pounds in weight.

Although most wolves in North America are a grizzled gray color, they can range in color from tan(棕褐色) to pure white or solid black. The gray wolf utilizes its acute senses to survive and flourish in the wilderness. Scientists have estimated that the wolf's sense of smell is up to one hundred times more sensitive than that of a human. One researcher found that a cow moose with twin calves was scented by wolves over four miles away. The sense of hearing is the next most acute of the wolf senses. Wolves can hear much higher frequencies than humans. While the upper auditory(听觉的) limit is 20 kHz for humans, wolves may detect frequencies as high as 80 kHz. Sight is perhaps the least developed of the wolf senses; however, it has been shown that the wolf's sight is at least as acute as that of humans.

The elk is the most abundant of the large mammals of Yellowstone National Park. Oftentimes reaching the size of a large horse, elk can be seen throughout Yellowstone year-around. The cows and calves oftentimes travel in large groups of a hundred or more during the summer months, while the males tend to travel by themselves or in very small groups, feeding on grasses and tree twigs. During the summer months, elk can usually be seen in the Lamar Valley and in the northwest sector of the Park. The Mammoth and Gardiner area, located at a relatively low elevation, provides forage for elk throughout the year and gives the visitor of any season ample viewing opportunities. Nonetheless, elk are liable to be spotted anywhere in the Park, and some of the mature males are majestic in stature with tremendous antler spreads. The mating season takes place in the fall, and the bulls lock antlers in fierce competition, their bugles ringing through the air, as they battle for a harem of cows. During the winter months, the weak and the young sometimes succumb(死于) to the harsh weather. Both grizzlies and wolves prey on elk, seeking the young and the weak that may lag behind the rest of the herd.

The Yellowstone bison story covers more than one hundred years of struggle and conflict surrounding one of America's most majestic animals. In the early 1800's, an estimated 65 million bison roamed throughout the continent of North America. However, market hunting and poaching (非法猎取) had a devastating effect on the bison population; and by 1890, fewer than 1,000 remained! Even with the establishment of Yellowstone National Park, protection and sanctuary for the bison did not occur until the U. S. Army arrived in 1886 to protect the park's resources.

Due to protection and manipulative management (transplanting bison to different parts of Yellowstone), there were approximately 1,500 bison in Yellowstone National Park by 1954. Beginning 1968, the manipulative management of Yellowstone bison was discontinued and the population was allowed to fluctuate based on environmental conditions (i. e. winter weather, food availability, etc). In the 1970's and 1980's, there were a series of cool, wet summers followed by mild winters. These conditions allowed for an abundance of grasses for the bison to feed on and a reduction in the winter mortality rate.

In addition, snowmobiling in the park has helped more bison survive winters since the groomed(修饰的) roads cut down on the amount of energy that a bison uses to travel when compared to traveling through deep snow. By the winter of 1996 to 1997, there were approximately 3,500 bison in Yellowstone National Park.

Today, the Yellowstone bison still faces conflict and struggle. The most recent modern issue is the disease calledbrucellosis which can cause abortions in domestic cattle. Due to a lawsuit filed by the State of Montana, a new bison management plan was developed. All bison entering Montana along Yellowstone's north boundary would be shot or shipped to slaughter. In addition, all bison entering Montana through the park's west boundary would be tested for brucellosis. Any bison testing positive for brucellosis would be shot or shipped to slaughter. Yellowstone National Park now has a new bison exhibit called "Where the Buffalo Roam". It opened on August 1, 1997 at the Canyon Visitor Center. The exhibit also presents issues that arise when the wild bison confronts the modern world. Although the Canyon Visitor Center is closed for the winter, it will be open again for the summer season.

Forest fires occur in the park each year; in the large forest fires of 1988, nearly one third of the park burned. Yellowstone has numerous recreational opportunities, including hiking, camping, boating, fishing and sightseeing. Paved roads provide close access to the major geothermal areas as well as some of the lakes and waterfalls. During the winter, visitors often access the park by way of guided tours that use either snow coaches(四轮大马车)or snowmobile.

Snowmobiling is a wonderful way to tour Yellowstone during the winter months. Because of environmental concerns, however, there are certain restrictions in place which govern the use of snowmobiles in the park.

From 200 – 250 machines typically enter the park on a given day, while the winter use management plan now in place allows up to 318 machines daily. All snowmobilers are required to be accompanied by a licensed guide.

Historically, nearly 800 machines entered Yellowstone daily, following introduction of snowmobiles into the park in the 1960s. The current winter use management plan ensures that resources are protected, gives visitors, employees and residents of the park gateway communities the information they want and need to plan for the near term, and helps minimize economic impacts on gateway communities. Snowmobiles must be rented from commercial operators in Yellowstone's gateway communities.

Notes

Ulysses Simpson Grant 尤利塞斯·辛普森·格兰特(1822 年 4 月 27 日—1885 年 7 月 23 日)美国第 18 任总统,生于俄亥俄州。总统任期:1869 年 3 月 4 日—1873 年 3 月 4 日;1873 年 3 月 4 日—1877 年 3 月 4 日。所属政党为共和党。格兰特是美国历史上第一位从美国军事院校(西点军校)毕业的军人总统。他在美国南北战争中屡建奇功,有“常胜将军”之称。但能征惯战并不等于善于理政,格兰特的平平政绩与他的赫赫战功成为鲜明对比。特别是在第二次总统任期内,他对南方奴隶主妥协让步以及对贪污腐化的属员采取姑息纵容态度,引起了选民的普遍不满。格兰特卸任后曾于 1879 年到世界游历。他也是第一位到达中国的美国总统。格兰特想在政治上东山再起,但未能如愿。晚年经商失败,抑郁病逝。

Wyoming 怀俄明州 美国西部洛矶山区的州。面积 25.33 万平方千米。人口 46 万(1991 年)。首府和最大城市是夏延。由于怀俄明州印第安人的抵抗,白人开拓较晚。1890 年建州。西部洛矶山脉占州面积的 2/3,高山集中于西北部黄石公园周围及其南面山地,最高的甘尼特峰海拔 4 200 米。偏南为怀俄明盆地,系半干燥高原。东部为大平原的一部分。平均海拔约 2 000 米,仅次于科罗拉多州。经济以矿业最盛,有煤、石油、天然气、铀等。煤藏量居全国首位,铀藏量居全国第二位。农业以畜牧为主,到处是牛、羊牧场。羊毛、羊肉生产仅次于得克萨斯州,居全国第二位。夏延、卡斯珀、拉腊光等为家畜贸易中心城市。工业以炼油和食品加工为主。旅游业发达。有著名的黄石国家公园和大蒂顿国家公园。

Montana 蒙大拿州 美国西北部的一个州。面积在美国名列第四大州,而人口相当稀少，人口密度也相当低，经济上以农牧为主，作物主要有燕麦、大麦和甜菜，还有重要的采矿和伐木业。蒙大拿州有两种非常壮观的自然地形：洛矶山脉的北端及大草原。草原覆盖了蒙大拿州3/5的面积，除了一些孤立的山脉和河谷之外，蒙大拿州的草原一直延伸到西边的洛矶山脉，该地区的气候非常不稳定。

Old Faithful Geyser 老忠实喷泉　美国黄石公园中最负盛名的景观。黄石最著名的间歇泉是老忠实喷泉，它不像其他喷泉那样没有规律地爆发，它每隔几十分钟就会喷发一次，从不叫旅客失望。它始终如一的忠实受到世人的喜爱。老忠实喷泉每隔56分钟喷出一次，每次历时约4分钟，喷得最高最美之时是前20秒，每次共喷出热水约1万加仑，高度达40～50米，水温摄氏93度。经冬历夏，老忠实喷泉都这样按照一定的规律不间歇地喷发着，遂得“老忠实”这样的美名。人们来到黄石公园，必到老忠实喷泉一睹为快。但见它不喷则已，一喷则如万马奔腾，更兼在阳光辉映下，水蒸气闪出七彩颜色，蔚为壮观。

Lewis and Clark Expedition 路易斯和克拉克远征　路易斯和克拉克是200年前美国的探险家。1803年美国从法国得到路易斯安纳领地，这块开阔的土地当时仍然是荒芜之地，他们奉杰弗逊总统的命令，组织探险队前往调查。

US Army 美国陆军　美国武装力量的组成部分之一，主要负责陆地上的作战。它组建于1775年6月14日，2002年有现役军人48万名，预备役军人55.5万名。预备役之中35万名编成为美国国民警卫队（ARNG），20.5万名编成为美国陆军预备役部队（USAR）。GI是美国陆军在第二次世界大战时的别称。在1918年之前，美国陆军还只是所谓的联邦军。在一战期间，美国国家级陆军才得以正式组建。在20世纪二三十年代，职业军人被称为所谓“正规军”，正规军中的空缺由募兵以及预备役军官补充。近几十年来，美国陆军被分为正规军和预备役。美国国民警卫队是美国各州政府指挥的武装力量。在21世纪之前，国民警卫队的队员是州政府的雇员，当他们被美国陆军征召时，他们转而成为美国陆军预备役部队的成员。现在，美国国民警卫队成员接受双重指挥：作为国民警卫队队员接受各州民兵司令官的指挥；作为预备役官兵接受美国陆军人力资源司令部的指挥。

Washburn Expedition 沃什伯恩－兰福德－多恩远征　1870年人类对黄石的一次最重要的造访——一次名为“沃什伯恩－兰福德－多恩”的探险行动开始了。黄石公园的别名“老忠实泉（Old Faithful）”就是这支探险队的人命名的。

Yellowstone River 黄石河　美国西北部河流。源自怀俄明州北部洛矶山区阿布萨罗卡岭，北流经黄石公园至蒙大拿州南部，折向东北到北达科他州西部汇入密苏里河。长约1080千米，流域面积18.2万平方千米。在黄石公园地区形成黄石湖，湖北面峡谷景色极佳。靠雪、雨水补给，春夏之交涨水。在蒙大拿州境内广泛用于灌溉。

Brucellosis 普鲁士菌病　一种可以导致牲畜胎儿夭折的疾病。允许猎杀是为了限制野牛的数量和活动区域。一些野牛携带病毒，所以一些牧场主担心，那些病菌可能传染到他们的牲畜身上。

Snowmobile 雪地车 一种在雪地、山地、沼泽地、水淹地、泥泞地或沙漠地执行任务的履带式后勤保障车辆。在北极边缘积雪区、山地或其他特种地带，气候寒冷、环境生疏、士兵穿着笨重、行动不便。雪地车是滑雪场用于救护、输送、娱乐的专用车辆。

Translation of Selected Sentences

1. Yellowstone was the first national park in the world, and is known for its wildlife and its many geothermal features, especially Old Faithful Geyser, one of the most popular features in the park.

黄石国家公园是世界上第一个国家公园，以特有的野生动物和许多的地热特征而闻名，特别是老忠实喷泉为公园内最受欢迎的景点。

2. The park is the centerpiece of the Greater Yellowstone Ecosystem, the largest remaining, nearly-intact ecosystem in the Earth's northern temperate zone.

黄石公园是整个大黄石生态系统的核心，是地球北温带最大的现存的近乎完整的生态系统。

3. Hundreds of structures have been built and are protected for their architectural and historical significance, and researchers have examined more than 1,000 archaeological sites.

数以百计的建筑已经建成，并因建筑或历史方面的重要性而得以保护，研究人员迄今为止已经检验了 1 000 多处的考古遗址。

4. In addition, snowmobiling in the park has helped more bison survive winters since the groomed roads cut down on the amount of energy that a bison uses to travel when compared to traveling through deep snow.

另外，公园里的雪地车可以帮助更多的水牛过冬，因为和深雪地相比，被修整过的道路减少了水牛行进时耗费的能量。

5. The current winter use management plan ensures that resources are protected, gives visitors, employees and residents of the park gateway communities the information they want and need to plan for the near term, and helps minimize economic impacts on gateway communities.

这个冬季使用管理计划确保资源保护，并且给游客、雇员、公园出入口社区的居民们信息，以便他们能够做短期计划，帮助减小出入口社区的经济影响。

Questions

1. What is the general condition of the Yellowstone National Park according to the passage?

2. Is the volcano in Yellowstone National Park still active recently and what is it like?

3. What is the Old Faithful Geyser and how is it formed?

4. What are the main animals in the Yellowstone National Park and what are the appearances and living habits?

5. What is the function of a snowmobile in Yellowstone National Park?

Chapter 2

Crater Lake National Park
火山口湖国家公园

【简介】

美国火山口湖国家公园位于美国俄勒冈州西南部的喀斯喀特山上,占地面积647平方千米。为保护火山口湖和周围的林木,美国政府于1902年将其连同附近地区划为国家公园,是美国第五个国家公园。火山口湖是公园的核心和主要部分,在美国湖泊中居首位。湖水呈现令人难忘的碧蓝和纯净,其浓淡随湖水深浅而变化。湖中央有座小岛,岛上青松翠柏、生机勃勃,景色极为秀丽。附近山头林立、湖畔岩石生长着松、红杉、铁杉,夏日杂花丛生,绿草铺地。园内多野生动物。

Crater Lake National Park is a United States National Park located in southern Oregon, whose primary feature is Crater Lake. This National Park was established on May 22, 1902, and it is the sixth oldest National Park in the U. S. This park encompasses(包围)the Crater Lake caldera, which rests in the remains of a destroyed volcano (eventually named Mount Mazama) and the surrounding forest land and hills. This is the only National Park in Oregon.

The lake was formed after the collapse of an ancient volcano, posthumously(于身后) named Mount Mazama. This volcano violently erupted approximately 7,700 years ago. That eruption was 42 times as powerful as the 1980 eruption of Mt. St. Helens. The basin or caldera(火山喷口) was formed after the top 5,000 feet of the volcano collapsed. Subsequent lava flows sealed the bottom, allowing the caldera to fill with approximately 4.6 trillion gallons of water from rainfall and snow melt, to create the seventh deepest lake in the world at 1,932 feet.

Rolling mountains, volcanic peaks and evergreen forests surround this enormous, high

Cascade Range Lake, recognized worldwide as a scenic wonder. On summer days, neither words nor photographs can capture Crater Lake's remarkable blueness. For much of the year, usually October to July at higher elevations(海拔), a thick blanket of snow encircles the lake. Snowfall provides most of the park's annual 66 inches of precipitation(沉淀物).

Crater Lake rarely freezes over completely; it last did in 1949. Heat from the summer sun stored in the immense body of water retards ice formation(冻结成冰)throughout the winter. On the earth clock, natural forces only recently constructed this landscape. Lava flows first formed a high plateau base on which explosive eruptions then built the Cascade volcanoes. Humans probably witnessed the cataclysmic(巨变的)eruption of Mount Mazama about 7,700 years ago.

One of the park's most outstanding features is its 33 mile "Rim Drive" which provides spectacular views of the lake and the surrounding cliffs and mountains as it completely circles the lake. The road, which is open for only a few months in the later summer because of the enormous amount of snow which accumulates on the rim during the winter, provides access to the lake for virtually(事实上)every visitor.

The park provides a number of other features in addition to the lake, many related to the vulcanism(火山作用) which has been the dominant force in shaping this area. There are a number of volcanic mountains along the rim of the crater, elsewhere in the park, and in the surrounding area. A barren desert composed of volcanic pumice(浮石), features composed of solidified(凝固的) lava, and strange volcanic formations can also be seen in various parts of Crater Lake National Park.

In addition to volcanic features, the park also features streams and creeks(小河), waterfalls, dense forests of red fir, mountain hemlock(铁杉) and whitebark pine(白皮巴尔干松). Much of the park area is wilderness and there are 140 miles of trails within the park boundaries.

Unquestionably, Crater Lake is one of the most spectacular physical features on the earth. It has created an unforgettable experience for the many people who have made their way there; the park itself has served as model for saving spectacular and precious geographical features all over the planet.

The clean, clear, cold lake water contained no fish until they were introduced by humans from 1888 to 1941. Today, rainbow trout(虹鳟鱼),kokanee(红大麻花鱼) and salmon(大马哈鱼) still survive in Crater Lake. Wildflowers bloom late and disappear early here, thriving(繁荣) in wet, open areas. Birds and other animals often seen are ravens(乌鸦), jays(松鸡), nutcrackers(胡桃夹子), deer, ground squirrels and chipmunks(花栗鼠). Present but seldom seen are elk, black bear, foxes, porcupines(豪猪), pine martens(松貂), chickaree squirrels and pikas(鼠兔).

Human beings seem to have inhabited the general area of the park for as long as 12,000 years. It is believed, deriving from archaeological(考古学的)evidence gathered in the 1930's, that some humans may have witnessed the cataclysmic volcanic eruption which formed Crater Lake. This must have been one of the most amazing sights ever seen by human beings at a place or any time.

It is believed that some Indians held the lake to be sacred(神圣的) and were forbidden to gaze on its waters or to speak of it. In any case, while the native peoples may have visited the area from time to time, the severe nature of winter weather in the park and along the rim of the caldera, precluded year round habitation. Some evidence of visits by native peoples has been found, such as rock cairns, perhaps associated with "spirit quests".

For a variety of reasons, the lake was generally unknown to people of European descent for a long while, although legendary explorer John C. Fremont in the years 1843—1846 led reconnaissance(勘测)parties within sight of the lake's rim and apparently heard tales of a "great sunken(凹陷的)hole". Traders and trappers, although they visited the mountains and valleys of central Oregon and the Klamath Basin, all managed to miss the great lake in the crater

Gold was discovered in the area, at Daisy Creek, in 1852. The attraction of this strike led directly to the first whitemen to view the lake. These were members of a group of prospectors including John Wesley Hillman, Henry Klippel, and Isaac Skeeters, on June 12, 1853. They were understandably impressed by the sight and christened it "Deep Blue Lake". They saw the lake from a location near where Rim Village currently exists, now called "Discovery Point". However, this sighting was largely forgotten in the years that followed and had little discernable effect.

The Klamaths revered(崇敬) the lake and the surrounding area, keeping it undiscovered by white explorers until 1853. That year, on June 12, three gold prospectors, John Wesley Hillman, Henry Klippel and Isaac Skeeters, came upon a long, sloping(斜坡的) mountain. Upon reaching its highest point, a huge, awe-inspiring lake was visible. "This is the bluest lake we've ever seen," they reported, and named it Deep Blue Lake. But gold was more on the minds of settlers at the time and the discovery was soon forgotten.

Captain Clarence Dutton was the next man to make a discovery at Crater Lake. Dutton commanded a U.S. Geological Survey party which carried the Cleetwood, a half-ton survey boat, up the steep slopes(陡坡)of the mountain then lowered it to the lake. From the stern(严厉的) of the Cleetwood, a piece of pipe on the end of a spool of piano wire sounded the depth of the lake at 168 different points. Dutton's soundings of 1,996 feet were amazingly close to the sonar(声呐)

readings made in 1959 that established the lake's deepest point at 1,932 feet.

William Gladstone Steel, the father of Crater Lake National Park, devoted his life and fortune to the establishment and management of Crater Lake National Park. He had been preoccupied with(全神贯注于)Crater Lake since 1870 when he was a sixteen year old boy in Kansas. He learned of Crater Lake by reading a newspaper that was use to wrap(包裹)his lunch. Two years later he moved to Oregon and in 1885 he and a druggist(药剂师)named John Beck joined a group headed for Crater Lake.

When the two men, Steel and Beck finally spotted the lake, the water was so blue it startled them. "All ingenuity of nature seems to have been exerted to the fullest capacity to build a grand awe-inspiring temple the likes of which the world has never seen before," said Steel. His involvement with Crater Lake covered 49 years. After it was made a national park, he realized his work had just begun.

In his efforts to bring recognition to the park, he participated in lake surveys that provided scientific support. He named many of the lake's landmarks, including Wizard Island, LlaoRock, and Skell Head.

Steel's dream was realized on May 22, 1902 when President Theodore Roosevelt signed the bill giving Crater Lake national park status. And because of Steel's involvement, Crater Lake Lodge was opened in 1915 and the Rim Drive was completed in 1918.

Notes

Crater Lake 火山口湖　美国最深的湖。位于俄勒冈州西南部,喀斯喀特山脉南段。被冰川覆盖的古火山锥马札马火山于更新世晚期喷发时形成的破火山口,经风化和侵蚀后扩大积水而成。呈圆形,直径10千米,面积54平方千米。湖面海拔1882米。最大深度589米。湖面变动小,湖水清澈,呈深蓝色。由于火山多次喷发,形成若干火山锥,部分露出湖面即成为小岛。如最大的威扎德岛,高出湖水面213米,顶部有一火山口。湖周为高150~600米的熔岩峭壁,火山岩经风化后呈各种形状,加上湖区松、杉林茂密,夏季野花丛生,景色幽美,已于1902年辟为国家公园,有鹿、熊、鹰、猫头鹰等野生动物。湖中有鱼。

Mount Mazama 马札马山　一座位于美国俄勒冈州境内的火山,属于喀斯喀特山脉的一部分。约7 000年以前,马札马火山爆发、崩塌后,留下了火山口湖。熔岩冷却后封住了火山底,因而形成了一个大型的碗状火山口。时间一年年地过去,火山口内填满了超过400万兆加仑的降雨及融化的雪水,因而它也成为美国境内最深的湖泊。另外,这座湖也以其深蓝色湖水而知名。有关马札马火山的爆发,被发现在克拉马斯族印地安人(Klamath Indians)代代相传的故事中,他们是远古马卡拉克人(Makalak)的后裔。根据马卡拉克人的传说,山神劳(Llao)与天空之神史凯尔(Skell)之间有一场激烈战斗,当史凯尔击败劳后,原为山神居所的马札马火山,随即爆发并崩塌。火山口湖现在是一座美国国家公园,一年吸引大约 500 000 名游客来访。

Mt. St. Helens 圣海伦斯火山　北美洲近期喷发的活火山。位于美国西北部华盛顿州,北纬 49.20 度,西经 122.18 度,海拔 2549 米,属喀斯喀特山脉。在喀斯喀特山脉的众多火山中,圣海伦斯火山是一座相对年轻的火山,大约形成于 4 万年前。

Cascade Range Lake 喀斯喀特火山　美国喀斯喀特山脉的高峰。位于华盛顿州西部,海拔 4 391 米。圆锥型火山。基盘为花岗岩,火山体为安山岩。最近一次喷发在 1870 年,现仅喷少量蒸汽。附近有温泉。山麓和低坡生长着针叶林(冷杉、松等),海拔 2 600 ~ 2 800 米为高山草甸,更高为永久积雪和冰川。在喀斯喀特山脉南段有一个火山口湖,这是美国最深的湖泊,最大深度 589 米,湖的轮廓近似圆形,直径 10 千米,面积 54 平方千米。火山口湖原来是被冰川覆盖的古火山锥马扎马火山,更新世晚期火山喷发后山顶形成火山口,在长期风化和流水侵蚀的作用下,火山口逐渐扩大并且积水成湖。后来火山又先后出现了几次小的喷发,形成一些火山锥,这些火山锥露出湖面成为小岛。湖的四周围绕着熔岩峭壁,高约 150 ~ 600 米,岩壁形状奇特,多姿多彩,湖水清澈,碧蓝如玉,湖区松、杉林茂密,空气清新,景色迷人,1902 年被辟为国家公园。

Klamaths 克拉马斯　北美印第安部落,居住在美国俄勒冈州中南部及加利福尼亚州北部,操佩纽蒂语系两种相近的语言(或方言),并具有许多共同的文化特征。他们的土地位于喀斯喀特山脉(Cascade Range)南段一大槽形地带,约 160 千米长、45 千米宽,沼泽、湖泊、河流、溪流,星罗棋布于其间。当地人主要以捕鱼及猎捕水禽为生。

John Wesley Hillman 约翰·卫斯理·希尔曼(1832 年 3 月 29 日—1915 年 3 月 19 日)美国勘探者,欧裔美国人,在加州淘金热中首先发现俄勒冈州的火山口湖。希尔曼出生在纽约州奥尔巴尼。1848 年 16 岁的他随家人搬到路易斯安那州的新奥尔良。在父亲的陪同下,希尔曼远征加州。据说他是第一个在俄勒冈州看到“深蓝湖”的欧裔美国人。这个湖后来改名为火山口湖。

Clarence Dutton 克拉伦斯·达顿　美国地质学家,陆军军官。1841 年 5 月 15 日出生于康涅狄格州的沃林福德。他毕业于耶鲁大学,1860 年入伍,并参加了研究生课程。他研究了夏威夷、加利福尼亚和俄勒冈州的火山。1884 年他当选为国家科学院的成员。1886 年他曾率领美国地质勘探局成员到俄勒冈州做火山口湖的研究工作。

William Gladstone Steel 威廉姆·格莱斯顿·斯蒂尔(1854 年 9 月 7 日—1934 年 10 月 21 日)被称为“火山口湖之父”,火山口湖指的是在美国俄勒冈州的火山口湖国家公园。他是一位土生土长的美国俄亥俄州人,曾是一名邮递员。

Translation of Selected Sentences

1. This park encompasses the Crater Lake caldera, which rests in the remains of a destroyed volcano (eventually named Mount Mazama) and the surrounding forestland and hills.

公园环绕着火山口湖,而火山口湖位于被毁坏的火山(最终被命名为马札马山)和周围的林地和山丘上。

2. Subsequent lava flows sealed the bottom, allowing the caldera to fill with approximately 4.6 trillion gallons of water from rainfall and snow melt, to create the seventh deepest lake in the world at 1,932 feet.

随后的熔岩流使底部密封,使得火山口蓄积了约4.6万亿加仑的雨水和融化的雪水,深度达到1932英尺,在世界上最深的湖水排名中名列第七。

3. It has created an unforgettable experience for the many people who have made their way there; the park itself has served as model for saving spectacular and precious geographical features all over the planet.

很多到过火山湖口的人对这次经历都难以忘怀;公园保护了壮观、珍贵的地貌特征,为全世界的国家公园做出了典范。

4. For a variety of reasons, the lake was generally unknown to people of European descent for a long while, although legendary explorer John C. Fremont in the years 1843—1846 led reconnaissance parties within sight of the lake's rim and apparently heard tales of a "great sunken hole".

尽管具有传奇色彩的探险家约翰·弗里蒙特在1843—1846年带领勘探队到达湖的边缘,并且似乎听到了"大凹洞"的故事,但是由于种种原因,这个湖在很长时间并不为欧洲人的后裔所熟知。

5. All ingenuity of nature seems to have been exerted to the fullest capacity to build a grand awe-inspiring temple the likes of which the world has never seen before

这座神殿巧夺天工、气势恢宏、威风凛凛、匠心独具。

Questions

1. What is the general condition of Crater Lake National Park and when was it founded?
2. Why is "Rim Drive" one of the park's most outstanding features and what is it?
3. How was the lake formed and what was it like?
4. Who discovered the Crater Lake according to this passage?
5. How do people describe Crater Lake? And why?

Chapter 3

Mesa Verde National Park
弗德台地国家公园

【简介】

科罗拉多州的弗德台地海拔 2 600 米,有建于 6 ~ 12 世纪的阿拉萨斯印第安人居址,有 3 800 多处遗迹地点,其中包括平原村庄和悬崖上的房屋,其中的悬崖宫殿最为壮观。弗德台地国家公园是北美印第安人史前居住地遗迹保留地,又译为梅萨维德国家公园,位于美国科罗拉多州西南部蒙特苏马山谷和曼科斯山谷之间,占地面积达 21 074 公顷。1888 年 12 月发现,1906 年美国政府辟为国家公园,设专门管理保护机构。迄今已有 3 800 个史前遗迹点与世人见面,其中 600 多处属于圣胡安盆地的阿纳萨扎伊文化。

Mesa Verde National Park is a U. S. National Park and UNESCO World Heritage Site located in Montezuma County, Colorado, United States. It was created in 1906 to protect some of the best preserved cliff dwellings in the world. The park occupies 81.4 square miles (211 km^2) (211 square kilometers) near the Four Corners and features numerous ruins(遗迹) of homes and villages built by the Ancestral Puebloan people, sometimes called the Anasazi.

The Anasazi inhabited Mesa Verde anywhere between 550 to 1300 AD. These people were mainly subsistence(自然经济) farmers; they grew crops on nearby mesas(台地). Their primary crop was corn, which was also the major part of their diet. Men were also hunters which further increased their food supply. The women of the Anasazi were famous for the elegant basket weaving (编织). Anasazi pottery is just as famous as their baskets; their artifacts, even today, are highly prized. Since the Anasazi kept no written records, their artifacts are the only link to understanding their interesting culture.

By 750 AD, the people were building mesa-top villages made of adobe(土砖). By the late 12th century they began to build the cliff dwellings for which Mesa Verde is famous.

Mesa Verde is best known for cliff dwellings, which are structures built within caves and under outcropping in cliffs—including Cliff Palace, which is thought to be the largest cliff dwelling in North America. The Spanish term Mesa Verde translates into English as "green table". Approximately 600 of the over 4 700 archeological sites found in Mesa Verde National Park are cliff dwellings.

Mesa Verde National Park is located in the south-western corner of the state of Colorado. Elevations in the park range from about 6,100 to 8,400 feet (1,900 to 2,600 m). The terrain (地势) in much of the park is dominated by ridges(山脊) and valleys running roughly north and south. The northernmost point is 13. 2 miles (21. 2 kilometers) farther north than the southernmost; the westernmost point is 11. 9 miles (19. 2 kilometers) farther west than the easternmost.

Although explorers from Spain went through the general region in the 18th century, actual sight of the cliffs dwellings by outsiders seems to have first occurred in the latter half of the 19th century. The fame of Mesa Verde soon began to spread thanks to the Wetherill ranchers(大农场工人) and the archeological work of Gustaf Nordenskiöld. Vandalism(破坏行为) led to President Teddy Roosevelt's support of protecting the area as a national park in 1906.

Spanish explorers seeking a route from Santa Fe to California in the 1760s and 1770s were the first Europeans to reach the Mesa Verde (green table) region, which they named after its high, tree-covered plateaus. But they never got close enough, or into the needed angle, to see the ancient stone villages, which would remain a secret for another century.

Occasional trappers and prospectors visited, with one prospector, John Moss, making his observations known in 1873. The following year he ledeminent(有名的) photographer William Henry Jackson through Mancos Canyon, at the base of Mesa Verde. There Jackson both photographed and publicized a typical stone cliff dwelling. In 1875 geologist William H. Holmes retraced Jackson's route. Reports by both Jackson and Holmes were included in the 1876 report of the Hayden Survey, one of the four federally financed efforts to explore the American West. These and other publications led to proposals to systematically study Southwestern archaeological sites. They did not lead to action for some years.

As concern grew over the archaeological well being of Mesa Verde's ruins, and those in other nearby sites, the area was established as a national park on June 29, 1906. As with all historical areas administered by the National Park Service, the park was listed on the National Register of

Historic Places on October 15, 1966. It was designated a World Heritage Site on September 6th, 1978. The park was named with the Spanish term for green table because of its forests of juniper (杜松) and piñon (矮松) trees.

A set of six buildings built by the National Park Service in 1921, the Mesa Verde Administrative District, was designated a National Historic Landmark on May 29, 1987. It consists of the first buildings constructed by the National Park Service which are based on cultural traditions represented in the park area. The principal designer believed that structures could be used for interpretive purposes to explain the construction of prehistoric dwellings in the Park, and be compatible (与和谐相处) with their natural and cultural setting. There have been 600 cliff dwellings that have been recorded within Mesa Verde National Park.

In the summers of 1996, 2000 (twice), 2001, 2002 and 2003, the park, which is covered with pinyon pine and Utah juniperforests, suffered from a large number of wildland fires; parts of it were closed. All areas of the park have since re-opened, but some areas show significant damage from the fires.

Mesa Verde's park entrance is about 9 miles (15 kilometers) east of the community of Cortez. The visitor center is 15 miles (24 kilometers) from the entrance, and Chapin Mesa (the most popular area) is another 6 miles (10 kilometers) beyond the visitor center.

The park's Chapin Mesa Archeological Museum provides information about the Ancient Puebloan civilization and displays findings and artwork.

Three of the cliff dwellings on Chapin Mesa are open to the public. Spruce Tree House is open all year, weather permitting. Balcony House and Cliff Palace are open except in the winter; visitors may tour them only on ranger-guided tours. The cliff dwellings on Wetherill Mesa, including Long House and Step House, can be reached via a 12 - mile (19.2 kilometers) long mountain road leading southwest from the park visitor center. Many other dwellings are visible from the road but not open to tourists.

In addition to the cliff dwellings, Mesa Verde boasts a number of mesa-top ruins. Examples open to public access include the Far View Complex, Cedar Tree Tower and the Sun Temple, all on Chapin Mesa, and Badger House Community, on Wetherill Mesa.

Also in the park are hiking trails, a campground and facilities for food, fuel, and lodging; these are unavailable in the winter.

Mesa Verde is best known for a large number of well preservedcliff dwellings, houses built in shallow caves and under rock overhangs along the canyon walls. The structures contained within these alcoves (凹室) were mostly blocks of hard sandstone, held together and plastered with (涂上灰泥) adobe mortar. Specific constructions had many similarities, but were generally unique in form due to the individual topography of different alcoves along the canyon walls. In marked contrast to earlier constructions and villages on top of the mesas, the cliff dwellings of Mesa Verde reflected a region-wide trend towards the aggregation(聚合)of growing regional populations into close, highly defensible(可防御的)quarters during the 13th century.

While much of the construction in these sites conforms to common Pueblo(印第安村落) architectural forms, including Kivas(基瓦会堂),towers, and pit-houses, the space constrictions of these alcoves necessitated what seems to have been a far denser concentration of their populations. Mug House, a typical cliff dwelling of the period, was home to around 100 people who shared 94 small rooms and eight kivas built right up against each other and sharing many of their walls; builders in these areas maximized space in any way they could and no areas were considered off-limits to construction.

Not all of the people in the region lived in cliff dwellings; many colonized the canyon rims and slopes in multi-family structures that grew to unprecedented size as populations swelled(膨胀). Decorative motifs for these sandstone/mortar constructions, both cliff dwellings and non-, included T-shaped windows and doors. This has been taken by some archaeologists, such as Stephen Lekson (1999), as evidence of the continuing reach of the Chaco Canyon elite system, which had seemingly collapsed around a century before. Other researchers see these motifs(图案)as part of a more generalized Puebloan style and spiritual significance, rather than evidence of a continuing specific elite socioeconomic system.

For most of the 12th and 13th centuries, known archaeologically as the Classic Period, the Ancient Puebloan Native Americans lived in the cliff dwellings. The reason for their sudden departure about 1 275 remains unexplained; theories range from crop failures due to droughts to an intrusion of foreign tribes from the North.

Cliff Palace is the largest and best-known of the cliff dwellings in Mesa Verde. The site has 150 identified(被鉴定的)rooms and 23 kivas(大地穴). Although this and other Mesa Verde sites are large and well constructed, they demonstrate a long history of occupation and their architectural design is an aggregation of dwellings and storage spaces that developed slowly and randomly. Accurate archaeological information from this site has been limited due to several decades of digging and collecting at the turn of the twentieth century.

Mug House situated on Wetherill Mesa was professionally excavated(挖掘)in the late 1960s by archaeologist Arthur Rohn. The structure contains 94 rooms, in four levels, including a large kiva, with simple vertical walls and masonry(石工)pilasters(壁柱). This ceremonial(仪式的) structure has a keyhole(锁眼)shape, due to a recess(凹处)behind the fireplace and a deflector (导流板), which is considered an element of the Mesa Verde style. The rooms clustered around the kiva formed part of the courtyard, indicating the kiva would have been roofed.

Spruce Tree House Located on Chapin Mesa, this cliff dwelling is easily accessible and well preserved. The ruins include a kiva with a restored roof which visitors can enter. Excavations indicate that this structure, like many other dwellings in Mesa Verde, was probably occupied for less than a century.

Square Tower House that gives this site its name is the tallest structure in Mesa Verde. This cliff dwelling was occupied between AD 1200 and 1300.

Mesa Verde Reservoirs built by the Ancient Puebloans, were named a National Civil Engineering Historic Landmark on September 26, 2004.

Records indicate that the Balcony House was probably first rediscovered by an excavator, S. E. Osborn, sometime in 1884, his name and the date "March 20, 1884" being found in a dwelling nearby. Furthermore, Osborn published a newspaper account of dwellings, including one that matches the description of Balcony House, in 1886. The site was excavated by archaeologist Jesse Nusbaum in 1910.

In February 2008, the Colorado Historical Society has decided to invest a part of its US $7 million budget into a culturally modified trees project in the National Park.

Notes

World Heritage Site 世界遗产　被联合国教科文组织和世界遗产委员会确认的人类罕见的、目前无法替代的财富，是全人类公认的具有突出意义和普遍价值的文物古迹及自然景观。狭义的世界遗产包括"世界文化遗产""世界自然遗产""世界文化与自然遗产"和"文化景观"四类。广义概念，根据形态和性质，世界遗产分为文化遗产、自然遗产、文化和自然双重遗产、记忆遗产、人类口述和非物质遗产（简称非物质文化遗产）、文化景观遗产。

Montezuma County 蒙提祖马县　美国科罗拉多州西南角的一个县，顺时针与新墨西哥州、亚利桑那州、犹他州相连。面积为 5 284 平方千米。根据美国 2000 年人口普查，共有人口 23 830 人。县治科特斯。成立于 1893 年 4 月 16 日。县名来自阿斯特克人最著名的统治者蒙提祖马二世。

Pueblo 普韦布洛　史前阿纳萨齐人，阿纳萨齐文化（Anasazi culture）的有史时期的后裔。住在亚利桑那东北和新墨西哥西北的若干地点，密集定居，名为普韦布洛（西班牙语意为村或镇）。史前的阿纳萨齐人有很多的地区性歧异，普韦布洛印第安人同样彼此差异很大，不管是在文化上还是在语言上都是如此。当代普韦布洛人分为东西两支。全体普韦布洛人都集体猎捕野兔，妇女采集野生植物。西部的普韦布洛人因气候干旱，农作物较少，霍皮人的情况更是如此。现代普韦布洛人的社会生活以村落为中心（村落亦为政治单位），不过他们基本上仍是神权政治。西普韦布洛人组成氏族和家系;氏族设有秘密会社,会社定期举行仪式以祈福或求雨。有一种全部落性的卡奇纳(kachina)崇拜仪式是为追念远祖举行的,而全部男子会社都负责保卫工作和参加丰产仪式。在格兰德河地区,每村一分为二,号称夏民及冬民,分别负责村内各项活动;该地的秘密会社主要负责医疗仪式,卡奇纳崇拜仪式则较少。地方性艺术品及手工艺品的生产,在霍皮人中间尤为活跃,纺织和编筐迄今仍很盛行,19 世纪 90 年代霍皮－特瓦人恢复了陶器制造。大多数村庄普遍制作银器及绿松石饰物,但银器制造工艺并非原住民技术。

Anasazi 阿纳萨齐　美国印第安部落,是早期欧洲白人在西部开发垦荒时代出现的一些相当有规模但是仍然被遗忘的群落。一般相信现今普韦布洛印第安人是阿纳萨齐人的后裔。遗址主要分部在科罗拉多、新墨西哥、亚利桑那和犹他四州的交界地带。遗址通常在岩壁巨大的洞穴下,其中经常能够发现编织精美的篮子。

the American West 美国西部　泛指美国西部各州。由于美国自建国以来疆域多次向西扩展,因此美国西部的定义也随着时代而变化,一般多以密西西比河作为美国东西部的分界线。如果采用广义的定义,美国西部占去了美国一半以上的领土。美国西部包括了自路易斯安那购地中获得的土地、1818 年自英国获得的土地、得克萨斯加入美国时的土地、1846 年自英国获得的土地、1848 年墨西哥割让给美国的土地及盖兹登购地中获得的土地。美国西部本身又可再细分为不同的地域。亚利桑那州、科罗拉多州、加利福尼亚州、新墨西哥州、内华达州、俄克拉荷马州、得克萨斯州及犹他州又被视为美国西南部。而蒙大拿州、怀俄明州、爱达荷州、俄勒冈州和华盛顿州则是美国西北部。华盛顿州、俄勒冈州和加利福尼亚州三州位于太平洋沿岸,又被称为美国西岸。

Administrative District 行政区划　行政区域划分的简称,指国家对辖境进行的行政分区。或者说,它是指国家为了便于行政管理,根据政治、经济、民族、历史等各种因素的不同,把领土划分成大小不同、层次不等的区域,并在此基础上建立相应的政权机关进行社会管理的制度。简言之,若行政区划按三级制,则各级为一级行政区即省级行政区、县级行政区即第二级行政区和乡级行政区即第三级行政区;若行政区划按四级制,则各级为省级、地级、县级和乡级。

Chaco Canyon 查科峡谷　位于美国新墨西哥西北的纳瓦约印第安人保留地。如今,这里气候干燥,草木不生。峡谷两旁耸立着 15 ~ 40 米高的峭壁,峭壁之间是时断时续的查科河冲积而成的沙石河床。大约 1000 年前,这里曾经是北美洲史前史上最辉煌、最成熟的文明中心之一。公元 1000 年前后,查科兴旺起来,成为阿那萨吉文化的重要中心,1130—1150 年人们遗弃了它,遗弃它的主要原因是干旱。

Translation of Selected Sentences

1. Mesa Verde is best known for cliff dwellings, which are structures built within caves and under outcropping in cliffs—including Cliff Palace, which is thought to be the largest cliff dwelling in North America.

梅萨维德国家公园以悬崖屋而著称。悬崖屋是建立在山洞和悬崖凸出部分的下面——包括悬崖宫殿,它被认为是北美最大的悬崖屋。

2. Spanish explorers seeking a route from Santa Fe to California in the 1760s and 1770s were the first Europeans to reach the Mesa Verde (green table) region, which they named after its high, tree-covered plateaus.

18 世纪六七十年代,寻求从圣菲到加利福尼亚州之间道路的西班牙探险家是最早到达梅萨维德(绿色平地)地区的欧洲人,他们按照高耸的悬崖、绿色覆盖的平原这些特征给它起了名字。

3. The principal designer believed that structures could be used for interpretive purposes to explain the construction of prehistoric dwellings in the Park, and be compatible with their natural and cultural setting.

主要设计师坚信这些结构可以用来解释公园内史前居住的建筑,并且符合他们的自然和文化背景。

4. In marked contrast to earlier constructions and villages on top of the mesas, the cliff dwellings of Mesa Verde reflected a region-wide trend towards the aggregation of growing regional populations into close, highly defensible quarters during the 13th century.

和早期台地顶端的建筑和村庄形成鲜明对比的是,梅萨维德的悬崖民居反应了大范围内地区人口不断增长的趋向。在 13 世纪时形成集中高密度的居住形式。

5. Other researchers see these motifs as part of a more generalized Puebloan style and spiritual significance, rather than evidence of a continuing specific elite socioeconomic system.

其他研究人员把这些图案看作更加广泛的古代普布罗人的风格和精神意义,而不是连续具体的精英社会经济系统的证据。

Questions

1. What is the general condition of Mesa Verde National Park?
2. What is Mesa Verde National Park famous for?
3. Who is Anasazi and can you tell the story of Ansazi?
4. When did actual sight of the cliffs dwellings by outsiders seem to have first occurred?
5. How should we describe the cliff?

Chapter 4

Glacier National Park
冰川国家公园

【简介】

冰川国家公园建于1910年,位于美国蒙大拿州北部,与加拿大的英属哥伦比亚省和阿尔亚伯塔省接壤。因这里有50余座冰川,故得名冰川国家公园。冰川是由降落在高山的大量积雪在重力作用下凝为沿地面运动的巨大冰体形成。陡峭的纵壁、有棱有角的外形是冰川的特征。冰川国家公园由多山的湖区组成,位于南纬49度以南,阿根廷圣克鲁斯省西南部的边远地区。它包括南安第斯山的一个被大雪覆盖的地区,以及许多发源于巴塔哥尼亚冰原的冰川。

Glacier National Park is located in the U. S. state of Montana, bordering the Canadian provinces of Alberta and British Columbia. The park encompasses(包含) over 1,000,000 acres (4 000 km^2) and includes parts of two mountain ranges (sub-ranges of the Rocky Mountains), over 130 named lakes, more than 1,000 different species of plants and hundreds of species of animals. This vast pristine(原始的) ecosystem is the centerpiece(核心) of what has been referred to as the "Crown of the Continent Ecosystem", a region of protected land encompassing 16,000 square miles (41 000 km^2).

The region that became Glacier National Park was first inhabited by Native Americans and upon the arrival of European explorers, was dominated by the Blackfeet in the east and the Flathead in the western regions. Soon after the establishment of the park on May 11, 1910, a number of hotels and chalets(小屋) were constructed by the Great Northern Railway. These historic hotels and chalets are listed as National Historic Landmarks, and a total of 350 locations are on the National Register of Historic Places. By 1932, work was completed on the Going-to-the-Sun Road, later designated a National Historic Civil Engineering Landmark, which provided greater accessibility for automobiles into the heart of the park.

After the creation of the park, the growing staff of park rangers(公园管理员) needed housing and offices to help protect the new park. The increasing number of park visitors made the need for roads, trails, and hotels urgent. A typical visit to Glacier involved a train ride to the park, followed by a multi-day journey on horseback. Each day after a long ride in the mountains, guests would stay at a different hotel or chalet. The lack of roads meant that, to see the interior of the park, visitors had to hike or ride a horse. Eventually, the demand for a road across the mountains led to the building of the Going-to-the-Sun Road.

The construction of the Going-to-the-Sun Road was a huge undertaking(事业). Even today, visitors to the park marvel at how such a road could have been built. The final section of the Going-to-the-Sun Road, over Logan Pass, was completed in 1932 after 11 years of work. The road is considered as an engineering feat and is a National Historic Landmark. It is one of the most scenic roads in North America. The construction of the road forever changed the way visitors would experience Glacier National Park. Future visitors would drive over sections of the park that previously had taken days of horseback riding to see.

While much has changed since the first visitors came to Glacier, it is possible to relive some of Glacier early history. You can take a horseback ride like an early visitor. Miles of hiking trails follow routes first used by trappers in the early 1800's. Several hotels and chalets, built by the Great Northern Railway in the early 1900's, house summer guests to the park. A visit to Glacier National Park is still a great adventure!

The mountains of Glacier National Park began forming 170 million years ago when ancient rocks were forced eastward up and over much younger rock strata. Known as the Lewis Overthrust, these sedimentary rocks are considered to have some of the finest fossilized examples of extremely early life found anywhere on Earth. Of the estimated 150 glaciers which existed in the park in the mid-19th century, only 25 active glaciers remained by 2010. Scientists studying the glaciers in the park have estimated that all the glaciers may disappear by 2030 if the current climate patterns persist.

Glacier National Park has almost all its original endemic(地方性的) plant and animal species. Mammals such as the grizzly and mountain goat as well as less common ones such as the wolverine(貂熊) and lynx(山猫) are known to inhabit the park. Hundreds of species of birds, more than a dozen fish species and even a few reptiles(爬行动物) and amphibian(两栖类) species have been documented. The park has numerous ecosystems ranging from prairie(大草原)

to tundra(苔原) and the easternmost forests of red cedar(红雪松) and hemlock normally found in large numbers closer to the Pacific Ocean. Though larger forest fires are uncommon in the park, in 2003 over 10% of the park was impacted by fires.

Glacier National Park borders Waterton Lakes National Park in Canada—the two parks are known as the Waterton-Glacier International Peace Park, and were designated as the world's first International Peace Park in 1932. Both parks were designated by the United Nations as Biosphere (生物圈) Reserves in 1976, and in 1995 as World Heritage sites.

The park is bordered on the north by Waterton Lakes National Park in Alberta, and the Flathead Provincial Forest and Akamina-Kishinena Provincial Park in British Columbia. To the west, the north fork of the Flathead River forms the western boundary, while its middle fork is part of the southern boundary. The Blackfeet Indian Reservation provides most of the eastern boundary, and the Lewis and Clark and the Flathead National Forests form the southern and western boundary. The remote Bob Marshall Wilderness Complex is located in the two forests immediately to the south.

The park contains a dozen large lakes and 700 smaller ones, but only 131 lakes have been named. Lake McDonald on the western side of the park is the longest (9.4 miles (15.1 km)), largest in area (6,823 acres (27.61 km^2)) and deepest (464 feet (141 m)) of these lakes. Numerous smaller lakes, known as tarns(山中小湖), are located in cirques(圆环) formed by glacial erosion(冰河侵蚀). Some of these lakes, like Avalanche Lake and Cracker Lake, are colored an opaque(不透明的) turquoise(蓝绿色) by suspended glacial silt(淤泥), which also causes a number of streams to run milky white. The lakes of Glacier National Park remain cold year round, with temperatures rarely above 50 °F (10 °C) at their surface. Cold water lakes such as these support little plankton(浮游生物) growth, ensuring that the lake waters are remarkably clear. The lack of plankton, however, lowers the rate of pollution filtration (过滤), and pollutants have a tendency to linger longer. Consequently, the lakes are considered environmental bellwethers as they can be quickly affected by even minor increases in pollutants.

Two hundred waterfalls are scattered throughout the park, however, during drier times of the year, many of these are reduced to a trickle(细流). The largest falls include those in the Two Medicine region, McDonald Falls in the McDonald Valley and Swiftcurrent Falls in the Many Glacier area, which is easily observable and close to the Many Glacier Hotel. One of the tallest waterfalls is Bird Woman Falls, which drops 492 feet (150 m) from a hanging valley beneath the north slope of Mount Oberlin.

Notes

Alberta 阿尔伯达省　加拿大西部省份，北纬 49°~60°之间，南北长 1 217 千米，东西宽 293~650 千米不等，面积约 66 万平方千米。西部为落矶山脉，东部为平原。2004 年人口约 318 万，其中 2/3 住在首府埃德蒙顿及最大城市卡加利。冬季寒冷，白天平均气温 -10 ℃；夏季暖和，白天平均气温约 25℃。阿尔伯达省是加拿大最富裕的省份之一，经济支柱为石油开采和加工。除此之外，它与相邻的萨斯喀彻温及曼尼托巴经常被合称为“草原三省”，是加拿大国家重要的粮仓。该地区畜牧业与农业发达，草原地区养有超过 500 万头牛，加拿大将近一半的牛肉产自阿尔伯达省。

British Columbia 不列颠哥伦比亚　位于加拿大西岸，临太平洋，海岸线长达 8850 千米，沿岸风景优美，海产丰富，全省面积 94.78 万平方千米，人口 414.13 万，仅次于魁北克和安大略省，是加拿大第三大省。该省南部与美国华盛顿州、爱达荷州及蒙大拿州接壤，是加拿大通往亚太地区的门户。不列颠哥伦比亚省气候温和，风景秀丽，环境保护良好，依山傍水，省内大部分面积是森林地带，是世界著名的旅游胜地。

Rocky Mountains 洛矶山脉　也被译为落矶山脉，位于北美洲西部，是美洲科迪勒拉山系在北美的主干，由许多小山脉组成，被称为北美洲的“脊骨”，从加拿大横越美国西部、直到新墨西哥州，绵延超过 4 800 千米。整个洛矶山脉由众多小山脉组成，其中有名称的就有 39 条。这条巨大的山脉南北狭长，北至加拿大西部，南达美国西南部的得克萨斯州一带，几乎纵贯美国全境。除圣劳伦斯河外，北美几乎所有大河都源于洛矶山脉，是大陆重要分水岭。山脉的最高峰是埃尔伯特峰，位于科罗拉多州境内，高度有 4 401 米。罗布森峰则是加拿大境内的最高峰，高度为 3 954 米。

Native Americans 美洲原住民　对美洲所有原住民的总称。美洲原住民中的绝大多数为印第安人，剩下的则是主要位于北美洲北部的爱斯基摩人。传统上将美洲原住民划归蒙古人种美洲支系。当欧洲人首次来到美洲时，美洲原住民早已遍布南北美洲各地。美洲原住民所说的语言众多，目前仍然存在的美洲原住民语言约有 350 个，分属十几个语系，至今没有公认的语言分类。

Blackfeet 黑脚族人　以凶猛勇悍而著称。黑脚族人靠着马和枪的优势，不停地与周边部落开战。他们在战争中压倒了周边的弱小部族。他们对进入这片领土的白人捕猎者毫不留情。整个洛矶山区几乎成了他们的天下。捕猎河狸的白人进入这片地区总是心惊胆颤，甚至一听到黑脚族的名字就吓得发抖。长期以来，没有白人敢踏入黑脚人的圣地。

Flathead 弗拉塞德人　操萨利什语的印第安部落，原住于美国蒙大拿州西部地区，其疆土自比特鲁特山脉之脊至洛矶山分水岭，以哥伦比亚河的克拉克支流上游为中心，通常被认为是高原文化区最东端的部落，但具有很多洛基山东部平原印第安人的特征。历史上他们曾拥有大量马匹,每年会骑马远征平原,捕获野牛,因此经常与平原部落发生战争。这种活动通常需要严密的政治组织,虽然平原区的典型村落组织由一地区群组及其首领与议会构成,足以处理地区范围内的日常事务,但大规模的战争与捕猎野牛则需要若干村落结合为真正的部落联盟,选出最高领袖负责指挥各村落的协同行动。平原圆锥形的帐篷是平常居所,但高原地带的A字形草席覆盖住所也是该族村落中较为常见的。西弗拉塞德人会使用独木舟,但东弗拉塞德人仅仅使用简陋的野牛皮制作的皮舟,这是平原印第安人的特色之一。渔业在生产生活中占有重要的地位。弗拉塞德人信仰保护精灵,有自己的巫医。

Going-to-the-Sun Road 向阳大道　冰河公园最值得一看的地方以及最好的路线,便是沿着向阳大道前行。向阳大道全长约80千米,途中会经过路易斯山脉(LewisRange),还会穿越高6 680米的罗根山隘(LoganPass)。由于这条路段冬天会被冰封,所以只限6月中旬至10月中旬才开放。

the Pacific Ocean 太平洋　世界最大的海洋。包括属海的面积为18 134.4万平方千米,不包括属海的面积为16 624.1万平方千米,约占地球总面积的1/3。从南极大陆海岸延伸至白令海峡,跨越纬度135°,南北最宽1.55万千米。在太平洋水系中,最主要的是中国及东南亚的河流。

Waterton Lakes National Park 沃特顿湖国家公园　位于阿尔伯塔省最南端,邻接美国冰河国家公园。建于1895年,占地525平方千米。园内有超过250种珍惜鸟类,111种珍贵植物。1932年在扶轮社力促之下,将瓦特顿与紧邻的蒙大拿州冰河国家公园(Glacier National Park)合称为沃特顿-冰河国际和平公园(Waterton-Glacier International Peace Park),希望两国不分你我,共同享用与保护自然资源。这个和平公园的设立,开创了全球先例,至今仍是世界上最大的国际和平公园。1995年联合国教科文组织又指定这两座国家公园为世界传承地(World Heritage Site),另外,这里也被该组织人类与自然环境计划指定为生态保育区,希望能为人类与大地的依存关系设立良好的典范。

Indian Reservation 印第安保留地　一片受美国内政部印第安事务局管辖的土地。由于美国政府对印第安部落区的管理被限制,导致部落地区的法律不同于周边地区。例如,某些保留地引来游客的原因是它的法律可以允许赌博。在美国大约有310个印第安人保留区,这意味着不是所有拥有550人以上的村落被视为可拥有保留区的部落。有的部落占有一个保留区,有的多个或者是没有。此外,由于过去的土地分配问题,导致一些保留地被销售给非印第安人,有一些保留地受到严重侵蚀。每一个单独的部落,或私人持有的土地就像是一个国中之国。这种私人的和集体所有土地的混杂造成了极大的行政上的困难。

Translation of Selected Sentences

1. The region that became Glacier National Park was first inhabited by Native Americans and upon the arrival of European explorers, was dominated by the Blackfeet in the east and the Flathead in the western regions.

欧洲探险家到来之前,冰川国家公园的原址最先居住着美国原住居民,东部地区由黑脚族人占据,而西部地区主要由弗拉塞德人占据。

2. Known as the Lewis Overthrust, these sedimentary rocks are considered to have some of the finest fossilized examples of extremely early life found anywhere on Earth.

这些沉积岩——也被称为刘易斯上冲断层——被认为是在地球上发现的最好的早期生命的石化样本。

3. The park has numerous ecosystems ranging from prairie to tundra and the easternmost forests of red cedar and hemlock normally found in large numbers closer to the Pacific Ocean.

公园里有很多生态系统,从草原生态系统到冻土地带,最东部的红雪松和铁杉通常会在邻近太平洋地区大量生长。

4. Glacier National Park borders Waterton Lakes National Park in Canada—the two parks are known as the Waterton-Glacier International Peace Park, and were designated as the world's first International Peace Park in 1932.

冰川国家公园和加拿大的沃特顿湖国家公园接壤——这两个公园合起来被认为是沃特顿 - 冰川国际和平公园,1932 年被指定为世界上第一个国际和平公园。

Questions

1. Where is Glacier National Park located?
2. State briefly the significant scenes in the Glacier National Park.
3. What kind of animalsdoes Glacier National Park have?
4. How do you describe the mountains in Glacier National Park?
5. Who inhabited the region that became Glacier National Park?

Chapter 5

Rocky Mountain National Park
落矶山国家公园

【简介】

落矶山国家公园位于美国科罗拉多州，成立于1915年，占地1 075平方千米，1/3的园区位于林线以上，包括了78个海拔超过3 670米的山峰。其中最高的是久久峰，海拔4 345米。公园自成一个完整的生态体系，不论海拔高低，到处可见野生动物的踪迹，如麋鹿、长耳鹿、土狼、角鹿和大角羊等。此外贯穿大峡谷的科罗拉多河也是在此地发源的。落矶山公园最大的宝藏就是绵延的山脉和清澈的高山湖泊。它也是北美洲的大分水岭，是一处山水兼具的旅游景点。

Rocky Mountain National Park is a national park located in the north-central region of the U. S. state of Colorado. It features majestic mountain views, a variety of wildlife, varied climates and environments—from wooded forests to mountain tundra—and easy access to back-country trails and campsites. The park is located northwest of Boulder, Colorado, in the Rockies, and includes the Continental Divide and the headwaters of the Colorado River.

The park has five visitor centers. The park headquarters, Beaver Meadows Visitor Center, is a National Historic Landmark, designed by the Frank Lloyd Wright School of Architecture at Taliesin West.

The park may be accessed by three roads: U. S. Highway 34, 36, and State Highway 7. Highway 7 enters the park for less than a mile, where it provides access to the Lily Lake Visitor Center. Farther south, spurs from route 7 lead to campgrounds and trail heads around Longs Peak and Wild Basin. Highway 36 enters the park on the east side, where it terminates after a few miles at Highway 34. Highway 34, known as Trail Ridge Road through the park, runs from the town of Estes Park on the east to Grand Lake on the southwest. The road reaches an elevation of 12,183 feet (3,713 m), and is closed by snow in winter.

The park is surrounded by Roosevelt National Forest on the north and east, Routt National Forest on the northwest, and Arapaho National Forest on the southwest.

Rocky Mountain National Park encompasses approximately 265,770 acres (1,076 km^2) of land in Colorado's northern Front Range. The park is split(分离) by the Continental Divide, which gives the eastern and western portions of the park a different character. The east side of the park tends to be drier, with heavily glaciated peaks and cirques. The west side of the park is wetter and more lush(苍翠繁茂的), with deep forests dominating.

The park contains 359 miles (578 km) of trails, 150 lakes, and 450 miles (720 km) of streams. The park contains over 60 named peaks higher than 12,000 feet (3,700 m), and over one fourth of the park resides above tree line. The highest point of the park is Longs Peak, which rises to 14,259 feet (4,346 m; surveys before 2002 show 14,255 feet (4,345 m)) above sea level. Longs Peak is the only fourteen thousand foot peak in the park.

Several small glaciers and permanent snowfields are found in the high mountain cirques, including Andrews Glacier, Sprague Glacier, Tyndall Glacier, Taylor Glacier, Rowe Glacier, Mills Glacier, and Moomaw Glacier.

The lowest elevations in the park aremontane(山区的)forests and grassland. The ponderosa (北美黄松)pine, which prefers drier areas, dominates, though at higher elevations douglas fir (花旗松)trees are found. Above 9,000 feet (2,700 m) the montane forests give way to the subalpine forest. Engelmann Spruce(恩格尔曼氏云杉)and Subalpine Fir trees are common in this zone. These forests tend to have more moisture than the montane and tend to be denser. Above tree line, at approximately 11,500 feet (3,500 m), trees disappear and the vast alpine tundra takes over. Due to harsh winds and weather, the plants in the tundra are short with very limited growing seasons. Streams have created lush riparian wetlands across the park.

July and August are the warmest months in the park, where temperatures can reach the 80s although it is not uncommon to drop to below freezing at night. Thunderstorms often appear in the afternoons, and visitors should plan on staying below tree line when they occur. Heavy winter snows begin around mid-October, and last into May. While the snow can melt away from the lowest elevations of the park, deep snow is found above 9,000 feet (2,700 m) in the winter, causing the closure of Trail Ridge and Fall River roads during the winter and spring. Most of the trails are under snow this time of the year, and snowshoeing(穿着雪鞋走)and skiing become popular. Springs tend to be wet, alternating between rain and possibly heavy snows. These snows can occur as late as July. The west side of the park typically receives more precipitation than the drier east side.

The park is dominated by Longs Peak, which is visible from many vantage points(有利位置), and has an elevation of 14,259 feet (4,346 m). Each year thousands of people attempt to scale it. The easiest route is the Keyhole Route, impassable(不能通过的) to regular hikers in all but the hottest summer months due to snow and ice. This eight-mile (13 km) one-way hike has an elevation gain of 4,850 ft (1,480 m). The vast east face, including the area known as The Diamond, is home to many classic big wall rock climbing routes.

Not all leave Longs Peak alive and safe. There is a stone gazebo(露台) at the Keyhole formation with a plaque memorializing Agnes Vaille, a well-known climber in the 1920s. In January 1925, Vaille fell 100 ft (30 m) while descending the North Face. Vaille survived the fall with minor injuries, but was unable to walk. Her climbing partner, professional mountaineering guide Walter Kiener, went for help; but when rescuers arrived, Vaille had died of fatigue and hypothermia(低温).

Bear Lake, in the heart of the park, is a popular destination and trailhead(小道的起点). The lake lies below Hallett Peak and the Continental Divide. Several trails start from the lake, ranging from easy strolls to strenuous(费力的) hikes. Bear Lake Road is open year round, though it may close temporarily due to bad weather.

Trail Ridge Road connects the town of Estes Park in the east with Grand Lake in the west. The road reaches an altitude of 12,183 feet (3,713 m), with long stretches above tree line. It passes the Alpine Visitors' Center, a popular destination, and crosses the Continental Divide at Milner Pass. Numerous short interpretive trails and pullouts along the road serve to educate the visitor on the history, geography, and ecology of the park.

The southern area of the park is Wild Basin, a wild and remote region. Several trails cross the area and backpacking(背包旅行) it is popular.

The Mummy Range is a short mountain range in the north of the park. The Mummies tend to be gentler and more forested than the other peaks in the park, though some slopes are rugged and heavily glaciated, particularly around Ypsilon Mountain and Mummy Mountain.

The snow-capped(积雪覆盖的) Never Summer Mountains are found in the west side of the park. Here the south-trending Continental Divide takes a brief sharp northward loop, which creates an interesting reverse scenario, where the Pacific Basin is on the east side of the divide and the Atlantic Basin on the west. The mountains themselves, the result of volcanic activity, are craggy(崎岖的) and, more often than not, covered in deep snow. This area saw the most extensive mining in the park, and trails lead past old mines and ghost towns.

Paradise Park is hidden in the peaks above Grand Lake. This rugged and wild area has no trails penetrating it.

Evidence has shown that Native Americans have visited the area of the park for the last 10,000 years. Their influence in the region was limited, however, and their visits are often transitory(短暂的). The Ute Tribe visited the west side of the park, particularly around Grand Lake. The Arapaho visited and hunted in the Estes Park region.

The Long Expedition, led byStephen H. Long, for whom Longs Peak was named, visited the area in 1820, though they never entered the mountains.

Joel Estes and his son stumbled across the meadows that eventually became Estes Park in 1859 while on a hunting expedition. He moved his family there in 1860 and raised cattle. He stayed only until 1866, forced out by long, harsh winters. In the next years, settlers and homesteaders staked their claims in the Estes Park region. Tourists, particularly those interested in climbing the high peaks of the region, appeared after this time.

In 1880 a small mining rush began in the Never Summer Mountains. The mining town of Lulu City was established with great fanfare(炫耀) and promotion by the media, particularly by Fort Collins newspapers. The ore mined, however, was low grade; by 1883 the rush went bust, and most of the residents moved on. A satellite town, Dutchtown, was abandoned by 1884.

Enos Mills, then a 14 year old boy, moved to Estes Park in 1884. He explored the mountains of the area and wrote many books and articles describing the region. He later supported the creation of Rocky Mountain National Park, and he split his time between the mountains he loved and the cities of the eastern United States, where he lobbied for the legislation to create the park. The legislation was drafted by James Grafton Rogers, a Denver lawyer and avid(热心的) outdoorsman(常在野外活动的人). Mills' original proposal for park boundaries went from Wyoming all the way down to the Mount Evans area, including areas such as the Indian Peaks Wilderness. Much of the land was favored for mining, logging, and other operations, however, so the proposed park was reduced to an area approximating the current park borders. The bill passed Congress and was signed by President Woodrow Wilson on January 26, 1915. A formal dedication ceremony was held on September 4, 1915 in Horseshoe Park. The park has expanded over the years, with the largest parcel—the Never Summer Range—added in 1929.

The 1920s saw a boom in building lodges and roads in the park, culminating with the construction of Trail Ridge Road between 1929 and 1933. The Civilian Conservation Corps handled several building projects during the Great Depression and remnants(残余) of their camps can be found in the park today.

On June 24, 2010, awildfire burned over 1,500 acres (6.1 km^2) of the park in Larimer County near Estes Park. Estes Park Fire Department believed that lightning may have started the fire.

Notes

Continental Divide 大陆分水岭　北美西部由一系列北 - 南向山峰形成的连绵不断的山岭,将该大陆主要水系分为东向(在加拿大注入哈得逊湾或在美国主要汇入密西西比河)和西向(注入太平洋)两部分。分水岭大部分沿落矶山的浪峰走,经过加拿大的不列颠哥伦比亚省,并沿着不列颠哥伦比亚 - 阿尔伯塔省边界下来,穿过美国蒙大拿、怀俄明、科罗拉多和新墨西哥州,继续往南到达墨西哥和中美洲,大致和西马德雷山脉、南马德雷山脉以及其在中美洲相关的山脉平行。

Colorado River 科罗拉多河　北美洲主要河流之一,有"西南部的生命线"之称。它发源于美国西部科罗拉多州中北部洛矶山脉,由大面积厚雪融化提供水源,是美国西南方、墨西哥西北方的河流,其曲折蜿蜒的水道长度大约有 2 333 千米,有一部分在美国洛矶山脉区域西边消失。科罗拉多河的干、支流多峡谷,以干流的大峡谷最著为著名。河水含泥沙量高,使河水呈红色,"科罗拉多"即为西班牙语"红色的"之意。另一个著名景观就是 1936 年建成的胡佛水坝,是美国最大的人工湖。

Longs Peak 朗斯峰　美国洛矶山支脉弗朗特山脉的高峰,位于科罗拉多州北部。海拔 4 344.7 米,是洛矶山国家公园地区的最高峰。于 1820 年被发现,山峰上覆盖着万年冰雪。自山顶向东可眺望大平原,向西俯瞰洛矶山脉群山。

Grand River 格兰德河　北美洲南部河流。源自洛矶山,最初向东,继而向南流,最后沿美国、墨西哥国界作东南流向,注入墨西哥湾。格兰德河长 3 030 千米,流域面积 57 万平方千米。在墨西哥境内被称为布拉沃河。下游可通航,但根据国际协定,航运关闭。河水大部分流经干燥地区,植被稀疏。夏多暴雨,河水含沙量高,常形成水灾。在肥沃三角洲地区有蔬菜农场和柑橘园林。沿河主要城市有美国的阿尔布开克、帕索和墨西哥的马塔莫罗斯等。

Satellite Town 卫星城　在大城市外围建立的既有就业岗位,又有较完善的住宅和公共设施的城镇,是在大城市郊区或其以外附近地区,为分散中心城市的人口和工业而新建或扩建的具有相对独立性的城镇。因其围绕中心城市象卫星一样,故名卫星城,旨在控制大城市的过度扩展,疏散过分集中的人口和工业。卫星城虽有一定的独立性,但是在行政管理、经济、文化以及生活上同它所依托的大城市有较密切的联系,与母城之间保持一定的距离,一般以农田或绿化带隔离,但有便捷的交通联系。卫星城概念产生于英国,美国的泰勒正式提出并使用"卫星城"这一形象性的概念。

Woodrow Wilson 伍德罗·威尔逊(1856 年12 月28 日—1924 年2 月3 日) 美国第28 任总统。作为进步主义时代的一位领袖级知识分子,他曾先后担任普林斯顿大学校长、新泽西州州长等职。在1912 年总统大选中,由于西奥多·罗斯福和威廉·塔夫脱的竞争分散了共和党选票,伍德罗·威尔逊以民主党人身份当选总统。迄今为止,他是唯一一名拥有哲学博士头衔的美国总统(法学博士衔除外),也是唯一一名任总统以前曾在新泽西州担任公职的美国总统。

Civilian Conservation Corps 民间资源保护队 罗斯福新政期间从事资源保护工作的重要组织。它的成立是由保护自然资源的愿望和为城市失业青年提供有报酬的工作的需要所决定的;同时它也是罗斯福具有进步主义内涵和特征的社会改革和资源保护思想与实践发展的逻辑结果。民间资源保护队在美国现代历史上占有非常重要的地位,它不仅在一定程度上缓解了美国当时所面临的“环境”危机和社会危机,而且扩展了资源保护运动的范围和群众基础,为战后美国的环境运动埋下了伏笔。

Translation of Selected Sentences

1. It features majesticmountain views, a variety of wildlife, varied climates and environments—from wooded forests to mountain tundra—and easy access to back-country trails and campsites.

美国洛矶山国家公园的突出特点是宏伟的山峰景色、各种各样的野生动物、多样化的气候和环境——从茂密的森林到高山苔原——从这里很容易去乡村野游和露营。

2. While the snow can melt away from the lowest elevations of the park, deep snow is found above 9,000 feet (2,700 m) in the winter, causing the closure of Trail Ridge and Fall River roads during the winter and spring.

虽然在公园低海拔地区的积雪可以融化,在冬季还是可以见到9 000 英尺(2 700 米)的深雪,这会导致在冬季和春季径山和落河公路的关闭。

3. Numerous short interpretive trails and pullouts along the road serve to educate the visitor on the history, geography, and ecology of the park.

在道路两旁可以看到很多简短的的路径说明和大张海报,供游客了解美国洛矶山国家公园的历史、地理和生态环境。

4. The Mummies tend to be gentler and more forested than the other peaks in the park, though some slopes are rugged and heavily glaciated, particularly around Ypsilon Mountain and Mummy Mountain.

虽然母亲山脉的一些斜坡较为崎岖,并被厚厚的冰雪覆盖着,尤其是在亚普斯林和母亲山的周围地区,母亲山脉较之公园里的其他山峰来说更趋于和缓、繁茂。

5. The 1920s saw a boom in building lodges and roads in the park, culminating with the construction of Trail Ridge Road between 1929 and 1933.

20 世纪 20 年代见证了修建公园胜地旅舍和公路的高峰，1929 年至 1933 年间建造的径山公路的修建大道能到达顶峰。

Questions

1. State briefly the general conditions of Rocky Mountain National Park.
2. How was Rocky Mountain National Park founded?
3. What does the climate like there?
4. What are the main mountains and lakes in the Rocky Mountain National Park?
5. How do people protect the Rochy Mountain National Park?

Chapter 6

Hawaii Volcanoes National Park
夏威夷(大岛)火山国家公园

【简介】

夏威夷群岛位于北太平洋的中央,由东南至西北的130多个岛屿组成。它是一个绵延伸展2 400千米的群岛整体。夏威夷火山国家公园位于美国夏威夷州的夏威夷岛上,面积929平方千米,主要包括冒纳罗业和基拉韦厄两座现代活火山。这里有茂密的热带雨林,还有蝙蝠、夏威夷雁、大鹰、乌鸦、夏威夷白腹水鸟等动物。基拉韦厄和冒纳罗亚两座现代活火山是这个公园的主要组成部分,同时它们也是夏威夷火山国家公园闻名遐迩的显著性标志。公园是1961年根据美国国会法令建立的,在1980年成为教科文组织生物圈保护区。

Hawaii Volcanoes National Park contains two of the most active volcanoes in the world, Mauna Loa (4,170 m high) and Kilauea (1,250 m high), both of which tower over(高出)the Pacific Ocean. Volcanic eruptions have created a constantly changing landscape, and the lava flows reveal surprising geological formations. Rare birds and endemic species(特有物种) can be found there, as well as forests of giant ferns(蕨类植物).

Through the process of shield-building volcanism, the park's landscape is one of relatively constant, dynamic change. This property is a unique example of significant island building through ongoing volcanic processes. It represents the most recent activity in the continuing process of the geologic origin and change of the Hawaiian Archipelago(群岛). The volcano Mauna Loa, measured from the ocean floor, is the greatest volcanic mass on earth.

Volcanoes are monuments to earth's origin, evidence that its primordial(原始的)forces are still at work. During a volcanic eruption, we are reminded that our planet is an ever changing environment whose basic processes are beyond human control. As much as we have altered the face of the earth to suite our needs, we can only stand in awe before the power of an eruption.

Volcanoes are also prodigious(巨大的) land builders—they have created the Hawaiian Island chain. Kilauea and mauna Loa, two of the world's most active volcanoes, are still adding land to the island of Hawaii. Mauna Loa is the most massive mountain on the earth, occupying an area of 10,000 cubic miles. Measured from its base on the seafloor(海底), it rises 30,000 feet, approximately 1,000 feet higher than Mount Everest. In contrast to the explosive continental volcanoes, the more fluid(流动的) and less gaseous(气态的) eruptions of Kilauea and Mauna Loa produce fiery(炽热的) fountains and rivers of molten(熔化的) lava. These flows, added layer upon layer, produced a barren volcanic landscape that served as a foundation for life. Hundreds of species of plant and animals found their way across the vast Pacific on wind, water, and the wings of birds. A few survived, adapted, and prospered during this time of isolation. The arrival of humans, first Polynesians, then Europeans, and the plants and animals that they brought with them drastically altered this evolutionary showcase, this grand experiment.

Today Hawaii Volcanoes National Park displays the results of 70 million years of volcanism, migration, and evolution—processes that thrust a bare land from the sea and clothed it with complex and unique ecosystems and a distinct human culture. Created to preserve the natural setting of Kilauea and Mauna Loa, the park is also a refuge for the island's native plants and animals and a link to its human past. Park managers work to protect the resources and promote understanding and appreciation of the park by visitors. Research by scientists at the Hawaiian Volcano Observatory has made Kilauea one of the best understood volcanoes in the world, shedding light on the birth of the Hawaiian Islands and the beginning of planet Earth.

Volcanoes attest to(证实) the dynamic nature of the earth. Divided into rigid plates, the outermost layer of the earth drifts slowly over the more plastic mantle(地幔) beneath. Most volcanic activity occurs along the edge of these plates, forming a "ring of fire". The series of volcanoes that include Washington's Mount Saint Helens, Alaska's Katmai(卡特迈火山), Japan's Mount Fuji and the Philippines' Mount Pinatubo define the margins of the Pacific plate.

Why are there volcanoes in Hawaii, which is located in the middle of the Pacific plate? Plumes of magma(岩浆) rise from a "hot spot" deep within the mantle. This fluid charged with gas, melts and pushes its way to the surface, erupting on the ocean floor to create a seamount(海底山). After several hundred thousand years and countless eruptions, the volcano rises above sea level to form an island. The volcano continues to grow until movement of the Pacific plate carries the island off the hot spot.

During the last 70 million years the Pacific plate has acted as a conveyor belt(传送带),

moving the islands northwest off the hot spot at the rate of about four inches a year. The park's active volcanoes, Kilauea and Mauna Loa, continue the island building process that formed the 3, 500 mile Emperor Seamount-Hawaiian Island chain. But they are not the last; to the southeast, Lo'ihi seamount is rising from the ocean floor.

Lies in the south-east part of the island of Hawaii (Big Island), the easternmost island of the State of Hawaii, and includes the summit and south-east slope of Mauna Loa and the summit and south-western, southern, and south-eastern slopes of the Kilauea Volcano. The park extends from the southern coast, with its volcanic sea cliff headlands(岬) to the summit calderas of Kilauea, the most active volcano in the world (with more than 50 recorded eruptions in the last 33 years up to 1985) and Mauna Loa volcanoes. The latter is a massive, flat-domed(平顶的) shield volcano (盾状火山) built by lava flow layers and is considered to be the best example of its type in the world, extending from 5,581 m below sea level to 4,169 m above.

The Hawaii Volcanoes National Park is greatest and the calderas of Kilauea are one of most studied in the world: in fact there is a geologic station dating from 1912.

The park contains a high diversity of plant communities with striking life-form and physiognomic differences; 23 distinct vegetation types have been described for the park, ranging from the very diverse tropical rainforest of Ola'a to the scrub(矮树) and grassland of Ka'u and the alpine tundra of Mauna Loa, grouped into five major ecosystems: subalpine, montane seasonal, montane rainforest, submontane seasonal and coastal lowlands. The Ola'a Forest tract, over 4,000 ha in size, is probably the largest remaining tract of virgin ohia and fern forest in the Hawaiian Islands. The native flora(植物群) numbering 41 species, with a further 40 listed as rare and warranting special attention.

The park is rich, in archaeological remains particularly along the coast with native villages, temples, graves, paved trails, canoe(独木舟) landings, petroglyphs(岩石雕刻), shelter caves and agricultural areas. Following the arrival of Captain James Cook in 1778 – 79, Christian influences started in or around 1823, with churches and schools built and the introduction of cattle, goat and pulu (tree-fern product) harvesting and the encouragement of visitors. Extensive ruins of stone structures dating back to the time of Pa'ao (a high priest) in 1 275 are present.

Like all natural areas in Hawaii, the park has been subject to considerable biological alteration since man's arrival. Direct removal or alteration of native forest for growing sugar and pineapple plantations, ranching(经营大牧场) and logging(砍伐), has altered the native biota (生物区) of the forest habitats, particularly at low and middle elevations. Except for a single species of bat Hawaiian hoary bat(灰蝙蝠), native mammalian(哺乳动物) forms are absent. Little is known to date about invertebrate(无脊椎动物) forms. Avian(鸟类) forms present interesting and significant examples of adaptive radiation (适应辐射) of extinction. Most endemic (地方性的) avian species are rare or endangered. Ranching activities and the introduction of species such as the pig, goat and mongoose (獴狐猴) have had serious biological consequences, including destruction of native ecosystems and widespread extinction of endemic species. Pockets of standing water, created by the wallowing of the feral (野生的) pigs, provide breeding places for mosquitoes, resulting in serious avian malaria.

The control programs undertaken on the basis of solid scientific research have been a model for removal of alien animals and plants (including illegal narcotics) on oceanic islands.

Hawaii National Park, created on 1 August 1916 by Act of the US Congress (39 Stat. 432), consisted of two units each on different islands, one on Hawaii Island and the other on Maui Island. The area of the park was more than doubled as a result of Congressional authorisation in 1922 (45 Stat. 503), in 1928 (45 Stat. 424) and in 1938 (52 Stat. 781). The Ola'a Forest Tract was donated in 1951 and 1953 (Executive Order #1640). The park was split into Hawaii Volcanoes National Park (on Hawaii Island) and Haleakala National Park (on Maui Island) in 1961 (75 Stat. 577). Hawaii Volcanoes National Park is protected under 16 U. S. C. 1 (National Park Service Organic Act) and under the terms establishing the park as set out in 16 U. S. C. 395b, and under several sections of 16 U. S. C. 391 – 396a. Hawaii Volcanoes and Haleakala national parks were accepted as MAB Biosphere Reserves in 1980, and combined to form one cluster, Hawaiian Islands Biosphere Reserve, in April 1983. Hawaii Volcanoes National Park is under consideration for inscription on the World Heritage List.

Notes

Mauna Loa 莫纳罗亚山　地球上最大的火山,自有记载的首次喷发以来,该火山已经喷发过 33 次,不过从 1984 年以来,这座火山相对沉寂了下来。美国地质调查局的艾切尔博格说莫纳罗亚火山的沉寂是因为太平洋地壳下沉于大陆板块之下,这导致大量的水涌进地幔的炽热区域。地幔导致岩石融化,引发火山喷发。熔岩不如固态岩石紧密,易涌出地表。比如在夏威夷莫纳罗亚山,涌出的岩浆流到火山附近区域,令周边深受其苦。

Kilauea 基拉韦厄火山　位于美国夏威夷岛东南部,北纬 19.43°,西经 155.29°,海拔 1 222 米。基拉韦厄火山是世界上活动力旺盛的活火山,至今仍经常喷发。在它 23 年的活动期里,总共发生了 55 次喷发。在世界众多火山中,基拉韦厄火山是非常独特的。它不仅是世界上最活跃的火山之一,而且它每年喷出的熔岩也是地球上所有火山中最多的。1960 年基拉韦厄火山大爆发时,熔岩流从高处奔腾下泻,涌入大海,在海边填造了一块约 2 平方千米的新陆地。此后,这座火山仍经常活动。在最近一次开始于 1983 年 1 月的爆发过程中,基拉韦厄火山释放出大量的熔岩。据统计,如果使用喷发的熔岩铺路的话,道路长度超过地球周长的3倍。

Polynesians 波利尼西亚人　大洋洲东部波利尼西亚群岛的民族集团,包括毛利人、萨摩亚人、汤加人、图瓦卢人、夏威夷人、塔希提人、托克劳人、库克岛人、瓦利斯人、纽埃人、复活节岛人等10多个支系,共有90多万人(1978年)。属南方蒙古人种和澳大利亚人种的混合类型,身材中等偏高,皮肤为浅褐色,体毛较少,头发为宽幅波状黑发,面部宽大,略呈颌突。使用多种语言和方言,彼此差别不大,同属南岛语系波利尼西亚语族。原崇拜多神,迷信巫术,现多改信基督教和天主教。

Mount Saint Helens 圣海伦斯火山　位于美国西北部华盛顿州,是旅游者熟悉的景点之一,圆锥形的峰峦及其颇有特色的雪冠,隆起在一片美丽的森林景观之上。

Mount Fuji 日本富士山　日本第一高峰,横跨静冈县和山梨县的休眠火山,位于东京西南方约80千米处,主峰海拔3 776米。2002年8月(平成14年),经日本国土地理院重新测量后,为3 775.63米,接近太平洋岸,东京西南方约100千米。富士山是日本国内的最高峰,也是世界上最大的活火山之一,目前处于休眠状态,但地质学家仍然把它列入活火山之列。

Mount Pinatubo 皮纳图博火山　位于菲律宾吕宋岛,东经120.35°,北纬15.13°,海拔1 486米。

James Cook 詹姆斯库克　一位因进行了3次探险航行而闻名于世的伟大探险家。他给人们关于大洋——特别是太平洋的地理学知识增添了新的内容。他还被认为在通过改善船员的饮食——包括增加水果和蔬菜等来预防长期航行中出现的坏血病方面也有所贡献。

Executive Order 行政命令　行政主体依法要求相对人进行一定的作为或不作为的意思表示。行政命令具有强制力,它包括两类:一类是要求相对人进行一定作为的命令,如命令纳税、命令外国人出境;另一类是要求相对人履行一定的不作为的命令,称为禁(止)令,如因修建马路禁止通行,禁止携带危险品的旅客上车等。

Translation of Selected Sentences

1. In contrast to the explosive continental volcanoes, the more fluid and less gaseous eruptions of Kilauea and Mauna Loa produce fiery fountains and rivers of molten lava.

和爆炸性的大陆性火山相比,基拉韦厄火山和莫纳罗亚火山流动性更强、气态喷发更少,会产生出炽热的火山喷泉和火山岩浆流。

2. The arrival of humans, first Polynesians, then Europeans, and the plants and animals that they brought with them drastically altered this evolutionary showcase, this grand experiment.

人类的到来,首先是波利尼西亚人,然后欧洲人,以及这些人所带来的动植物都极大地改变了这里的发展规律,这个伟大的尝试。

3. Today Hawaii Volcanoes National Park displays the results of 70 million years of volcanism, migration and evolution—processes that thrust a bare land from the sea and clothed it with complex and unique ecosystems and a distinct human culture.

今天的夏威夷火山国家公园展示了7000万年的火山形成、移民和生态的进化——这是由海洋迸发出的并且逐渐形成被复杂独特的生态系统和别致的人类文化包围着陆地的过程。

4. Research by scientists at the Hawaiian Volcano Observatory has made Kilauea one of the best understood volcanoes in the world, shedding light on the birth of the Hawai'ian Islands and the beginning of planet Earth.

由夏威夷火山观测台的科学家们所做的调查已经使得基拉韦厄火山变成世界上最为人们所了解掌握的火山之一,人们充分了解了夏威夷岛的起源和地球的形成开端。

5. The park is rich, in archaeological remains particularly along the coast with native villages, temples, graves, paved trails, canoe landings, petroglyphs, shelter caves and agricultural areas.

夏威夷火山国家公园有很多的考古遗迹,尤其在沿着海岸附近小路可以看到当地的村落、寺庙、墓地、独木舟着陆地点、岩石雕刻、岩洞和农业地区。

6. Direct removal or alteration of native forest for growing sugar and pineapple plantations, ranching and logging, has altered the native biota of the forest habitats, particularly at low and middle elevations.

为了生产糖类、建造菠萝种植园,大量的天然树林被直接砍伐和改建;经营大牧场和砍伐已经改变了森林栖息地的原始生物区,尤其是在较低或者是中级海拔高度。

Questions

1. What is the general condition of Hawaii Volcanoes National Park?
2. How do you describe the two famous volcanoes?
3. What animals does Hawaii Volcanoes National Park have?
4. How is the climate in Hawaii Volcanoes National Park?
5. Why was the place altered and how did people take actions to protect it?

Chapter 7

Lassen Volcanic National Park
拉森火山国家公园

【简介】

拉森火山国家公园位于美国加利福尼亚州的北部，以美国大陆部分唯一的活火山拉森峰而得名。1916 年建成，占地 433 平方千米。整个公园中最主要的景观就是拉森火山，是世界上最大的穹顶火山，也是喀斯喀山脉中最南端的火山。拉森火山国家公园是世界上少数几个能找到所有火山类型的地区，包括穹顶、盾状火山、火山渣锥及层状火山。园内有 40 多处湖泊，生长着 20 多种茂密的针叶树和灌木类植物，包括松、柏、杉等，还有多种野生动物。

Lassen Volcanic National Park is a United States National Park in northeastern California. The dominant feature of the park is Lassen Peak; the largest plug dome(岩颈丘)volcano in the world and the southern-most volcano in the Cascade Range(喀斯喀特山脉). Lassen Volcanic National Park started as two separate national monuments designated by President Theodore Roosevelt in 1907: Cinder Cone National Monument and Lassen Peak National Monument.

The source of heat for volcanism in the Lassen area is subduction(俯冲)off the Northern California coast of the Gorda Plate diving below the North American Plate. The area surrounding Lassen Peak is still active with boiling mud pots(泥温泉), stinking(发恶臭的)fumaroles(喷气孔), and churning(搅拌)hot springs. Lassen Volcanic National Park is one of the few areas in the world where all four types of volcano can be found (plug dome(岩颈丘), shield, cinder cone, and strato).

The park is accessible via State Routes SR 89 and SR 44. SR 89 passes north-south through the park, beginning at SR 36 to the south and ending at SR 44 to the north. SR 89 passes immediately adjacent(邻近的) the base of Lassen Peak.

There are a total of five vehicle entrances to the park: the north and south entrances of SR 89, and unpaved roads entering at Drakesbad and Juniper Lake in the south, and Butte Lake in the northeast. The Park can also be accessed by trails leading in from Caribou Wilderness to the east, as well as the Pacific Crest Trail, and two smaller trails leading in from Willow Lake and Little Willow Lake to the south.

A large lodge (the Lassen Chalet) with concession facilities formerly was located near the south-west entrance, but was demolished in 2005. A new, full-service(全方位服务) visitor center was constructed in the same location, and opened to the public in 2008. Near the old lodge location was also located Lassen Ski Area, which ceased operation in 1992; all infrastructure(基础设施) has been removed.

Native Americans have inhabited the area since long before white settlers first saw Lassen. The natives knew that the peak was full of fire and water and thought that it would one day blow itself apart.

White immigrants in the mid-19th century used Lassen Peak as a landmark on their trek(艰苦跋涉) to the fertile Sacramento Valley. One of the guides to these immigrants was a Danish blacksmith named Peter Lassen, who settled in Northern California in the 1830s. Lassen Peak was named after him. Nobles Emigrant Trail was later cut through the park area and passed Cinder Cone and the Fantastic Lava Beds.

As late as 1859 a witness reported seeing fire in the sky from a distance, attributing it to an eruption. Early geologists and volcanologists(火山学家) who studied the Cinder Cone concluded the last eruption occurred between 1675 and 1700. After the 1980 eruption of Mount St. Helens, the United States Geological Survey (USGS) began reassessing the potential risk of other active volcanic areas in the Cascade Range. Further study of Cinder Cone estimated the last eruption occurred between 1630 and 1670.

Starting in May 1914 and lasting until 1921, a series of minor to major eruptions occurred on Lassen. These events created a new crater, and releasedlava and a great deal of ash. Fortunately, because of warnings, no one was killed, but several houses along area creeks were destroyed. Because of the eruptive activity, which continued through 1917, and the area's stark volcanic beauty, Lassen Peak, Cinder Cone and the area surrounding were declared a National Park on August 9, 1916.

The 29 – mile (47 km) Main Park Road was constructed between 1925 and 1931, just 10 years after Lassen Peak erupted. Near Lassen Peak the road reaches 8,512 feet (2,594 m), making it the highest road in the Cascade Mountains. It is not unusual for 40 ft (12 m) of snow to accumulate on the road near Lake Helen and for patches(块)of snow to last into July.

In October 1972, a portion of the park was designated as Lassen Volcanic Wilderness by theUS Congress (Public Law 92 – 511). The National Park Service seeks to manage the wilderness in keeping with the Wilderness Act of 1964, with minimal developed facilities, signage (引导标示), and trails. The management plan of 2003 adds that, "The wilderness experience offers a moderate to high degree of challenge and adventure."

In 1974 the National Park Service took the advice of the USGS and closed the visitor center and accommodations(住宿)at Manzanita Lake. The Survey stated that these buildings would be in the way of a rockslide(崩落的岩石)from Chaos Crags if an earthquake or eruption occurred in the area. An aging seismograph(地震仪)station remains. However, a campground, store, and museum dedicated to Benjamin F. Loomis stands near Manzanita Lake, welcoming visitors who enter the park from the northwest entrance.

After the Mount St. Helens eruption, the USGS intensified its monitoring of active and potentially active volcanoes in the Cascade Range. Monitoring of the Lassen area includes periodic measurements of ground deformation (变形) and volcanic-gas emissions and continuous transmission of data from a local network of nine seismometers to USGS offices in Menlo Park, California. Should indications of a significant increase in volcanic activity be detected, the USGS will immediately deploy scientists and specially designed portable monitoring instruments to evaluate the threat. In addition, the National Park Service (NPS) has developed an emergency response plan that would be activated to protect the public in the event of an impending(即将发生的)eruption.

Notes

California 加利福尼亚州　通常简称为加州，是美国西部太平洋岸边的一个州，在面积上是全美第三大州，人口上是全美第一大州。加利福尼亚无论是在地理、地貌、物产还是人口构成上都十分多样化。由于早年的淘金热，加州有一个别名叫做金州（The Golden State），邮政缩写是 CA。面积 41.1 万平方千米。人口 3387 万（2000 年），城市人口占 91.3%，主要集中在太平洋沿岸地区，其中 50% 聚居在洛杉矶和圣弗朗西斯科（旧金山）一带。拥有全国 50% 以上的华裔和墨西哥裔美国人（契卡诺人）。圣弗朗西斯科的沙加缅度是州政府所在地。唐人街是华人最密集的居住区。首府是萨克拉门托（又名沙加缅度）。

Lassen Peak 拉森峰　劫后余生的拉森峰曾经是一座名为泰哈马山体（Mount Tehama）的一部分，它是在 60 万年前由地下深层涌出的熔岩（岩浆）开始形成的。经过漫长的岁月，熔岩逐渐形成了高达 3 353 米的山体，底部山体宽度超过 18 千米。最后，这座巨型火山逐渐崩塌，由此在其半径内诞生了许多较小的山体。拉森峰就是其中之一。

Cascade Range 喀斯喀特山脉　北美洲环太平洋海岸山脉的一部分。位于美国西部。南接内华达山脉，北与加拿大海岸山脉相连，从美国加利福尼亚州向北经俄勒冈州和华盛顿州延伸到加拿大，长 1 127 千米。海拔 1 800 ~ 2 500 米。山体自南向北逐渐增高。形成于中生代内华达造山运动，大部分由喷出岩组成。北段地势最高，主要由古生代花岗岩和较老的变质岩构成，表现为一处切割的高地或高原，顶部有花岗岩蚀余山；中段和南段均覆盖第三纪熔岩，尤以南段广泛而深厚。山脉第四纪时有大量火山喷发，形成许多火山堆，以南段最密集，多为死火山。部分还在活动中，如圣海伦斯火山，海拔多在 3 000 米以上。雷尼尔山海拔 4 392 米，为山脉最高峰。山区有森林、水力、旅游资源，建有多处公园。

Cinder cone 火山渣锥　玄武岩碎片堆积而成的山丘，喷出气体携带熔岩滴进入大气，然后在火山口附近降落，形成火山锥。熔岩滴通常在飞行中间，降落地面之前已经是固态的或部分固态的，称之为火山弹。如果气体压力下降，最后阶段的可能是熔岩流冲出渣锥的底部，如果熔融岩浆中具有充裕的水，它们的相互作用将形成玛珥湖（低平火山口）而不是火山渣锥。喷发时间越长，火山锥就越高。一些火山锥只有几米高，而一些如墨西哥的帕里库廷火山（Paricutin），从 1943 至 1952 年连续喷发，火山锥高达 610 米，随着火山碎屑活动，熔岩流从其底部流出，摧毁了帕里库廷村。火山渣锥可以单独存在，也可以组成小的或大的群，或火山场。

Sacramento 萨克拉门托　加利福尼亚州首府。位于州中部、萨克拉门托河下游，西南距旧金山 137 千米。人口 36.9 万（1990），大市区由萨克拉门托等县组成，面积 2 582 平方千米。1839 年为瑞典移民定居点；1848 年其东北 56 千米处发现金矿，成为淘金者的物资供应中心。1854 年成为州府；1863 年设市；1869 年铺设北美第一条横贯大陆的铁路。金矿枯竭后，它作为农产品贸易中心而继续繁荣。第二次世界大战后，工业趋向多样化，是谷地农产品集散和贸易中心。食品加工是最大工业部门。第二次世界大战后建立的导弹和火箭发动机制造业也占重要地位，此外还有运输设备、木树加工、化学等工业。它是纵横加利福尼亚的高速公路和铁路的交叉点，重要河港，有深水航道，并与圣弗朗西斯科湾诸港相连，市中心在萨克拉门托河西岸；商业集中在市区，工业主要分布于城郊。市内有加利福尼亚大学、克罗克美术馆、加利福尼亚州立图书馆和州立萨特堡历史公园等文化设施。近郊有空军基地。

Peter Lassen 彼得 · 拉森　丹麦裔美国人（1800 年 10 月 31 日 – 1859 年 4 月 26 日）铁匠，牧场主，勘探者，互济会成员。拉森火山以他的名字而命名。

United States Geological Survey 美国地质勘探局　美国政府内政部的一个下属机构。此机构的科学家主要研究美国的地形、自然资源和自然灾害与应付方法（如地震、火山喷发、泥石流等）。该部门负责四大科学范畴：物理学、地理学、地质学和水文学。

Menlo Park 门洛帕克　美国加利福尼亚州圣马刁县东南部的一座城市。根据美国人口调查局2000年的人口普查，该市共有30 787人。值得一提的是，“加利福尼亚州门洛帕克”的名称最早见于汤玛斯·爱迪生在新泽西州门洛帕克所创的发明品。

Translation of Selected Sentences

1. Because of the eruptive activity, which continued through 1917, and the area's stark volcanic beauty, Lassen Peak, Cinder Cone and the area surrounding were declared a National Park on August 9, 1916.

因为火山喷发活动一直持续到1917年，这个地区突显的火山美景，拉森峰，火山渣锥和地区周围的环境于1916年8月9日被纳入国家公园。

2. The National Park Service seeks to manage the wilderness in keeping with the *Wilderness Act* of 1964, with minimal developed facilities, signage, and trails.

美国国家公园管理局努力管理这片荒野，使用最小的发展设施，引导标示和游径，使之符合1964年的《荒野保护法案》。

3. The Survey stated that these buildings would be in the way of a rockslide from Chaos Crags if an earthquake or eruption occurred in the area.

调查显示如果在这片地区内发生地震或者火山爆发，那么这些建筑就会挡在从卡尔斯峭壁上崩落的岩石中间。

4. However, a campground, store, and museum dedicated to Benjamin F. Loomis stands near Manzanita Lake, welcoming visitors who enter the park from the northwest entrance.

然而，在蒙撒尼莎河周围的露营地、商店、博物馆和本杰明·卢米斯雕像一起欢迎从西北入口进来参观公园的游客。

5. Monitoring of the Lassen area includes periodic measurements of ground deformation and volcanic-gas emissions and continuous transmission of data from a local network of nine seismometers to USGS offices in Menlo Park, California.

拉森地区的监管包括定期的地面变形测量，火山喷发气体排放和从当地9个地震检查器网络持续传送到位于加利福尼亚州门罗帕克的美国地震勘探局的数据。

6. In addition, the National Park Service (NPS) has developed an emergency response plan that would be activated to protect the public in the event of an impending eruption.

另外，美国国家公园管理局已经展开了紧急响应的规划，可以积极地投入到保护公众以应对以后可能会发生的火山喷发事件。

Questions

1. Why is the volcano called Lassen Peak?
2. What is the general condition of Lassen Volcanic National Park?
3. Who have inhabited the area since white settlers first saw Lassen?
4. When was Lassen area first protected?
5. What did the National Park Service do in 1974?

Chapter 8

North Cascades National Park
北瀑布国家公园

【简介】

北瀑布国家公园(一译北卡斯卡德山国家公园)位于美国华盛顿州北部。1919 年 2 月 26 日国会批准其升级为国家公园,占地面积 2 768 平方千米。其地形包括山地、海岸线、丛林以及少数淡水湖泊。公园里山峦连绵起伏,葱郁浓密,有迪亚波罗(Diablo)和罗斯(Ross)两个大湖,青山绿水相映衬,煞是好看。不过山路崎岖,转弯又多又急,常常要紧盯着路标才不至于迷路。

North Cascades National Park is a U. S. National Park located in the state of Washington. The park is the largest of the three National Park Service units that comprise the North Cascades National Park Service Complex. Several national wilderness areas and British Columbia parkland (公用场地) adjoin the National Park. The park features rugged(崎岖的) mountain peaks.

North Cascades National Park came into being, officially, on 02 Oct 1968, when Lyndon B. Johnson signed the North Cascades Act. This act created the North Cascades National Park Service Complex which comprised of 684,000 acres of wild land. It also included the park's north and south units, as well as Ross Lake and Lake Chelan national recreation areas. This same act created the adjacent Pasayten Wilderness of 550,000 acres and enlarged the Glacier Peak Wilderness to 464,000 acres. In 1988 Congress designated approximately 93% of the three areas as the Stephen Mather Wilderness to provide additional legislated protection. The Cascades were among the world's greatest mountain ranges. Extending from Canada's Fraser River south beyond Oregon, they contribute greatly to shaping the Pacific Northwest's climate and vegetation(植被).

Fur traders, traveling on foot and by canoe(独木舟), were among the first Euro-Americans to venture into the North Cascades wilderness in the late 1 700 s.

Many of the early settlers trapped to supplement(补充) their income. Trapping was primarily a winter activity, the most difficult season to be afield(去野外) in the mountains. The Weaver brothers came to Stehekin primarily to trap animals for a living. They were so successful that they opened a taxidermy business across the river at what is now called Weaver Point. John McMillan, a miner, ran traplines along Big Beaver Creek and the upper Skagit River in the late nineteenth century. Beaver, bear, cougar, wolf, lynx(山猫), fisher, marten and fox were all sought by trappers in the North Cascades.

The earliest recorded crossing of the North Cascades by a Euro-American occurred in 1814. Alexander Ross, a fur trader, crossed Twisp Pass and descended Bridge Creek to the Stehekin River, which he then followed upstream. Finally crossing Cascade Pass, he traced the Cascade River downstream to its confluence with the Skagit River. Maps of Washington Territory in 1860 show large areas still labeled "unexplored". The handful of explorers who followed Ross, also commented on the region's rugged, isolated nature.

Miners prospected for(勘探) gold, lead, zinc(锌), and platinum(铂) here from 1880 to 1910. They recorded moderate strikes, but transportation proved to be arduous and profits so limited that mining was abandoned. Some logging and homesteading occurred around 1900. The electricity generating potential of the Skagit River was early recognized. Between 1924 and 1961, Seattle City Light built three dams on the river. Mountains do not stop at the park boundaries. The three areas are flanked on the south, east and west by national forest lands and on the north by provincial lands of British Columbia, Canada. The national forest lands encompass a number of outstanding federal wilderness areas, including the Glacier Peak Wilderness on the Baker-Snoquaimie and Wenatchee National Forests. Only an invisible boundary separates the first two national park units from the two national recreation areas and the adjoining national forest lands.

Evidence of Indian use of the Cascades is widespread, but little is know about it. History has touched little of this area. Today, readily reached areas are heavily visited, but some remote locations have yet to feel the boots of today's backcountry(偏远地区) traveler. Forest giants of western red-cedar and Douglas-fir dot the deep valleys. Off the trail, tangled growth's of alder, vine maple, stinging nettles(荨麻), and devil's club still defy crosscountry(横贯全国的) hikers. Glaciers scored by crevasses, permanent snowfields, sheer-walled cliffs, spires and pinnacles challenge mountaineers. From North Cascades Highway, on clear days, you may get glimpses of alpine wonders that lie just beyond.

The Cascade Mountain range runs 500 miles from Northern California to British Columbia, but it is not until it reaches Northwest Washington that the mountains are at their most breathtaking. The Cascades are higher in other parts of the range, but nowhere are they as dramatic. Jagged, rocky peaks of up to 10,000 feet give way to near-sea level valleys; glaciers cling dizzyingly to the sides of foreboding slopes; everywhere waterfalls tumble down from the mountains, the characteristic that gave the Cascades their name.

Many millions of years ago, before it bumped into North America, this region was an itinerant(流动的)land mass drifting in the Pacific Ocean. The mountain-building forces at work before and since that unification—accumulation of sediment from pre-historic seas, colliding tectonic plates and volcanic activity—have combined to create one of the fastest growing mountain ranges in the world. Indeed, the North Cascades would be even taller if the counteracting forces of water and glaciers did not conspire to keep the mountains at more modest heights.

Still, the elevational distance from valleys to summits throughout the North Cascades can exceed 5,000 feet—a relief as great as any other range in the United States. The steep and imposing(壮观)North Cascades presented a formidable barrier to early white explorers and the names they gave some of these mountains betray their dread: Mt Terror, Mt Challenger, Mt Fury, Mt Despair, Mt Torment, Desolation Peak.

Water is the essence of the North Cascades. Small trickles(细流)and rivulets(小溪)tumble from glaciers and snowfields to join together as rushing creeks. These tributary(支流)streams merge into the mighty rivers of the western slope. The east side of the range receives less rainfall and therefore has fewer glaciers, creeks(小河)and rivers.

Hundreds of small lakes, isolated jewels accessible only by arduous cross-country hiking, are scattered throughout the North Cascades. Many of these lakes are tarns, remnants of the alpine glaciers that once covered the North Cascades. Mountain lakes are rich reservoirs of life surrounded by marshes and meadows. Voracious dragonfly nymphs are common in streams and lakes, as are caddisfly and mayfly larvae. Rainbow and cutthroat trout have been introduced into many high lakes, where they feed upon abundant aquatic(水上的)insects.

The larger lakes, including Ross Lake and Lake Chelan, host breeding populations of osprey (鱼鹰)and other fish-eating birds. Common mergansers, large diving ducks, are frequently seen along rivers, as are great blue herons(大蓝鹭)and spotted sandpipers. Osprey fish the rivers and lakes. They dive from hundreds of feet above the water, then rise with a struggling trout clutched tightly in their talons.

Hiking trails wind through most of the major river valleys. Travel off-trail is difficult at best and nearly impossible in many places. River valleys are nearly impenetrable jungles of vegetation. Creeks and streambank habitats are choked with armies of inhospitable brush, including spiny devil's club, prickly(多刺的)currant(红醋栗), salmonberry(美洲大树莓), thimbleberry and willow. There are tales of mountaineers who spent days trying to bushwhack cross-country out of the Picket Range—crawling through brush that yielded less than a mile's passage for each day's effort.

Lowland river valleys contain magnificent forests dominated by western hemlock, western red cedar and Douglas-fir trees. While some forests on public and private land are managed for timber harvest and some are managed for multiple uses, others are designated wilderness areas. Old-growth stands (also called ancient forests) on public and private lands in the North Cascades are those areas of virgin, uncut forest that contain large, old trees together with trees of different ages. The diversity of age classes in an old-growth forest provides a rich understory and canopy of vegetation. The different levels of canopy allow sunlight to touch the forest floor in places. Old-growth forests also contain snags—standing dead trees—and "nurse" logs that provide shelter for a variety of plant and animal life. Eventually, the snags and logs decompose(分解)and become new soil. Some of the last extensive stands of old-growth forest left in the United States are the lowland forests of the Cascade Mountains.

The greatest concentration of old-growth western red cedar in the North Cascades is found along the lower reaches of Big Beaver Creek below 2,000 feet. Western red cedars are a long-lived species; the oldest are estimated to be over 1 000 years old. Red cedars grow in moist habitats. Thus, they are ideally suited to the North Cascades, where summers are cloudy and winters are mild. Tapering from heavily buttressed bases often over 10 feet in diameter, they rise 200 feet above the forest floor, balanced on a shallow, wide-spreading root system that provides stability in the wet valley soils. They are distinguished by their stringy, fibrous bark, which peels away in thin strips. Lacy sprays of flat needles give western red cedars a delicate appearance despite their great size. Native peoples living on the west side of the Cascades and along the coast used all parts of the "tree of life", making clothing and blankets with the inner bark and totems (图腾), dugouts, cooking utensils and lodging with the wood.

Another dominant tree of the river valleys is western hemlock. Both western red cedar and western hemlock are shade-tolerant; their seedlings thrive in the dense shade of lowland forests. Western hemlocks in an undisturbed area eventually out-compete less shade tolerant trees like Douglas-fir to form the climax community of the low-elevation forests of the North Cascades. Growing in scattered, open glades throughout the hemlock forest are red alder and bigleaf maple (大叶枫). These two deciduous trees must have some scattered sunlight for their seedlings to grow. They appear along trails and streams and in openings created when a forest giant falls during a windstorm. Red alder is an important pioneer species that adds nitrogen to the soil through nodules of symbiotic bacteria (mycorrhizae) on its roots. Bigleaf maple is easy to identify with its giant five-lobed(有叶的)leaves spanning 16 inches or more.

When walking along a forest trail, occasionally climbing over fallen trees and ducking under low-hanging branches, we see a spectrum of the many shades of green. Huckleberries tantalize the hiker along the trail. In drier locations, salal and Oregon grape, both evergreen shrubs, dominate the understory, while sword fern and its many relatives grow throughout the forest. In springtime the forest floor is thick with wild flowers. Trillium, bleeding heart, yellow violet, calypso orchid(兰花), twin flower and arrow-leaf balsam root display their delicate colors and brighten the forest shade. Mosses, lichens(地衣)and fungi(蘑菇)grow thickly over and under the ground and on exposed rocks and fallen trees.

As dusk settles over the forest, they come, navigating by sonar, silent flyers on leathery wings. Bats: creatures of awe and mystery. There are over 40 species of bats in North America but we know little about them because they are nocturnal(夜间的). Bats come out when people go in. These small mammals eat more insects by body weight than any other predator, snatching their prey on the wing. Bats roost upside-down in large groups, sleeping all day and hibernating (过冬) all winter. Some large species native to the Cascades migrate south in the coldest months. Mother bats nurse their young constantly in the first few weeks after birth, when they seldom leave their upside-down perch inside tree cavities or caves, even to feed. Old-growth forest, with snags, tree cavities and loose bark offer important roosting and nesting habitat(栖息地)for bats in the Cascades. Bats have extremely sensitive hearing. They use echolocation(回声定位)to locate prey and navigate, changing the speed and focus of their high frequency "blips" for precise aerial hunting.

Ten thousand years ago, beavers the size of black bears roamed the continent, and today beavers are still the largest rodent(啮齿的)in North America. Beaver sightings are rare, but evidence of their handiwork is common throughout the Cascades: beaver dams, ponds, mounds, and beaver-gnawed trees. Beavers need only hear the sound of running water and the urge to build a dam takes control. Poles, saplings, and mud are dragged and piled into place to form a very efficient structure. Smaller mounds, or lodges, are constructed in the pond as living quarters. Hollow inside, with the entrance underwater, these lodges provide a safe haven from predators. In areas where lodges aren't feasible, beavers construct burrows in river banks. After their pond freezes in winter, the beavers let some water out through the dam, lowering the water level of the pond and providing a "breathing space" between the water's surface and the bottom of the ice. Then they go happily about their business: feeding on poles stashed(躲藏)underwater in the summer, grooming, playing, and sleeping. Hunted and trapped almost to extinction for their pelts, beavers have made a comeback. Persistent and adaptable, they can even be found trying to construct dams near central Seattle and Portland.

The flying squirrel(鼯鼠) is seldom seen because of its crepuscular(黄昏的) life style (active in the hours before dawn and after dusk), but it is quite common. Distinguished by huge eyes and velvet-soft fur, this fragile squirrel doesn't really fly, but glides gracefully through the forest, stretching broad flaps of skin between its front and rear legs. In old-growth forests flying squirrels seek out old woodpecker cavities for nesting. Here the babies are safe until their eyes open at one month of age. Unlike most members of the rodent family, these squirrels like meat, but their main food in summer and fall is truffles. Truffles are underground fungi that provide vital nutrients to coniferous trees through a symbiotic relationship. Flying squirrels have an important role in the ecosystem, that of spreading truffle spores throughout the forests. On quiet evenings, you may hear a soft chirp(唧唧声) and the thump of a landing as they go about their nightly business of finding food.

Notes

Washington 华盛顿州　位于美国西北太平洋沿岸的一个州,北接加拿大不列颠哥伦比亚省,南接俄勒冈州,东临爱达荷州,西邻太平洋。为纪念美国首任总统,乔治·华盛顿而成立,也是唯一以总统名称命名的州。于 1889 年 11 月 11 日加入联邦成为美国的第 42 个州。

Fraser River 弗雷泽河　北美洲西部的一条大河,发源于加拿大不列颠哥伦比亚省东部洛矶山脉耶洛黑德(Yellowhead)山口附近海拔 1 109 米的耶洛黑德湖,干流先向北缓慢流过底部平坦的洛矶山谷,然后转向西北,至乔治王子(Prince George)城再转向南,流经不列颠哥伦比亚省中部的高原和山谷。

North America 北亚美利加洲　简称北美洲,位于西半球北部。东濒大西洋,西临太平洋,北濒北冰洋,南以巴拿马运河为界与南美洲相分隔。面积为 2 422.8 万平方千米(包括附近岛屿),约占世界陆地总面积的 16.2%,是世界第三大洲。通用英语、西班牙语,其次是法语、荷兰语、印第安语等。

Western Hemlock 异叶铁杉　松科铁杉属,为常绿针叶乔木。通常生长在美国阿拉斯加的基奈半岛(Kenai Peninusla)至加州西北部的太平洋沿岸混合林段中,也生长在内陆地区,是重要的经济林带的木材,木材是优于其他建筑用铁杉,这种树木在针叶木中较为坚硬,是强度较高的树种。它是架构和建筑用材,主要用于嵌条、木工成品和镶板。

Western Red Cedar 红雪松　又称西部红柏,是北美西北部太平洋沿岸,特别是加拿大不列颠哥伦比亚省(BC 省)内陆雨林中的常见树种。

Douglas-fir 花旗松　常绿大乔木，高达100米，胸径达12米。树皮厚，深裂成鳞状。叶条形。长1.5～3厘米，先端钝或微尖，上面深绿色，下面色较淡，有两条灰绿色气孔带。球果呈椭圆状卵圆形，长约8厘米，褐色，有光泽；种鳞斜方形或近菱形；苞鳞长于种鳞，中裂片窄长渐尖，两侧裂片较宽而短。树皮呈金钱豹花纹状，心材淡红色，边材淡黄色而有树脂，材质坚韧，富有弹力，保存期长，是良好的建筑及器具用材。分布范围东西自洛矶山脉至太平洋沿岸，南北自墨西哥至加拿大哥伦比亚省，纽西兰、英国、澳洲等地则有栽植。一般木材可分为产自奥瑞冈州、华盛顿州及加州沿岸，为海岸型(Coast Type)，产自洛矶山脉各州者为洛矶山脉型(Rocky Mountain Type)或内陆型(Interior Type)。

Translation of Selected Sentences

1. The three areas are flanked on the south, east and west by national forest lands and on the north by provincial lands of British Columbia, Canada.

这三个地区的南侧、东侧和西侧都被国家林地包围着，北侧和加拿大不列颠哥伦比亚省接壤。

2. Only aninvisible boundary separates the first two national park units from the two national recreation areas and the adjoining national forest lands.

只有一条无形的分界线将原先这两个国家公园从两个国家娱乐区域和相邻的国家林地分离出来。

3. The Cascade Mountain range runs 500 miles from Northern California to British Columbia, but it is not until it reaches Northwest Washington that the mountains are at their most breathtaking.

瀑布山脉从北加利福尼亚州到不列颠哥伦比亚省绵延了500英里，但是直到华盛顿西北部时，山峰才到达最高点。

4. Jagged, rocky peaks of up to 10,000 feet give way to near-sea level valleys; glaciers cling dizzyingly to the sides of foreboding slopes; everywhere waterfalls tumble down from the mountains, the characteristic that gave the Cascades their name.

最高可达1万英尺的参差不齐的岩石山峰让路于接近海平面的山谷；冰川令人晕眩地紧贴着令人毛骨悚然的斜坡的侧面；四周的瀑布从山上悬垂下来，正是这些特点形成了现在我们熟知的国家瀑布。

5. Voracious dragonfly nymphs are common in streams and lakes, as are caddisfly and mayfly larvae. Rainbow and cutthroat trout have been introduced into many high lakes, where they feed upon abundant aquatic insects.

贪婪的蜻蜓在溪流和湖泊随处可见，石娥、蜉蝣、幼虫也一样。彩虹鱼和切喉鳟生长在许多高压地带的湖泊里，以捕捉大量的水面上的昆虫为生。

6. Tapering from heavily buttressed bases often over 10 feet in diameter, they rise 200 feet above the forest floor, balanced on a shallow, wide-spreading root system that provides stability in the wet valley soils.

拱壁根基通常直径可达 10 英尺多，从下往上逐渐变细，可达 200 英尺高，与森林地面相平，在浅浅的但延展很广的，又可以在湿润的山谷土壤里提高稳定性的根系上保持平衡。

7. Distinguished by huge eyes and velvet-soft fur, this fragile squirrel doesn't really fly, but glides gracefully through the forest, stretching broad flaps of skin between its front and rear legs.

以超大的眼睛和天鹅绒般的柔软皮毛为特征，这只脆弱的松鼠并不是真的会飞，但因其在前腿和后退之间有着可以延展的宽大的皮毛，从而可以优雅地在森林中滑翔。

Questions

1. How was the North Cascades National Park built?
2. Describe the main features of the North Cascades National Park.
3. What special plants does the North Cascades National Park have?
4. What kind of animals does the North Cascades National Park have?
5. How about the water landscape in North Cascades National Park?

Chapter 9

Death Valley National Park
死谷国家公园

【简介】

美国死谷国家公园主要位于美国加利福尼亚州东南部，一小角延伸入内华达州的西部。1933 年建成。谷底原有的大湖泊由于缺乏水分、干涸后成为沙漠。公园里有盐碱地、沙丘、火山口、峡谷、雪山等，有丰富的地质地貌，还有地壳里各式各样的断层、峡谷往荒漠注入时形成大大小小的冲积扇平原等。1873 年这里发现硼砂矿，此后又发现铜、金、银、铝等矿，采矿人在此兴建过一些城镇。有少量耐盐碱植物生长在沼泽地，还有野兔、狐、山狼和沙漠大角羊等动物。

Death Valley National Park is a national park located east of the Sierra Nevada in the arid (干燥的) Great Basin of the United States. Parts of the park are in southern Inyo County and northern San Bernardino County in Eastern California, with a small extension into southwestern Nye County and extreme southern Esmeralda County in Nevada. In addition, there is an exclave (Devil's Hole) in southern Nye County. The park encompasses Saline Valley, a large part of Panamint Valley, almost all of Death Valley, and parts of several mountain ranges. Death Valley National Monument was declared a U. S. National Monument in 1933, placing the area under federal protection. In 1994, the monument was redesignated (重新指定) a national park, as well as being substantially (相当大的) expanded to include Saline and Eureka valleys.

It is the hottest and driest of the national parks in the United States. The second-lowest point in the Western Hemisphere is in Badwater Basin, which is 282 feet (86 m) below sea level. The park is home to many species of plants and animals that have adapted to this harsh desert environment. Some examples include creosote bush(木焦油灌木), Bighorn Sheep(大角羊), Coyote(北美小狼), and the Death Valley Pupfish(鳉鱼), a survivor of much wetter times. Approximately 95% of the park is designated as wilderness. Its wilderness area covers 4,774 square miles ($12,360\ km^2$), making it the largest in the Lower 48 states, and the sixth largest in the United States overall. Death Valley National Park is visited annually by more than 770,000 visitors who come to see its diverse geologic features, desert wildlife, historic sites, scenery, and clear night skies.

Mining was the primary activity in the area before it was protected. The first documented non-Native Americans to enter Death Valley did so in the winter of 1849, thinking they would save time by taking a shortcut to the gold fields of California. They were stuck for weeks and in the process gave the valley its name, even though only one of their groups died there. Several short-lived boom towns sprang up during the late 19th and early 20th centuries to exploit minor local bonanzas(富矿带) of gold. The only long-term profitable ore(矿) to be mined, however, was borax(硼砂), a mineral used to make soap and an important industrial compound. Today, borax is an essential component of high-temperature resistant boro-silicate(硼硅酸盐) glass products, for example Pyrex cookware(炊具). Twenty-mule teams were used to transport ore out of the valley; helping to make it famous and the subject of books, radio programs, television series, and movies.

The natural environment of the area has been shaped largely by its geology. The valley itself is actually a graben(地堑). The oldest rocks are extensively metamorphosed and at least 1.7 billion years old. Ancient warm, shallow seas deposited(浅海沉积) marine sediments(海洋沉积物) until rifting(裂谷作用) opened the Pacific Ocean. Additional sedimentation occurred until a subduction zone(俯冲带) formed off the coast. This uplifted the region out of the sea and created a line of volcanoes. Later the crust started to pull apart, creating the current Basin and Range landform. Valleys filled with sediment and, during the wet times of glacial periods(冰河时期), with lakes, such as Lake Manly.

Death Valley's outstanding natural beauty and scientific importance were first brought to the attention of the National Park Service in the late 20th. With the support of Stephen T. Mather, Director of the National Park Service, Death Valley's national significance was recognized, and it was proclaimed a national monument by President Hoover on February 11, 1933. With the passage of the Desert Protection Act of October 31, 1994, Death Valley grew by 1,200,000 acres and was designated a national park. Today Death Valley National Park is made up of 3,336,000 acres and contains more than 3,000,000 acres of wilderness.

Death Valley, the name is foreboding(预兆) and gloomy. Yet here in this valley, much of it below sea level, or in its surrounding mountains you can find spectacular wildflower displays, snow covered peaks, beautiful sand dunes(沙丘), abandoned mines and industrial structures, and the hottest place in North America.

Will the flowers be good this year?

Where can we go to see them?

What kinds/colors will there be?

When will the peak bloom be?

These questions are often asked by people planning spring visits to Death Valley. Although there are many variables involved in the desert wildflower shows, there are a few guidelines you can use to find answers to these questions.

Some years the desert is spectacular with wildflowers; other years the blossoms are almost nonexistent (but never totally absent). A good wildflower year depends on at least three things:

Well-spaced(间隔合适的)rainfall throughout the winter and early spring.

There are over 1000 plant species in Death Valley National Park, including 13 species of cactus(仙人掌)and 23 endemics(地方性的)(plants that are known to grow only in the Death Valley region). Most of the "showy"(艳丽的)plants are desert annuals, also referred to as ephemerals(短暂的)(shortlived). Colors range from white and yellow to purple, blue, red and bright magenta(品红).

The best time to see a spring floral display is in years when rainfall has been several times the Death Valley annual average of about 1.9 inches. In general, heavy rains in late October with no more rain through the winter months will not bring out the flowers as well as rains that are evenly-spaced throughout the winter and into the spring.

G. K. Gilbert, a geologist who worked in the area in the 1870's, noted that the rock formations were "beautifully delineated(描绘)on the slopes of the distant mountains, revealing at a glance relations that in a fertile country would appear only as the results of extended and laborious investigation". The rock layers that Gilbert noticed comprise a nearly complete record of the earth's past, but that record has been jumbled(搞乱)out of sequence. The reason is that the rock layers that form the mountains are very ancient, but only in recent geologic time have they risen.

Even as the mountains rose, erosion began to wear them down. An example of this is the formation of the alluvial fans(冲积扇). Intermittent(间歇的)streams, resulting mostly from the bursts of infrequent rains, rush down the steep canyons. It scours boulders, soil and other debris (碎片)and pushes and carries the whole mass with it. It then deposits it on the valley floor at the canyon's mouth.

The first documented(备有证明文件的) tourist facilities in Death Valley were a set of tent houses built in the 1920s where Stovepipe Wells is now located. People flocked to resorts built around natural springs thought to have curative(有治疗的) and restorative(滋补的) properties (性能). In 1927, Pacific Coast Borax turned the crew quarters of its Furnace Creek Ranch into a resort, creating the Furnace Creek Inn and resort. The spring at Furnace Creek was harnessed to develop the resort, and as the water was diverted, the surrounding marshes and wetlands started to shrink.

Soon the valley was a popular winter destination. Other facilities started off as private getaways but were later opened to the public. Most notable among these was Death Valley Ranch, better known as Scotty's Castle. This large ranch home built in the Spanish Revival style became a hotel in the late 1930s and, largely because of the fame of Death Valley Scotty, a tourist attraction(观光胜地). Death Valley Scotty, whose real name was Walter Scott, was a gold miner (淘金者) who pretended to be owner of "his castle", which he claimed to have built with profits from his gold mine. Neither claim was true, but the real owner, Chicago millionaire Albert Mussey Johnson, encouraged the myth. When asked by reporters what his connection was to Walter Scott's castle, Johnson replied that he was Mr. Scott's banker.

Death Valley National Park is a popular location for stargazing(眺望星星) as it has one of the darkest night skies in the United States. Despite Death Valley's remote location, its air quality and night visibility are threatened by civilization. In particular, light pollution is introduced by nearby Las Vegas. The darkest skies are, in general, located in the northwest of the park.

Notes

Sierra Nevada 内华达山脉　美国西南部花岗岩断块山,科迪勒拉山系西缘山地的组成部分。北起拉森峰,南至蒂哈查皮山口,长 640 千米,宽 80 ~ 130 千米,以大致南北走向绵亘于加利福尼亚州东部。山体形成于侏罗纪末至白垩纪初的内华达造山运动,一般海拔 1 800 ~ 3 000 米,山势自东向西倾斜;东坡断崖陡峻,拔起于东邻大盆地之上,高差达 1 500 ~ 3 000 米,植被稀疏,多为灌木草类;西坡平缓倾向加利福尼亚中央谷地,多河流切割成的深邃峡谷,森林茂密,迎太平洋湿润气流而雨雪丰沛,河流、湖泊众多,水力资源丰富,使中央谷地成为富饶的农业区。

Great Basin 大盆地　美国西部科迪勒拉山系中的高原性内陆盆地。西为内华达山脉和喀斯喀特山脉,东为瓦萨奇山脉,北面是哥伦比亚高原,南边为科罗拉多高原。由一系列南北走向的块状山脉(海拔 1 800 ~ 3 300 米)及其间许多盆地构成。盆地底部海拔 1 200 ~ 1 500 米,最低的死谷在海平面下 85 米。气候干燥,年降水量 100 ~ 350 毫米,植物稀少,以耐旱灌木为主。多咸水湖。大的湖泊有大盐湖、皮拉米德湖、塞维尔湖、犹他湖等。有丰富的铜、金、铅、锌、铁等矿藏。山麓地区有畜牧和灌溉农业。

San Bernardino County 圣贝纳迪诺县　美国加利福尼亚州东南部的一个县，东邻内华达州和亚利桑那州，东界科罗拉多河。面积 52 073 平方千米，是全国面积最大的县。根据美国 2000 年人口普查数字，共有人口 1 709 434 人，2005 年人口为 1 932 000 人。县治圣贝纳迪诺（San Bernardino）。成立于 1853 年。县名是为纪念天主教圣人锡耶纳的圣伯纳丁（Saint Bernardino of Siena）。

Nevada 内华达州　美国西南部内陆州，又称北美艾灌木丛州，以其发达的赌博业而著称。“内华达”来源于西班牙语，意思是“被雪覆盖”，原本是用来形容该州西部的内华达山的寒冷。内华达州是美国第七大州，也是美国山脉最多的州，农业和矿业是传统的支柱产业。

Exclave 飞地　一种特殊的人文地理现象，指隶属于某一行政区管辖但不与本区毗连的土地。通俗地讲，如果某一行政主体拥有一块飞地，那么它无法取道自己的行政区域到达该地，只能“飞”过其他行政主体的属地，才能到达自己的飞地。一般把本国境内包含的外国领土称为内飞地（enclave），外国境内的本国领土称为外飞地（exclave）。飞地的概念产生于中世纪，飞地的术语第一次出现于 1526 年签定的马德里条约的文件上。

Panamint Valley 帕纳明特岭　位于美国加利福尼亚州东部，其最高峰高度为 3 365 米。

Western Hemisphere 西半球　地球上西经 20°以西，东经 160°以东的半球。为了避免把欧洲和非洲的一些国家分在东西两个半球，习惯上，把西经 20°向西到东经 160°的半个地球称为西半球。西半球主要有南美洲、北美洲和南极洲的一部分。海陆比例东半球陆地占 19.1%；海洋占 62%。西半球的南美洲、北美洲的划分有两种方法：①以巴拿马运河为分界线，划分成为南美洲和北美洲；②以语言的语系为分界线，把美国及其北面的国家划分为北美洲，把美国南面的国家划分为拉丁美洲（因为这里的居民大都讲拉丁语系的语言）。

Badwater Basin 恶水盆地　加利福尼亚州因约县死亡谷国家公园的一个内流盆地，前身是全新世时期的曼利湖（Lake Manly）。海拔高度为-86 米，因而成为北美洲的最低点。盆地的命名由来是因为该内流盆地的水质盐度过高而不能饮用，故而称作“恶水”。不过盆地内也是有一些动植物生存，如植物海蓬子、小型水生昆虫和当地特有的恶水蜗牛（学名 Assiminea infirma）。该盆地并不会永久被水覆盖，而是反复蒸发循环，在反复循环下让盐分形成六角蜂窝状的结晶。恶水盆地常被误认为“西半球的最低点”，但实际上西半球最低点是位于阿根廷的德尔炭丽盆地（Laguna del Carbón），海拔高度为-105 米。

Las Vegas 拉斯维加斯　美国内华达州的最大城市，以赌博业为中心的庞大的旅游、购物、度假产业而著名，是世界知名的度假圣地之一。从一个巨型游乐场到一个真正有血有肉、活色生香的城市，拉斯维加斯在 10 年间脱胎换骨。每年来拉斯维加斯旅游的 3890 万旅客中，来购物和享受美食的占了大多数，专程来赌博的只占少数。内华达州这个曾经被人讽刺为"罪恶之城"的赌城,已经逐步成熟,成为一个真正的城市了。

Translation of Selected Sentences

1. Parts of the park are in southern Inyo County and northern San Bernardino County in Eastern California, with a small extension into southwestern Nye County and extreme southern Esmeralda County in Nevada.

公园的一部分是在东加利福尼亚南部的因约县,北部的圣贝纳迪诺县,还有一小部分延伸到了西南部的奈伊县和内华达州最南部的艾丝美拉达县。

2. Death Valley National Monument was declared a U. S. National Monument in 1933, placing the area under federal protection. In 1994, the monument was redesignated a national park, as well as being substantially expanded to include Saline and Eureka valleys.

死亡谷国家纪念碑于 1933 年被确定为美国国家纪念碑,处于联邦的保护之下。1994 年被重新指定为国家公园,同时还相当大范围的延伸到包括盐湖和尤里卡山谷。

3. The first documented non-Native Americans to enter Death Valley did so in the winter of 1849, thinking they would save time by taking a shortcut to the gold fields of California.

第一个记录在案的进入死亡谷的并非美国本土人,他们是在 1849 年的冬天,超近路以节省时间到达加利福尼亚的淘金地区的。

4. Yet here in this valley, much of it below sea level, or in its surrounding mountains you can find spectacular wildflower displays, snow covered peaks, beautiful sand dunes, abandoned mines and industrial structures, and the hottest place in North America.

但是在这个山谷中,许多山脉都是在海平面以下的,或者在周围的山上,你可以看到很特别的野花大范围的开着,积雪覆盖的山峰,美丽的沙丘,废弃的工厂工业建筑,这里是北美最炎热的地区。

5. In general, heavy rains in late October with no more rain through the winter months will not bring out the flowers as well as rains that are evenly-spaced throughout the winter and into the spring.

总的来说,冬春两季的雨水分布均衡,冬天雨水稀少,只有十月下旬的大雨不足以开出绚烂的花。

6. Death Valley Scotty, whose real name was Walter Scott, was a gold miner who pretended to be owner of "his castle", which he claimed to have built with profits from his gold mine.

真名为沃尔特斯考特的死谷斯考特是一个淘金者,自称是用淘金矿赚来的利润建造的"他的城堡"的主人。

Questions

1. Describe the general condition of Death Valley National Park.
2. What are the main plants in the Death Valley National Park?
3. What is the Death Valley National Park famous for?
4. Why is Death Valley National Park popular for stargazing?
5. Why is the valley a popular winter destination?

Chapter 10

Big Bend National Park
大转弯国家公园

【简介】

大转弯国家公园位于美国德克萨斯州西南部与墨西哥的交界处。建于 1949 年,景区面积3 237.6 平方千米。整个公园呈现为一片荒原和荒漠景观。公园的自然景观可以分为东部、中部和西部三部分。东部是一大片荒漠,到处都是仙人掌。公园西部虽然也是荒漠地带,可是这里充满了火山运动遗迹,曾经是一片地壳非常活跃的地方。中部以山为主。山上树林茂密,可是四周都是荒漠。因此山上的动植物世世代代被困在山上,与世隔绝地繁衍,形成一个生态岛屿。

Big Bend National Park is a national park located in the U. S. state of Texas. For more than 1,000 miles (1,600 km), the Rio Grande/Río Bravo forms the international boundary between Mexico and the United States, and Big Bend National Park administers(管理)approximately 244 miles (393 km) along that boundary. The park was named after the area, which is bounded by a large bend in the Texas-Mexico border.

Big Bend National Park has national significance as the largest protected area of Chihuahuan Desert topography and ecology in the United States, which includes more than 1,200 species of plants, more than 450 species of birds, 56 species of reptiles(爬行动物), and 75 species of mammals. The park covers 801,163 acres (1,252 sqmi; 3,242 km^2). Few areas exceed the park's value for the protection and study of geologic and paleontologic(古生物学的) resources. Cretaceous and Tertiary fossil organisms exist in variety and abundance. Archeologists have discovered artifacts(史前古器物)estimated to be 9,000 years old, and historic buildings and landscapes offer graphic illustration of life along the international border in the 19th century.

Because the Rio Grande serves as an international boundary, the park faces unusual constraints when administering and enforcing park rules, regulations, and policies. In accordance with the Treaty of Guadalupe Hidalgo, the park has jurisdiction only to the center of the deepest river channel as the river flowed in 1848. The rest of the land south of that channel, and the river, lies within Mexican territory.

The park exhibits dramatic contrasts; its climate may be characterized as one of extremes. Dry and hot late spring and summer days often exceed 100 °F (38 °C) in the lower elevations. Winters are normally mild throughout the park, but sub-freezing temperatures occasionally occur. Because of the range in altitude from approximately 1,800 feet (550 m) along the river to Emory Peak in the Chisos Mountains at 7,832 feet (2,387 m) a wide variation in available moisture and in temperature exists throughout the park. These variations contribute to an exceptional diversity in plant and animal habitats. Some species in the park, such as the Chisos Oak (Quercus gravesii), are found nowhere else in the United States.

The 118 miles (190 km) of river that form the southern park boundary include the spectacular canyons of Santa Elena, Mariscal, and Boquillas. The Rio Grande, meandering(漫步)through this portion of the Chihuahuan Desert, has cut deep canyons with nearly vertical walls through three uplifts made primarily of limestone. Throughout the open desert areas, the highly productive Rio Grande riparian(河边居民)zone includes numerous plant and animal species and significant cultural resources. The vegetative belt extends into the desert along creeks and arroyos.

South of the border lie the Mexican states of Chihuahua and Coahuila and the new protected areas for flora and fauna(动植物), which are regions known as the Maderas del Carmen and the Cañón de Santa Elena.

Cultural resourcesin the park range from the Paleo-Indian period 10,500 years ago through the historic period represented by Native American groups, such as the Chisos, Mescaleros, and Comanche. More recently, Spanish, Mexican, Anglo and Irish settlers farmed, ranched, and mined in the area.

Throughout the prehistoric period, humans found shelter and maintained open campsites throughout the park. The archeological record reveals an archaic(古代的)-period desert culture, whose inhabitants developed a nomadic(游牧的)hunting and gathering lifestyle that remained virtually unchanged for several thousand years.

The historiccultural landscape centers upon various subsistence or commercial land uses. The riparian and tributary environments were used for subsistence and irrigation farming(灌溉农业).

Transportation networks, irrigation structures, simple domestic residences and outbuildings, and planed and terraced(阶地的)farm land lining the stream banks characterize these landscapes.

Despite its harsh environment, Big Bend has an amazing variety and number of plant and animal species. It has more than 1,200 species of plants (including 60 different cacti species), more than 600 animal species, and about 3,600 insect species. The diversity of life is largely due to the diverse ecology and changes in elevation, ranging from the dry, hot desert to the cool mountains to the fertile river valley.

Most of the animals are not visible in the day, particularly in the desert. The park comes alive at night, with many of the animals foraging for(搜查) food. About 150 Cougar (Puma conocolor) sightings are reported per year, despite the fact that there are only a total of two dozen Cougars. Other species that inhabit the park include Black-tailed Jackrabbit (Lepus californicus), kangaroo rats (Dipodomys spp.), Greater Roadrunner (Geococcyx californianus), Golden Eagle (Aquila chrysaetos), Collared Peccary (Pecari tajacu), and Coyote (Canis latrans). Mexican Black Bears (Ursus americanus eremicus) are also present in the mountain areas.

The variety of cactus and other plant life add color to the Big Bend region. Cactus species in the park include prickly pear(仙人掌)(Opuntia spp.), Claret Cup (Echinocereus coccineus) and Pitaya (E. enneacanthus). In the spring, the wildflowers are in full bloom and the yucca(丝兰) flowers display bright colors. Bluebonnets (Lupinus spp.) are prevalent in Big Bend, and white and pink bluebonnets(矢车菊)are sometimes visible by the road. Other flowering plants such as the Desert Marigold (万寿菊)(Baileya multiradiata), Desert Willow (Chilopsis linearis), Ocotillo (Fouquieria splendens), Rock Nettle (Eucnide urens) and Lechuguilla (Agave lechuguilla) abound in(富于)Big Bend.

The first U.S. record of the Northern Tufted Flycatcher (Mitrephanes phaeocercus), a Central American species, was from this site in November 1991. Birders(养鸟人)also flock to the park as it is home to the only area in the United States within the breeding range of the Colima Warbler (Vermivora crissalis).

Big Bend is one of the largest, most remote, and least-visited national parks in the lower 48 United States. In recent years only 300,000 – 350,000 visitors have entered the park annually.

Big Bend's primary attraction is itshiking and backpacking trails. Particularly notable among these are the Chimneys Trail, which visits a rock formation in the desert, the Marufo Vega trail, a loop trail that passes through scenic canyons on the way to and from the Rio Grande, and the Outer Mountain Loop trail in the Chisos, which begins in the Chisos Basin, climbs into the high mountains, descends into the desert along the Dodson Trail, and then returns to the Chisos Basin, completing a thirty mile loop. Other notable locations include Santa Elena Canyon, Grapevine Hills, and the Mule Ears, two imposing(壮观的) rock towers in the middle of the desert. There are professional backpacking guide services that provide trips in the park.

The park administers 245 miles (394 km) of the Rio Grande for recreational use. There are professional river outfitters(旅行用品商) that provide tours of the river. Use of a personal boat is permitted but a free river float permit is required. By 2002 visitors often crossed the Rio Grande to visit the Mexican village of Boquillas, but the Department of Homeland Security closed the border crossing. In June 2009 the Department of Homeland Security began treating all float trips as trips that had left and reentered the country, and required participants to have an acceptable form of identification, such as a passport.

Another popular activity isbirdwatching with more than 450 species of birds recorded in the park. Many species stop in the park during their migrations.

There are five paved roadsin Big Bend. Persimmon Gap to Panther Junction is a 28-mile (45 km) road from the north entrance of the park to park headquarters at Panther Junction. Panther Junction to Rio Grande Village is a 21-mile (34 km) road that descends 2,000 feet (610 m) from the park headquarters at Panther Junction to the Rio Grande River. Maverick Entrance Station to Panther Junction is a 23-mile (37 km) route from the western entrance of the park to the park headquarters. Chisos Basin Road is 6 miles (10 km) long and climbs to 5,679 feet (1,731 m) above sea level at Panther Pass before descending into the Chisos Basin. The 30 miles (48 km) Ross Maxwell Scenic Drive leads to the Castolon Historic District and Santa Elena Canyon.

Notes

Texas 得克萨斯州　简称得州,是美国南方最大的一个州,也是全美第二大州,仅次于阿拉斯加州。州名来自于印地安 Hasinai 族中的 Caddoan 语“tejas”,意味着“朋友”或是“盟友”。西班牙的探险家在命名得州时将这个应为人称的词误以为地名,就由此沿用下来。

Chihuahuan Desert 奇瓦瓦沙漠　墨西哥面积最大的一个沙漠,东北部与美国接壤,面积 245 612 平方千米(具体数字众说纷纭),占墨西哥总体面积的 12.6%,与英国相当。奇瓦瓦沙漠的首府叫奇瓦瓦市,四面环山,是商业和加工业中心。奇瓦瓦州的最大城市是美墨边境地带的华雷斯市。在世人的眼中,奇瓦瓦沙漠充满神奇,无论是水晶洞、干尸,还是世间稀有的洒脱美酒、仙人掌,都为奇瓦瓦沙漠蒙上了一层神秘的面纱。

Cretaceous 白垩纪　一个地质时代,位于侏罗纪和古近纪之间,约 1 亿 4550 万年(误差值为 400 万年)前至 6550 万年前(误差值为 30 万年)。白垩纪是中生代的最后一个纪,长达8000万年,是显生宙最长的一个阶段。发生在白垩纪末的灭绝事件,是中生代与新生代的分界。白垩纪的缩写记为K,是德文白垩纪(Kreidezeit)的缩写。

Tertiary 第三纪　新生代的最老的一个纪(距今6500万年—180万年)。第三纪的重要生物类别是被子植物、哺乳动物、鸟类、真骨鱼类、双壳类、腹足类、有孔虫等,这与中生代的生物界面貌迥异,标志着"现代生物时代"的来临。

Coahuila 科阿韦拉州　墨西哥的一个州,位于该国北部。北临美国得克萨斯州。州名源于当地的原住民科阿韦拉特卡部落(Coahuilteca)。在墨西哥独立时,与得克萨斯州同为一州,称为科阿韦拉特亚斯(Coahuila y Tejas),后者在1836年独立成为得克萨斯共和国。

cultural resources 文化资源　人们从事文化生活和生产所必须的前提准备。从对人们的贡献力量来看,文化资源有广义和狭义之分。广义上的文化资源泛指人们从事一切与文化活动有关的生产和生活内容的总称。它以精神状态为主要存在形式。狭义上的文化资源是指对人们能够产生直接和间接经济利益的精神文化内容。文化资源的丰富程度和质量高低直接对当地文化经济的发展产生多重作用。

Native American 印第安人　对除爱斯基摩人外的所有美洲原住民的总称。美洲土著居民中的绝大多数为印第安人,分布于南北美洲各国,传统将其划归蒙古人种美洲支系。印第安人所说的语言一般总称为印第安语,或者称为美洲原住民语言。印第安人的族群及其语言的系属情况均十分复杂,至今没有公认的分类。印第安人是拉丁美洲的最早居民。他们之所以被称为"印第安人",主要是因为当年哥伦布等探险者以为他们到达的"新陆地"是印度,称当地居民为"印第安"人("印度"一词的英文发音)。由于英国殖民者、美国人、欧洲殖民者(西班牙人)迫害、杀戮印第安人达4 000万以上,毁灭印第安文化,致使现在残存的古代印第安文明已经不多,在美国建国后和西进运动中,原居住在今天美国东部的超过2 000万印第安人受到迫害、惨遭杀戮,其他的土著居民则被驱逐到西部荒凉地,现在在美国已经很难看到印第安人了。印第安人以前曾被称为红种人,因为他们的皮肤经常是红色的,后来才知道这些红色是由于习惯在面部涂红颜料而给人造成的错误认识。

cultural landscape 文化景观　地理学中,景观一般指地球表面各种地理现象的综合体,可以分为自然景观和文化景观两大类。自然景观指完全未受直接的人类活动影响或受这种影响的程度很小的自然综合体。文化景观则是指居住在其土地上的人的集团,为满足某种需要,利用自然界所提供的材料,有意识地在自然景观之上叠加了自己所创造的景观。

Department of Homeland Security 国土安全部　2002年11月25日美国总统小布什在白宫签署《2002年国土安全法》,宣布成立国土安全部。成立国土安全部是美国自1947年成立国防部以来规模最大的一次政府机构调整。

Translation of Selected Sentences

1. The archeological record reveals an Archaic-period desert culture, whose inhabitants developed a nomadic hunting and gathering lifestyle that remained virtually unchanged for several thousand years.

考古学的记录反映了古代时期的沙漠文化,那时的居民发展为一种游牧式的狩猎及采集的生活方式,事实上这种方式保持了几千年都没有改变过。

2. The diversity of life is largely due to the diverse ecology and changes in elevation, ranging from the dry, hot desert to the cool mountains to the fertile river valley.

生物的多样性很大部分是源于生态的多样性和海拔高度的变化,从干旱炎热的沙漠到凉爽的高山,再到富饶的河谷。

3. Particularly notable among these are the Chimneys Trail, which visits a rock formation in the desert, the Marufo Vega trail, a loop trail that passes through scenic canyons on the way to and from the Rio Grande, and the Outer Mountain Loop trail in the Chisos, which begins in the Chisos Basin, climbs into the high mountains, descends into the desert along the Dodson Trail, and then returns to the Chisos Basin, completing a thirty mile loop.

特别值得注意的是炊烟小径,一直延伸到沙漠里的岩层;玛如浮维珈小径,环形的小径穿过风景优美的峡谷一直延伸到格兰德河;位于迟疏斯的远山环径起源雨迟疏斯盆地,爬过高山,降到沿着得逊小径的沙漠,最后回到迟疏斯盆地,完成全程 30 英里的环形路程。

4. In June 2009 the Department of Homeland Security began treating all float trips as trips that had left and reentered the country, and required participants to have an acceptable form of identification, such as a passport.

2009 年 6 月国土安全部开始处理这些出入国家的旅行,要求参与者要有可以接受的身份鉴定形式,例如,护照。

Questions

1. How does the climate contribute to the uniqueplant and animal habitats?
2. How does the variety of cactus and other plant life add color to the Big Bend region?
3. Why do you think the Department of Homeland Security closed the border crossing?
4. Describe generally the conditions of the Big Bend National Park.
5. What are the five paved roads in Big Bend?

Chapter 11

Petrified Forest National Park
化石林国家公园

【简介】

化石林国家公园位于美国亚利桑那州北部阿达马那镇附近，是世界上最大、最绚丽的化石林集中地。1906年被辟为国家保护区，1962年起被指定为国家公园，占地面积381平方千米。国内分6片林区，最美丽的是“彩虹森林”，还有“碧玉森林”“水晶森林”“玛瑙森林”“黑森林”和“蓝森林”等。遍布五彩斑斓、犹如镶金叠玉的石化树木，年轮清晰，纹理斐然，阳光之下闪闪发光。园内有羚羊、山猫、郊狼、响尾蛇等野生动物以及丝兰花、百合、仙人掌、紫苑等植物。

Petrified Forest National Park is a United States national park in Navajo and Apache counties in northeastern Arizona. The park's headquarters are about 26 miles (42 km) east of Holbrook along Interstate 40 (I-40), which parallels a railroad line, the Puerco River, and historic U. S. Route 66, all crossing the park roughly east-west. Named for its large deposits of petrified wood, the park covers about 146 square miles (380 km^2), encompassing semi-desert shrub steppe as well as highly eroded and colorful badlands. The site, the northern part of which extends into the Painted Desert, was declared a National Monument in 1906 and a national park in 1962. About 600,000 people visit the park each year and take part in activities including sightseeing, photography, hiking, and backpacking.

Averaging about 5,400 feet (1,600 m) in elevation, the park has a dry windy climate with temperatures that vary from summer highs of about 100 °F (38 °C) to winter lows well below freezing. More than 400 species of plants, dominated by grasses such as bunchgrass, blue grama, and sacaton, are found in the park. Fauna include larger animals such as pronghorns(叉角羚), coyotes, and bobcats(短脚猫); many smaller animals such as deer mice(波士白足鼠), snakes, lizards, seven kinds of amphibians, and more than 200 species of birds, some of which are permanent residents and many of which are migratory(迁徙的). About half of the park is designated wilderness.

The Petrified Forest is known for its fossils, especially fallen trees that lived in the Late Triassic, about 225 million years ago. The sediments containing the fossil logs are part of the widespread and colorful Chinle Formation(钦尔组), from which the Painted Desert gets its name. Beginning about 60 million years ago, the Colorado Plateau, of which the park is part, was pushed upward by tectonic forces and exposed to increased erosion. All of the park's rock layers above the Chinle, except geologically recent ones found in parts of the park, have been removed by wind and water. In addition to petrified logs, fossils found in the park have included Late Triassic ferns, cycads(苏铁植物), ginkgoes(银杏), and many other plants as well as fauna including giant reptiles called phytosaurs, large amphibians(两栖动物), and early dinosaurs. Paleontologists have been unearthing(发掘) and studying the park's fossils since the early 20th century.

The park's earliest human inhabitants arrived at least 8,000 years ago. By about 2,000 years ago, they were growing corn in the area and shortly thereafter building pit houses in what would become the park. Later inhabitants built above-ground dwellings called pueblos. Although a changing climate caused the last of the park's pueblos to be abandoned by about 1,400 CE, more than 600 archeological sites, including petroglyphs(岩石雕刻), have been discovered in the park. In the 16th century, Spanish explorers visited the area, and by the mid-19th century a U.S. team had surveyed an east-west route through the park and noted the petrified wood. Later roads and a railway followed similar routes and gave rise to tourism and, before the park was protected, to private removal of the park's fossils. Theft of petrified wood remains a problem in the 21st century.

The colorful Chinle, which appears on the surface in many parts of the southwestern United States and from which the Painted Desert gets its name, is up to 800 feet (240 m) thick in the park. It consists of a variety of sedimentary rocks including beds of soft, fine-grained mudstone, siltstone(沙泥岩), and claystone(黏土石)—much of which is bentonite—as well as harder sandstone and conglomerate, and limestone. Exposed to wind and water, the Chinle usually erodes differentially into badlands made up of cliffs, gullies, mesas, buttes, and rounded hills. Its bentonite clay, which swells(膨胀) when wet and shrinks while drying, causes surface movement and cracking that discourages plant growth. Lack of plant cover makes the Chinle especially susceptible(易受影响的)to weathering(平安度过).

About 60 million years ago, tectonic movements of Earth's crust began to uplift the Colorado Plateau, of which the Painted Desert is part. Eventually parts of the plateau rose to 10,000 feet (3,000 m) above sea level. This warping(弯曲)of Earth's surface led to the gradual and continuing destruction of the plateau by erosion. An unconformity(不符)(break in the rock record) of about 200 million years occurs within the park, where erosion has removed all the rock layers above the Chinle except geologically recent ones. The Bidahochi Formation, laid down only 8 to 4 million years ago, rests directly atop the Chinle, and rocks laid down in the Jurassic, Cretaceous, and much of the Tertiary are absent.

During the period of the Bidahochideposition, a large lake basin covered much of northeastern Arizona. The older (lower) layers of the formation consist of fluvial and lacustrine (湖泊的)(lake-related) deposits of silt, sand, and clay. The younger (upper) Bidahochi contains ash and lava from volcanoes that erupted nearby and as far away as southwestern Nevada. Although much of the Bidahochi has since eroded, a small part of it outcrops(露头)in the northern part of the park—on Pilot Rock in the park's wilderness section and along the rim of the Painted Desert between Pintado and Tawa points. Exposed by erosion of the Bidahochi are volcanic landforms called maars(低平火山口)(flat-bottomed, roughly circular volcanic craters of explosive origin). A maar vent can be seen from the Pintado Point lookout.

During the Quaternary Period (1.8 million years ago to today), deposits of windblown(被风吹的)sand and alluvium covered much of the Chinle and Bidahochi. Older dunes range in age from 500,000 years at higher elevations in the northern part of the park to about 10,000 years in sandy drainage areas(排水面积)such as Lithodendron Wash. Stabilized by grasses and other vegetation, young dunes of about 1,000 years old are found throughout the park.

During the Late Triassic, downed trees accumulating in river channels in what became the park were buried periodically by sediment containing volcanic ash. Groundwater dissolved silica (silicon dioxide) from the ash and carried it into the logs, where it formed quartz crystals that gradually replaced the organic matter. Traces of iron oxide(氧化铁)and other substances combined with the silica to create varied colors in the petrified wood.

In Petrified Forest National Park, most of the logs in the park retained their original external form during petrification but lost their internal structure. However, a small fraction of the logs and most of the park's petrified animal bones have cells(小屋)and other spaces that are mineral-filled but still retain much of their original organic structure. With these permineralized fossils, it is possible to study the cellular make-up of the original organisms with the aid of a microscope. Other organic matter—typically leaves, seeds, cones, pollen grains(花粉粒), spores, small stems, and fish, insect, and animal remains—have been preserved in the park as compression fossils, flattened by the weight of the sediments above until only a thin film remains in the rock.

Much of the park's petrified wood is from Araucarioxylon arizonicum(南洋杉型木) trees, while some found in the northern part of the park is from Woodworthia arizonica and Schilderia adamanica trees. At least nine species of fossil trees from the park have been identified; all are extinct. The park has many other kinds of fossils besides trees. The Chinle, considered one of the richest Late Triassic fossil-plant deposits in the world, contains more than 200 fossil plant taxa(分类群). Plant groups represented in the park include lycopods, ferns, cycads, conifers, ginkgoes, as well as unclassified forms. The park has also produced many fossil vertebrates(脊椎动物)—including giant crocodile-like(鳄鱼) reptiles called phytosaurs, large salamander-like amphibians called Buettneria, and early dinosaurs—and invertebrates(无脊椎动物), including fresh water snails and clams(哈蚌).

Notes

Triassic 三叠纪　始于距今2.5亿年至2.03亿年,延续了约5 000万年。海西运动以后,许多地槽转化为山系,陆地面积扩大,地台区产生了一些内陆盆地。这种新的古地理条件导致沉积以及生物界的变化。在古气候方面,三叠纪初期继承了二叠纪末期干旱的特点;到中、晚期之后,气候向湿热过渡,由此出现了红色岩层含煤沉积、旱生性植物向湿热性植物发展的现象。植物地理区也同时出现了分异。

Painted Desert 彩色沙漠　位于美国亚利桑那州中部,面积约1.9万平方千米。大部分地区为海拔1 500米以上的沙漠荒野,地势起伏,点缀着孤山、台地、小丘和河谷。由于当地温度差别极大(41 ℃到零下30 ℃),风化作用十分强烈,在光秃的崖壁上刻蚀出各种奇特的造型,宛如人工雕刻一般。其中千变万化的沙漠色彩,像染上了各种颜色,在阳光的照耀下,形成了罕见的自然奇观。加上热、光和沙尘的影响,使沙漠呈现出蓝色、紫水晶色、黄褐色、红色、淡绿色和灰色,辉煌灿烂,五彩分呈。在热气蒸腾之下,这些色泽又会凝聚成各种颜色的烟雾,硕硕发光。随着阳光照射角度的改变,色彩也时时变化,色调有浓有淡;雨后天晴,颜色尤为明丽鲜艳。

Colorado Plateau 科罗拉多高原　美国唯一的一个沙漠高原,位于美国西南部,面积30多万平方千米。东起科罗拉多州(Colorado)和新墨西哥州(New Mexico)的西部,西迄内华达州(Nevadan)的南部,科罗拉多河贯穿整个高原。经科罗拉多河及其支流的冲蚀,科罗拉多高原形成多条深邃的峡谷。其名字来自于西班牙语"科罗拉多",理论上如此荒凉的高原是没有任何工业、农业发展潜力的,但是正是因为这里独特的冲积岩地貌,让乐于挑战极限的人们找到了理想的目的地。

bentonite 膨润土　由于有良好的物理化学性能，素有“万能”黏土之称，可做黏结剂、悬浮剂、触变剂、稳定剂、净化脱色剂、充填料、饲料、催化剂等，广泛用于石油开采、定向穿越、钢铁铸造、冶金球团、化工涂料、复合肥、浆纱、橡胶、塑料、造纸、净化水、吸潮剂、农药等领域。

Quaternary Period 第四纪　6500 万年前那次生物大灭绝后，地球进入了新生代。新生代是地球历史的最新阶段，而第四纪是新生代最后一个纪。第四纪还可以分为更新世、全新世等。这一时期形成的地层称第四纪。第四纪一名是法国学者 J. 德努瓦耶于 1829 年提出的。从第四纪开始，全球气候出现了明显的冰期和间冰期交替的模式。第四纪生物界的面貌已经很接近于现代。哺乳动物的进化在此阶段最为明显，而人类的出现与进化则更是第四纪最重要的事件之一。

quartz crystals 石英晶体　晶体属三方晶系的氧化物矿物，是石英族矿物中分布最广的一个矿物种。低温石英常呈带尖顶的六方柱状晶体状，柱面有横纹，类似于六方双锥状的尖顶实际上是由两个菱面体单形所形成的。石英集合体通常呈粒状、块状或晶簇、晶腺等。纯净的石英无色透明，玻璃光泽，贝壳状断口上具油脂光泽，无解理。受压或受热能产生电效应。

organic matter 有机质　泛指土壤中来源于生命的物质。包括：土壤微生物和土壤动物及其分泌物以及土体中植物残体和植物分泌物。

Translation of Selected Sentences

1. Fauna include larger animals such as pronghorns, coyotes, and bobcats; many smaller animals such as deer mice, snakes, lizards, seven kinds of amphibians, and more than 200 species of birds, some of which are permanent residents and many of which are migratory.

动物群包括例如叉角羚、土狼、短脚猫的大型动物；像波士白足鼠、蛇、蜥蜴等小型动物；7 种两栖动物；超过 200 多种鸟，其中的一些是永久地在这旦居住，另外一些是迁徙的。

2. In addition to petrified logs, fossils found in the park have included Late Triassic ferns, cycads, ginkgoes, and many other plants as well as fauna including giant reptiles called phytosaurs, large amphibians, and early dinosaurs.

除了让人瞠目的原木之外，公园里发现的化石还包括三叠纪晚期的蕨类植物、苏铁植物、银杏，还有其他许多包括大型爬行动物的植蜥类植物和动物群，大型的两栖动物和早期的恐龙。

3. It consists of a variety of sedimentary rocks including beds of soft, fine-grained mudstone, siltstone, and claystone—much of which is bentonite—as well as harder sandstone and conglomerate, and limestone.

它由各种各样的沉积岩组成，包括柔软的有细腻螺纹的泥石，沙泥岩，黏土石组成的河床——很大一部分是膨润土——还有更加坚硬的砂岩、砾岩和石灰岩。

4. Although much of the Bidahochi has since eroded, a small part of it outcrops in the northern part of the park—on Pilot Rock in the park's wilderness section and along the rim of the Painted Desert between Pintado and Tawa points.

虽然比达胡齐很多部分后来都被腐蚀了，还有一小部分在公园的北部开始露头——在公园荒地部分的百乐岩石上和在沿着宾塔图和塔瓦点之间的彩色沙漠的边缘上。

5. Other organic matter—typically leaves, seeds, cones, pollen grains, spores, small stems, and fish, insect, and animal remains—have been preserved in the park as compression fossils, flattened by the weight of the sediments above until only a thin film remains in the rock.

其他的有机质——有代表性的树叶、种子、锥细胞、花粉粒、孢子、小茎、鱼、昆虫和动物残骸——在公园里都被当成压缩化石保护着，由于在上面的沉积岩的重量被压平，所以在岩石里只存在了很小的薄膜。

Questions

1. What are the unique plants and animals in the Petrified Forest National Park?
2. What is the Petrified Forest known for, especially those in the Late Triassic?
3. How did Painted Desert get its name?
4. What other kinds of fossils does the have besides trees?
5. What is the origin ofthe Petrified Frest National Park's petrified wood?

Chapter 12

Redwood National Park
红杉树国家公园

【简介】

红杉树国家公园,所属美国北美洲。红杉树国家公园,位于美国西部加利福尼亚州西北部的太平洋沿岸,1980 年联合国教科文组织将红杉树国家公园作为自然遗产,列入《世界遗产名录》。红杉树国家公园拥有世界上最高大的植物——可长到 350 英尺的红杉树常青原始森林。国家公园内有世界上现存面积最大的红杉树林,其中百年以上的老林区有 170 多平方千米。这里靠近海洋,气候温和湿润,为红杉的生长创造了极为有利的条件。

Redwood National Park was established on 2 October 1968 and expanded on 27 March 1978. Redwood National and State Parks protect old growth coast redwoods, some of the world's tallest trees. Less well-known are the prairies and oak woodlands and the coastal and marine ecosystems. Three California state parks and the National Park Service unit represent a cooperative management effort of the National Park Service and the California Department of Parks and Recreation. Together these parks are a World Heritage Site and International Biosphere Reserve protecting resources cherished by citizens of many nations.

The coast redwood, whose scientific name is Sequoia sempervirens(北美红杉), is the tallest known plant species in the world. It is also the fastest growing conifer(针叶树), or cone-bearing tree, in North America.

Coast redwoods are among the oldest living things on earth. The oldest known specimen was logged in 1933. A count of the growth rings revealed that the tree was 2,200 years old. Most redwoods, however, live an average of six centuries.

The tallest known redwood (the "Tall Tree") is about 600 years old. It is located on Redwood Creek within Redwood National and State Parks, near Orick, California. The Tall Tree was measured at 367.8 feet by the National Geographic Society in 1963. The top of the Tall Tree broke off recently. New measurements by the National Geographic Society in 1995 revealed the "National Geographic Tree" measured in 1963 to be the third tallest tree in the world, is now the tallest. It, too, is located in Tall Trees Grove(树林).

From a seed no bigger than one from a tomato, California's coast redwood (Sequoia semperviren) may grow to a height of 367 feet (122 m) and have a width of 22 feet (7 m) at its base. Imagine a 35-story skyscraper in your city and you have an inkling(略知) of the trees' ability to arouse humility.

Some visitors envision(想象) dinosaurs rumbling through these forests in bygone eras. It turns out that this is a perfectly natural thought. Fossil records have shown that relatives of today's coast redwoods thrived in the Jurassic Era 160 million years ago. And while the fantastic creatures of that age have long since disappeared, the redwoods continue to thrive, in the right environment.

California's North Coast provides the only such environment in the world. A combination of longitude, climate, and elevation limits the redwoods' rangeto a few hundred coastal miles. The cool, moist air created by the Pacific Ocean keTIF the trees continually damp, even during summer droughts. These conditions have existed for some time, as the redwoods go back 20 million years in their present range.

Exactly why the redwoods grow so tall is a mystery. Theories continue to develop but proof remains elusive(难以捉摸). The trees can reach ages of 2,000 years and regularly reach 600 years.

Resistance to natural enemies such as insects and fire are built-in(固定的) features of a coast redwood. Diseases are virtually unknown and insect damage insignificant thanks to the high tannin(丹宁酸) content of the wood. Thick bark and foliage that rests high above the ground provides protection from all but the hottest fires.

The redwoods' unusual ability to regenerate(再生) also aids in their survival as a species. They do not rely solely upon sexual reproduction, as many other trees must. New sprouts(芽) may come directly from a stump(树桩) or downed tree's root system as a clone. Basal burls, hard, knotty(多节的) growths that form from dormant seedlings on a living tree, can sprout a new tree when the main trunk is damaged by fire, cutting, or toppling(倒塌).

Undoubtedly the most important environmental influence upon the coast redwood is its own biotic community. The complex soils on the forest floor contribute not only to the redwoods' growth, but also to a verdant(翠绿的)array(排列)of greenery(绿色植物), fungi(真菌), and other trees. A healthy redwood forest usually includes massive Douglas-firs, western hemlocks, tanoaks, madrones(浆果鹃), and other trees. Among the ferns(蕨类)and leafy(多叶的) redwood sorrels(栗色), mosses and mushrooms help to regenerate(使再生)the soils. And of course, the redwoods themselves eventually fall to the floor where they can be returned to the soil.

The coast redwood environment recycles naturally: because the 100-plus inches of annual rainfall leaves the soil with few nutrients, the trees rely on each other, living and dead for their vital nutrients. The trees need to decay naturally to fully participate in this cycle, so when logging (伐木工作)occurs, the natural recycling is interrupted.

There are three members of the redwood family: coast redwoods (Sequoia sempervirens) of the California coastal fog belt, giant sequoias(巨杉)(Sequoiadendron giganteum) of the Sierra Nevada, and dawn redwoods(水杉)(Metasequoia glyptostroboides) of central China. The names sequoia and redwood are often used interchangeably(可交换的), leading to confusion between the two California members of this family, but they are very different trees.

Almost all remaining giant sequoias are now safe from the axe(斧)and saw(锯); about one-third were cut before being protected. Old growth coast redwood continues to be cut for lumber today, with only a small part of its original uncut(未切的)acreage still remaining. Of the estimated 1,950,000 acres of coast redwood originally occurring in California, only about 4.4%, or 86,000 acres of old growth, remain. About 80,000 acres of this old growth is protected in parks and reserves.

The ancient forests of the Pacific Northwest, containing as many as 25 different species of conifers(松柏类), are dominated by only a few. Sitka spruce(西加云杉)dominates southeast Alaska and coastal British Columbia, while Douglas fir dominates in Oregon, Washington, and inland British Columbia. The stately coast redwood (Sequoia sempervirens), largest of all, dominates in northern California.

Old and younger trees grow together in a mixture of species. The larger trees often have wind-damaged tops and larger but relatively fewer branches. A thick growth of mosses and lichens (地衣)harbors many insects, birds, and small mammals. These forests are efficient at trapping large amounts of moisture. The uneven canopy can even retain moisture from fog during the drier seasons.

Dead snags(断枝)may remain standing for many decades. Insects and woodpeckers(啄木鸟)open up this dead wood, providing habitat(栖息地)for many other species. In turn, these species become food for larger predators(捕食者), such as the northern spotted owl, marten (貂), and black bear.

Fallen trees lie in random patterns in small streams. Since the water flow(水流) in these streams is not powerful enough to move large logs, they form semi-permanent "stairway"(阶梯) that hold woody debris in place and stabilize the streambed(河床).

Fish populations benefit by consuming insects harbored in the debris. They also benefit from the shelter provided by fallen logs. Recent research indicates that fish are heavily dependent on old growth forests. Their populations suffer as a consequence of old growth destruction.

A variety of wildlife species call Redwood National and State Parks home. The diversity of ecosystems in the parks means that creatures as different as black bears, sea stars(海星), and bald eagles(秃鹰) can be seen by a lucky visitor in a single day. In addition to the more common inhabitants, many threatened and endangered species rely on the parks' old-growth forests, open prairies, estuaries(河口), and the coastline for crucial havens of survival.

Marine mammals such as sea lions and gray whales are among the most visible wildlife in the parks. Visitors are also likely to see Roosevelt elk browsing in the prairies. Pelicans, ospreys, and gulls are frequently spotted along the coast. Of course, tidepool creatures aren't likely to run very far at your approach, so anenomes and crabs(蟹类) are easy to spot too.

While you're out watching wildlife, remember that all wildlife is protected in the parks. Never feed or approach animals such as bears, mountain lions(美洲狮), elk, chipmunks, racoons, or rabbits. Please treat the members of this natural community with respect.

The Marbled Murrelet or "Fog Lark" is a small ten-inch-long seabird about the size of a robin which nests in trees from Alaska to central California. Scientists believe they nest only in the high canopy of ancient coastal forests. They seek food in the nearby ocean. Marbled murrelets are rare, and became listed in 1992 as an endangered species by the California state government. Debates on the importance of protecting endangered species often include this species.

Each species contributes to the richness and diversity of life. No species exists separate from others; all are tied together in the interdependent web of life. Valuing one species above all others and denying the importance of a species to the whole system may profit some of us in the short term. Long-run consequences for all of us, however, are often costly. The extinction of one species may cause a domino effect leading to the extinction of others.

Seldom seen, black bear roam throughout the park. They favor recently logged Redwood Creek area whose re-vegetation fosters abundant food. Fond of acorns, bears travel far to harvest them. These bears have generally not lost their fear of humans. The best way to prevent wild bears from becoming problem bears is to keep human food away from them. Use sound food storage practices. Counter-balance all food, scented items (soap, toothpaste, lotion, etc), and garbage in a tree; 12 feet up and 10 feet out from the trunk(树干); and 5 feet down from the branch. Ask a ranger for proper techniques. Remember a fed bear is a dead bear.

Notes

marine ecosystems 海洋生态系　海洋中由生物群落及其环境相互作用所构成的自然系统,生态系(Ecosystem)一词是英国 A. G. 坦斯利于 1935 年提出的。在此之前,德国 K. A. 默比乌斯(1877)和美国 S. A. 福布斯(1887)曾分别用生物群落(Biocoenosis)和小宇宙(Microcosm)这两个词,记述了类似坦斯利所说的内容。广义而言,全球海洋是一个大生态系,其中包含许多不同等级的次级生态系。每个次级生态系占据一定的空间,由相互作用的生物和非生物,通过能量流和物质流形成具有一定结构和功能的统一体。海洋生态系分类,目前无定论,按海区划分,一般分为沿岸生态系、大洋生态系、上升流生态系等;按生物群落划分,一般分为红树林生态系、珊瑚礁生态系、藻类生态系等。海洋生态系研究开始于20 世纪 70 年代,一般涉及自然生态系和围隔实验生态系等领域。近几十年,以围隔(或受控)实验生态系研究为主,主要开展营养层次、海水中化学物质转移、污染物对海洋生物的影响、经济鱼类幼鱼的食物和生长等研究。

International Biosphere Reserve 国际生物圈保护区　联合国教科文组织在其"人与生物圈计划"(MAB)中提出的一项概念,指受到保护的陆地、海岸带或海洋生态系统的代表性区域。截至 2007 年,该组织已经认定了 109 个国家的 527 个地区为生物圈保护区。

National Geographic Society 国家地理学会　一个于 1888 年 1 月 27 日在美国正式创立的非营利科学与教育组织。该学会的创立是在当年的 1 月 13 日,由 33 位初期创会会员决议创设,目的在于"增进及普及地理知识",在经过两周左右的讨论后正式确定该学会的组织章程与营运计划,正式开始运作。一百多年来,学会秉持"增进并传播地理知识"的宗旨,赞助了 8000 项以上的研究探索计划,并用优秀的杂志、地图、书籍、电影和交互媒体产品鼓励人们保护地球。众多优秀的科学家、探险家、摄影师、电视工作者和作者队伍肩负学会使命,往来于世界各地,辛勤地工作,为学会赢得显赫荣誉著名的黄框标志成为令人向往的探险与优质标记,见于全球每个国家的越来越多的语言当中。

Jurassic Era 侏罗纪　一个地质时代,界于三叠纪和白垩纪之间,约 1 亿 9 960 万年前(误差值为 60 万年)到 1 亿 4 550 万年前(误差值为 400 万年)。侏罗纪是中生代的第二个纪,开始于三叠纪 - 侏罗纪灭绝事件。虽然这段时间的岩石标志非常明显和清晰,其开始和结束的准确时间却如同其他远古的地质时代,无法非常精确地被确定。侏罗纪的名称取自于德国、法国、瑞士边界的侏罗山。超级陆块盘古大陆此时真正开始分裂,大陆地壳上的缝生成了大西洋,非洲开始从南美洲裂开,而印度则准备移向亚洲。

sexual reproduction 有性繁殖　一般繁殖多用此法,不仅有大量种子产生可以繁殖较多的新苗,而今日所有名种名花,也多是利用有性繁殖的改良育种而来,不似无性繁殖法所产生的新个体完全与母体一样而无变化，即便是以相同的亲本再次交配，因为因子的结合机会不同，所产生的子代也多与亲代及前次交配产生的子苗不一样而富有极多的变化，为了满足人的好奇心而喜爱新种有变化进步的心理，有性繁殖法实优于无性繁殖法。而所谓的育种也绝不是授粉、采蒴果、播种、移植等这么简单，或者是任意而为，育种者本身应有学术性，严肃性与神圣性，并应对亲本花有足够的认识，不但如此也要追溯到亲本的兄弟株，并详实的统计分析，做出育种交配的评估，并按照既定的目标计划，对其育出的子代再做出比较与评估，这样才能造就优良的子代。

clone 克隆　指生物体通过体细胞进行的无性繁殖，以及由无性繁殖形成的基因型完全相同的后代个体组成的种群。通常是利用生物技术由无性生殖产生与原个体有完全相同基因组织后代的过程。

Alaska 阿拉斯加　一个位于美国西北太平洋东岸的州，第 49 个加入美利坚合众国的州，其也是美国最大的州、世界最大的飞地地区，该州的邮政缩写是 AK。“阿拉斯加”一名可能源于阿留申语的“Alyeska”，意思是“很大的陆地”，阿拉斯加州东接加拿大的育空、不列颠哥伦比亚，南邻阿拉斯加湾、太平洋，西濒白令海、白令海峡、楚科奇海，北临北冰洋。原属俄罗斯，1867 年被俄国人以 720 万美元的价格卖给了美国。

Domino effect 多米诺效应　第一棵树的砍伐，最后导致了森林的消失；一日的荒废，可能是一生荒废的开始；第一场强权战争的出现，可能是使整个世界文明化为灰烬的力量。这些预言或许有些危言耸听，但是在未来我们可能不得不承认它们的准确性，或许我们惟一难以预见的是从第一块骨牌到最后一块骨牌的传递过程会有多久。有些可预见的事件最终出现要经历一个世纪或者两个世纪的漫长时间，但它的变化已经从我们没有注意到的地方开始了。从电视画面可看出，骨牌瞬间依次倒下的场面蔚为壮观，其间显示的图案丰富多彩，令人惊叹，其中蕴含着一定的科学道理——多米诺骨牌效应。该效应产生的能量是十分巨大的。这种效应的物理道理是：骨牌竖着时，重心较高，倒下时重心下降，倒下过程中，将其重力势能转化为动能，它倒在第二张牌上，这个动能就转移到第二张牌上，第二张牌将第一张牌转移来的动能和自己倒下过程中由本身具有的重力势能转化来的动能之和，再传到第三张牌上……所以每张牌倒下的时候，具有的动能都比前一块牌大，因此它们的速度一个比一个快，也就是说，它们依次推倒的能量一个比一个大。

Translation of Selected Sentences

1. Redwood National and State Parks protect old growth coast redwoods, some of the world's tallest trees.

红杉树国家公园保护了沿海岸生长的古老的红杉树，它们是世界上最高的树。

2. The coast redwood environment recycles naturally: because the 100-plus inches of annual

rainfall leaves the soil with few nutrients, the trees rely on each other, living and dead for their vital nutrients.

沿海岸生长的红杉树林形成了自我给养补给的循环系统,因为这里年均降水量只有100英寸,导致土壤缺乏养分,所以树木们只有依靠同伴生存或死亡化作珍贵的养料。

3. Almost all remaining giant sequoias are now safe from the axe and saw; about one-third were cut before being protected.

几乎所有留存下来的巨型红杉树目前都远离了被斧子或电锯砍伐的威胁,在此之前大概有三分之一的树木被砍伐。

4. Recent research indicates that fish are heavily dependent on old growth forests. Their populations suffer as a consequence of old growth destruction.

近期研究表明,鱼类的生存非常依赖原始森林。鱼类种群的增减反映了原始森林遭破坏的程度。

5. Marbled murrelets are rare, and became listed in 1992 as an endangered species by the California state government.

带大理石条纹的小海鸭的数量不断减少,1992年被加利福尼亚政府列为濒危物种。

Questions

1. Why was Redwood National Park established?
2. Describe the features of the coast redwood.
3. Were there dinosaurs rumbling through these forests in bygone eras?
4. Why do the old and younger trees grow together in a mixture of species?
5. Why may the extinction of one species cause a domino effect leading to the extinction of others?

Chapter 13

Grand Canyon National Park
大峡谷国家公园

【简介】

大峡谷国家公园位于美国西部亚利桑那州凯巴布高原,面积 2 724.7 平方千米,也被称为"科罗拉多大峡谷"。大峡谷国家公园建立于 1919 年,1979 年纳入联合国教科文组织的世界遗产(World Heritage)名单。大峡谷广阔无垠,尽管全球已有数百万人游览过大峡谷,但至今无人见过它的全貌。大峡谷的壮观,不仅在于其千姿百态的奇峰异石和峭壁石柱,还在于其色彩的变幻。由于峡谷两壁的岩石性质、所含矿物质的不同,在阳光照射下,呈现不同的色彩。

Grand Canyon National Park is one of the United States' oldest national parks and is located in Arizona. Within the park lies the Grand Canyon, a gorge(峡谷)of the Colorado River, considered to be one of Seven Natural Wonders of the World. The park covers 1 902 mi^2(4 927 km^2) of unincorporated area in Coconino County and Mohave County.

Most visitors to the park come to the South Rim, arriving on Arizona State Route 64. The Highway enters the park through the South Entrance, near Tusayan, Arizona, and heads eastward, leaving the park through the East Entrance. All park accommodations are operated by the Xanterra corporation. Park headquarters are at Grand Canyon Village, a short distance from the South Entrance, being also the center of the most popular viewpoints. Some thirty miles of the South Rim are accessible by road. A much smaller venue for tourists is found on the North Rim, accessed by Arizona State Route 67. There is no road connection between the two within Arizona except via the Navajo Bridge, near Page, Arizona, entailing(使必须)a five-hour drive.

Otherwise, the two rims of the Canyon are connected via Las Vegas, Nevada, and the Hoover Dam. The rest of the Grand Canyon is extremely rugged and remote, although many places are accessible by pack trail and backcountry(偏远地区)roads.

Grand Canyon National Park became a national park in 1919. So famous is this landmark to modern Americans that it seems surprising that it took more than thirty years for it to become a national park. President Theodore Roosevelt visited the rim in 1903 and exclaimed: "The Grand Canyon fills me with awe. It is beyond comparison—beyond description; absolutely unparalleled throughout the wide world. Let this great wonder of nature remain as it now is. Do nothing to mar its grandeur(壮丽), sublimity(庄严)and loveliness. You cannot improve on(改进) it. But what you can do is to keep it for your children, your children's children, and all who come after you, as the one great sight which every American should see."

Despite Roosevelt's enthusiasm and his strong interest in preserving land for public use, the Grand Canyon was not immediately designated as a national park. The first bill to create Grand Canyon National Park had been introduced in 1882 and again in 1883 and 1886 by Senator Benjamin Harrison. As President, Harrison established the Grand Canyon Forest Reserve in 1893. Theodore Roosevelt created the Grand Canyon Game Preserve by proclamation in 1906 and Grand Canyon National Monument in 1908. Senate bills to establish a national park were introduced and defeated in 1910 and 1911; the Grand Canyon National Park Act was finally signed by President Woodrow Wilson in 1919. The National Park Service, which had been established in 1916, assumed administration of the park.

The creation of the park was an early success of the environmental conservation movement; its National Park status may have helped thwart(横跨)proposals to dam the Colorado River within its boundaries (Lack of this fame may have enabled Glen Canyon Dam to be built upriver, flooding Glen Canyon and creating Lake Powell). In 1975, the former Marble Canyon National Monument, which followed the Colorado River northeast from the Grand Canyon to Lee's Ferry, was made part of Grand Canyon National Park. In 1979, UNESCO declared it as a World Heritage Site.

The Grand Canyon itself, including its extensive system of tributary canyons, is valued for the combination of large size, depth, and the exposed layering of colorful rocks dating back to Precambrian times. It was created through the incision(切口) of the Colorado River and its tributaries after the Colorado Plateau was uplifted and the Colorado River system developed along its present path.

The South Rim is easier to visit than the North Rim. Access is available from Las Vegas, California and the west coast and from Flagstaff and points east, i. e., New Mexico or east of the

Mississippi River. For areas to the north, U. S. 89 connects Utah and Colorado and the North Rim of the canyon to the south rim.

Carved out by the Colorado River, the Grand Canyon (nearly 1,500 m deep) is the most spectacular gorge in the world. Located in the state of Arizona, it cuts across the Grand Canyon National Park. Its horizontal strata retracethe geological history of the past 2 billion years. There are also prehistoric traces of human adaptation to a particularly harsh environment.

The Grand Canyon is among the earth's greatest on-going geological spectacles. Its vastness is stunning, and theevidence it reveals about the earth's history is invaluable(无价的). The 1.5-kilometer (0.9 mile) deep gorge ranges in width from 500 m to 30 km (0.3 mile to 18.6 miles). It twists and turns 445 km (276.5 miles) and was formed during 6 million years of geologic activity and erosion by the Colorado River on the upraised earth's crust(外壳). The buttes(孤峰), spires (尖顶山), mesas(平顶山) and temples in the canyon are in fact mountains looked down upon from the rims. Horizontal strata(层) exposed in the canyon retrace geological history over 2 billion years and represent the four major geologic eras.

Widely known for its exceptional natural beauty and considered one of the world's most visually powerful landscapes, the Grand Canyon is celebrated for its plunging depths; temple-like buttes; and vast, multihued, labyrinthine (复杂的) topography. Scenic wonders within park boundaries include high plateaus, plains, deserts, forests, cinder cones, lava flows, streams, waterfalls, and one of America's great whitewater(急流) rivers.

Within park boundaries, the geologic record spans all four eras of the earth's evolutionary history, from the Precambrian to the Cenozoic. The Precambrian and Paleozoic portions of this record are particularly well exposed in canyon walls and include a rich fossil assemblage. Numerous caves shelter fossils and animal remains that extend the paleontological(古生物学的) record into the Pleistocene.

Grand Canyon is an exceptional example of biological environments at different elevations that evolved as the river cut deeper portraying five of North America's seven life zones within canyon walls. Flora and fauna species overlap in many of the zones and are found throughout the canyon.

The park's diverse topography has resulted in equally diverse ecosystems. The five life zones within the canyon are represented in a remarkably small geographic area. Grand Canyon National Park is an ecological refuge, with relatively undisturbed remnants of dwindling ecosystems, and numerous endemic, rare or endangered plants and animal species.

Altitudinal range provides a variety of climates and habitats, ranging from desert to mountain conditions. The canyon is a vast biological museum stretching through five different life and vegetation zones: Hudsonian on the North Rim plateau with Colorado blue spruce and Rocky Mountain maple; Canadian near the North Rim, with aspen and Ponderosa and forests of Douglas fir, white fir and aspen at 2,500 m; high-altitude Transition (Ponderosa) forests of Pinus ponderosa (北美黄松) and gambel oak; upper Sonoran on and below the South Rim with Utah juniper(杜松), pinon(矮松) pine and sagebrush(蒿属植物); and the Lower Sonoran down the canyon and at the bottom (desert cacti, rabbitbrush, mesquite, Morman tea and Manzanita).

Over 1,000 plant species have so far been identified from the park. There are 11 plant species listed as threatened in the United States statutes in the park, including Palmer amsonia (水甘草属), goldenweed, plains cactus, scouler catchfly(捕虫草), wild buckwheats(荞麦), primrose(报春花) and clute penstemon (钓钟柳属). In addition, 15 plant species are recommended for consideration as threatened species under the Endangered Species Act.

Concerning the fauna, 76 mammal, 299 bird and 41 reptile and amphibian species have been identified from the park and some 16 fish species inhabit the Colorado River and its tributaries. Rare or threatened birds are listed under the United States Endangered Species Act of 1973.

Created a national park on 26 February 1919 by an act of Congress. First protected in 1893 as a forest reserve in which mining, lumbering and hunting continued to be allowed; upgraded to a game reserve in 1906, giving protection to the wildlife; redesignated a national monument in 1908, accepted as a World Heritage site on 24 October 1979.

Notes

Seven Natural Wonders of the World 世界七大自然奇迹　1. 美国的科罗拉多大峡谷;2. 澳大利亚的大堡礁;3. 尼泊尔中国边界的珠穆朗玛峰;4. 赞比亚津巴布韦边界的维多利亚大瀑布; 5. 两极极光;6. 墨西哥的帕里库廷火山——世界最年轻的火山;7. 巴西里约热内卢的港口——天然港口。

Las Vegas 拉斯维加斯　美国内华达州的最大城市,以赌博业为中心的庞大的旅游、购物、度假产业而著名,是世界知名的度假圣地之一。从一个巨型游乐场到一个真正有血有肉、活色生香的城市,拉斯维加斯在10年间脱胎换骨。每年来拉斯维加斯旅游的3 890万旅客中,来购物和享受美食的占了大多数,专程来赌博的只占少数。内华达州这个曾经被人讽刺为"罪恶之城"的赌城,已经逐步成熟,成为一个真正的城市了。

Hoover Dam 胡佛水坝　美国综合开发科罗拉多河(Colorado)水资源的一项关键性工程,位于内华达州和亚利桑那州交界之处的黑峡(Black Canyon),具有防洪、灌溉、发电、航运、供水等综合效益。大坝系混凝土重力拱坝,坝高221.4米,大坝形成的水库叫米德湖

(Mead),总库容348.5亿立方米,水电站装机容量原为134万千瓦,现已扩容到208万千瓦,计划达到245.2万千瓦。1931年4月开始动工兴建,1936年3月建成,1936年10月第一台机组正式发电。

New Mexico 新墨西哥州　美国西南部四州之一。北接科罗拉多州,西接亚利桑那州,东北邻俄克拉何马州,东部和南部与得克萨斯州毗连,西南与墨西哥的奇瓦瓦州接壤。新墨西哥的景致迷人,有红岩峭壁、沙漠、仙人掌等。但是在这些景观的背后有着古老的文化和历史。

Mississippi River 密西西比河　世界第四长河,也是北美洲流程最长、流域面积最广、水量最大的河流。密西西比河位于北美洲中南部,注入墨西哥湾。“密西西比”是英文“mississippi”的音译,来源于印第安人阿耳冈昆族语言,“密西”(misi)和“西比(sipi)分别是“大、老”和“水”的意思,“密西西比”即“大河”或“老人河”。干流发源于苏必利尔湖以西,美国明尼苏达州西北部海拔501米的、小小的艾塔斯卡湖,向南流经中部平原,注入墨西哥湾。

Translation of Selected Sentences

1. The Grand Canyon fills me with awe. It is beyond comparison—beyond description; absolutely unparalleled throughout the wide world... Let this great wonder of nature remain as it now is. Do nothing to mar its grandeur, sublimity and loveliness. You cannot improve on it. But what you can do is to keep it for your children, your children's children, and all who come after you, as the one great sight which every American should see.

我对大峡谷怀着一种敬畏感。她的美无与伦比——根本无法用文字来形容;绝对是在这个世界上空前绝后的……一定要让这个伟大的自然奇迹一直保持现在这个样子。决定不要毁坏了她的壮丽,庄严和魅力。根本就不用对她再进行改进。但是你所能做的是把她保存下来留给你们的孩子,孩子的孩子,和你后代的所有人,让所有美国人都能看到这个宏伟的景象。

2. The Grand Canyon itself, including its extensive system of tributary canyons, is valued for the combination of large size, depth, and the exposed layering of colorful rocks dating back to Precambrian times.

大峡谷自身包括成广阔体系的附带峡谷,因其庞大的规模和厚度而备受重视,暴露的彩岩层可以追溯到前寒武纪时期。

3. The buttes, spires, mesas and temples in the canyon are in fact mountains looked down upon from the rims. Horizontal strata exposed in the canyon retrace geological history over 2 billion years and represent the four major geologic eras.

峡谷的孤峰、尖顶山、平顶山和庙宇事实上是从峡谷边缘向下看到的山。峡谷中暴露的垂直层可以追溯到20亿年前的地质历史，并且是四个主要地质时期的代表。

4. Grand Canyon is an exceptional example of biological environments at different elevations that evolved as the river cut deeper portraying five of North America's seven life zones within canyon walls.

大峡谷是个特殊的例子，生物环境在不同的纬度下逐渐的形成，河水流的更加深远；在北美7个生物带中峡谷内竟然就占据了五个之多。

5. Grand Canyon National Park is an ecological refuge, with relatively undisturbed remnants of dwindling ecosystems, and numerous endemic, rare or endangered plants and animal species.

大峡谷国家公园就是一个生态庇护，在现存的越来越小的生态系统里是相对不被打扰的残存的，许多地方性的，有着稀有的和濒临灭绝的植物和动物种类。

Questions

1. Why do you think the park has been one of the Seven Natural Wonders of the World?
2. Why has the park taken more than thirty years to become a national park?
3. How did President Theodore Roosevelt describe the park?
4. Why was the creation of the park an early success of the environmental conservation movement?
5. What is the specialized plant in the park?

Chapter 14

Everglades National Park
大沼泽地国家公园

【简介】

大沼泽地国家公园建于公元1974年，现在已经覆盖140万英亩。被称为“一条被草覆盖的河从内地缓缓流向海洋”的大沼泽地国家公园位于佛罗里达州的南端。6英寸深、50英里宽的淡水河缓缓流过广袤的平原，因而造就了这种独特的大沼泽地环境。这里是美国本土上最大的亚热带野生动物保护地。这里水中的生活环境为无数的鸟类和爬行动物以及海牛一类的濒危动物提供了很好的避难场所。辽阔的沼泽地、壮观的松树林和星罗棋布的红树林为无数野生动物提供了安居之地。

On 30 May 1934, an Act was passed authorizing a park to be acquired through public donations. Everglade National Park was to be “... wilderness, (where) no development ... or plan for the entertainment of visitors shall be undertaken which will interfere with the preservation intact of the unique flora and fauna of historic values the essential primitive natural conditions now prevailing in this area.” This mandate(命令) to preserve wilderness and its biota is one of the strongest in the legislative history of the National Park System. Thirteen years later, through a combination of(…的组合) federal, state and private lands, a vast wetland teeming with life was dedicated as a national park. Everglades was the first national park preserved primarily for its abundance and variety of life, rather than for scenic or historic values.

“Here are no lofty peaks seeking the sky, no mighty glaciers or rushing streams wearing away(磨损) the uplifted land. Here is land, tranquil in its quiet beauty, serving not as the source of water, but as the receiver of it. To its natural abundance we owe the spectacular plant and animal life that distinguishes this place from all others in our country." With these words, President Harry S. Truman formally dedicated Everglades National Park on 6 December 1947 in a ceremony held at Everglades City. This event culminated(到绝顶)years of effort by a dedicated group of conservationists to make a national park in the Florida Everglades a reality. The Visitor Center near the main park entrance is dedicated to one of the foremost of these far-sighted individuals, Ernest "Tom" Coe.

Ernest F. Coe, a Yale-educated landscape architect, made the Everglades park project his life work shortly after moving to Miami in 1925. In 1928, Coe and others organized the Tropical Everglades Park Association (later known as the Tropical Everglades National Park Association), devoted solely to the creation of a national park in south Florida. Dr. David Fairchild, the former head of the U. S. Department of Agriculture's Bureau of Plant Exploration, was the association's first president; Coe was named executive secretary. The following year, the Florida legislature authorized the Tropical Everglades National Park Commission to take over (接管) the responsibilities of the Tropic Everglades National Park Association and with the power to acquire land by purchase, gift, bequest (遗赠) or condemnation (征用). Ernest Coe was the commission's executive chairman. Also in 1929, the U. S. Congress authorized an investigation into the feasibility(可行性)of a national park project in South Florida. A special committee of the National Parks Association, which included NPS Director Horace Albright, Assistant Director Arno B. Cammerer and Yellowstone superintendent(监督人)Roger Toll, toured the area by auto, boat, and Goodyear blimp, with local park advocates, including Ernest Coe, David Fairchild, Dr. T. Gilbert Pearson, president of the National Association of Audubon Societies, and Ruth Bryan Owen, the U. S. representative to Congress from Miami. Upon their return to Washington, D.C., the committee reported favorably to Congress on the proposed park.

Everglades National Park is open year round. Highest visitation is from December through April, and the lowest visitation is May through November. Walking and canoe trails, boat tours and tram tours are excellent for viewing wildlife, including alligators(短吻鳄)and a multitude of tropical and temperate birds.

Given present trends, the endangered wood storks may no longer nest in South Florida by the year 2000. The wood stork has declined from 6,000 nesting birds to just 500 since the 1960's. Their feeding behavior explains their predicament(窘况). Wood storks feed not by sight, but by touch—"tacto-location"—in shallow and often muddy water full of plants. Fish can't be seen in those conditions. Walking slowly forward the stork sweTIF its submerged bill from side to side. Touching prey, mostly small fish, the bill snaps shut with a 25-second millisecond reflex action, the fastest know for vertebrates. Only seasonally drying wetlands (mostly in drying ponds) concentrate enough fish to provide the 440 pounds a pair of these big birds requires in a breeding season. When natural wetlands cycles are upset by human water management, wood storks fail to nest successfully. The wood stork, which stands over 3 feet tall, and has a 5-foot wing spread, and weighs 4 to 7 pounds, was placed on the endangered species list in 1984.

Native trees, such as mangroves and cypress, are being replaced by exotic (introduced) species from other countries. Florida largemouth bass share their nesting beds with tilapia(罗非鱼) and oscars(地图鱼), fish imported from Africa and South America. As the Everglades yield to human introduced plants and fish, native species diminish.

Solutions are underway, but the fate of the Everglades still hangs in the balance. In one of the world's largest ecosystem restoration projects, Congress has extended the park boundaries to protect the eastern Shark River Slough. Historically it hosted higher concentrations of wading bird (涉水鸟) nesting populations than any other area in the park location. The enlargement should help turn around the 93 percent decline these species have suffered by restoring critical, suitable habitat. The National Park and the State of Florida have agreed to be partners in enforcing the existing water quality regulations to address water quality problems. The Park Service is working with the US Army Corps of Engineers and other water management jurisdictions to adopt natural rainfall models of manipulating water supplies.

The park was established to save the 'Glades, but real problems continue to beset this landscape. Nothing is saved for good; the Everglades' fate remains our mandate.

Threatened, endangered and extinct are words that have become all too common in our 20th century vocabulary. The natural process of species evolution, taking hundreds and thousands of years, has accelerated rapidly since the turn of the century. Today because of man's desire for land and raw materials, his continued pollution and indiscriminate(任意的) hunting many plant and wildlife species are on the brink of extinction.

Drainage of wetlands, alteration of overland (陆上的) water flow and hunting have all contributed to species decline. The Everglades, once known for its abundant bird life, has seen its wading bird population decline drastically since the turn of the century. The Florida Panther, once common throughout the state, today is on the verge of extinction. Within the four National Park areas of Everglades National Park, Biscayne National Park, Big Cypress National Preserve and Fort Jefferson National Monument there are 16 endangered and 6 threatened wildlife species. The mere physical boundaries of a National Park do not guarantee a species survival.

Maintaining harmony between "20th century progress" and wilderness areas requires research, legislation and public awareness. For the last decade the South Florida Research Center, Everglades National Park, has been studying how changes occurring outside the parks influence the fragile areas within their boundaries. Research going on today may lead to a brighter future for many species.

Legislation such as the Endangered Species Act of 1973 has also afforded some measure of protection for wildlife. The Act provided for the classification of wildlife species as "endangered" or "threatened", and mandated legal protection for species so listed. In justification for such protection, the Act also recognized that the various species of fish, wildlife and plants have aesthetic, educational, historical and scientific value.

Public support is also vital for species preservation. "What can I do?" you might ask. You can:

· Become informed on the status of plants and wildlife in your state.

· Do not purchase products that you suspect come from endangered or threatened plant or wildlife species.

· Report those people who are known dealers in endangered or threatened plants and wildlife.

· Support conservation legislation.

Today it is not enough to merely appreciate nature; we have to actively work to protect it. What we do today toward that goal is the legacy we leave our children and their children. The extinction of a species is forever... and the decision is ours.

Notes

Harry S. Truman 哈利·斯·杜鲁门(1884年5月8日—1972年12月26日) 1945年接替因病逝世的富兰克林·德拉诺·罗斯福总统,成为第33任美国总统(1945—1953年)。1949—1953年连任总统。1945年对日本使用原子弹而使第二次世界大战很快结束。1947年提出"杜鲁门主义",要求国会拨款4亿美元援助希腊和土耳其,镇压当地人民革命运动。1948年批准以控制欧洲为目的的"马歇尔计划"。1953年卸任回乡。1972年12月16日在堪萨斯城病故。著有回忆录《决定的年代》《试验和希望的年代》和《公民先生》。

Miami 迈阿密　美国佛罗里达州东南部滨海城市,运输业、商业中心,海滨旅游胜地。市区面积88平方千米。人口约38.1万。市区沿比斯坎湾延伸19.2千米,宽13千米。以工商业为主,主要工业部门有电子、电器、服装、食品、塑料制品、装饰品等,为全国第三大服装中心,也是葡萄、水果等集散地。渔业亦盛。陆军、海军基地。是拉美航空交通的中继站。也是重要的金融中心,有100余家本国和外国银行。城西的迈阿密国际机场辟有国内外航线60余条。道奇岛等地建有大型客运码头,已替代纽约成为美国最大的客运港。市区与郊区间有高架铁路连接。旅游业为经济支柱。全市有豪华旅馆400多家,可同时接待20

万游客。雅兹卡亚艺术馆是宏伟的别墅式建筑。市区还设有蛇类展出馆、水族馆、费尔柴尔德热带公园等。建有迈阿密大学等6所高等院校和博物馆、图书馆等。海洋和气象学科研机构闻名世界。

Department of Agriculture's Bureau 美国农业部　联邦政府内阁13个部门之一，是重要的经济管理部门。美国农业部是按照法律设置的，机构稳定，职能明确。美国农业部的职能，可以用一句话来概括，那就是"从田间到餐桌"。

Africa 非洲　位于亚洲的西南面。东濒印度洋，西临大西洋，北隔地中海与欧洲相望，东北角习惯上以苏伊士运河为非洲和亚洲的分界。大陆东至哈丰角（东经51度24分，北纬10度27分），南至厄加勒斯角（东经20度02分，南纬34度51分），西至佛得角（西经17度33分、北纬34度45分），北部至吉兰角（东经9度50分、北纬37度21分）。面积约3 020万平方千米（包括附近岛屿），南北约长8 000千米，东西约长7 403千米。约占世界陆地总面积的20.2%，仅次于亚洲，为世界第二大洲。

South America 南美洲　南美洲位于西半球的南部，东濒大西洋，西临太平洋，北濒加勒比海，南隔德雷克海峡与南极洲相望。一般以巴拿马运河为界同北美洲相分隔，包括哥伦比亚、委内瑞拉、圭亚那、苏里南、厄瓜多尔、秘鲁、巴西、玻利维亚、智利、巴拉圭、乌拉圭、阿根廷、法属圭亚那等13个国家和地区。

Biscayne National Park 比斯坎国家公园　95%是水域。它根本上是一个海洋公园。整个比斯坎国家公园成立的目的是在保护园区，主要是比斯坎湾（Biscayne Bay）里的海洋生态。比斯坎湾位于佛罗里达东南岸和基拉戈岛（Key Largo）以北延伸的一串行屿之间，受到列屿的保护，没有太大的风浪。这个湾区的水非常浅，水也格外清澄。这个海域也是美国最大的珊瑚礁生长地区，种种得天独厚的条件，使这里成了浮潜与潜水者的天堂。另一个特别的生态是海岸边的红树林。其实红树林在全佛罗里达南边的海岸与列屿（Keys）随处可见。它在当地生态系扮演重要角色。红树（Mangrove）有很强的耐力，能在咸水环境生存。它多分支的根部强而有力，能伸入水下并固定树干。多亏它的根，岸边的小石块及泥沙得以被固定而不致被潮水冲失。也因此很多小岛得以存在而不致流失。

Big Cypress National Preserve 大落羽衫国家保留地　坐落在南佛罗里达的美国国家保护区，距迈阿密以西45英里，是美国国家公园体系最早的国家保护区。

Fort Jefferson National Monument 杰弗逊堡国家纪念地　1846年美国开始建造杰弗逊堡，虽然没有建成，最多的时候能容下2 000人。在过去的150年中，这个位于佛罗里达岛以西70英里的群岛一直是监狱。1935年富兰克·林罗斯福把这个地方叫做杰弗逊堡国家纪念地。1970年被列入国家历史名录，1992年被建成为国家公园。

Translation of Selected Sentences

1. Everglade National Park was to be "... wilderness, (where) no development... or plan for the entertainment of visitors shall be undertaken which will interfere with the preservation intact of the unique flora and fauna of historic values the essential primitive natural conditions now prevailing in this area."

美国大沼泽地国家公园是茫茫荒原,一切破坏有历史价值的稀有动植物的完整性保存的发展或者游客娱乐计划都不能实施,现如今基本的原始自然条件在这个地区占主要位置。

2. "Here are no lofty peaks seeking the sky, no mighty glaciers or rushing streams wearing away the uplifted land. Here is land, tranquil in its quiet beauty, serving not as the source of water, but as the receiver of it. To its natural abundance we owe the spectacular plant and animal life that distinguishes this place from all others in our country."

这里没有直上云霄的高峰,没有巨型的冰山和激流冲刷抬高的地面。这里有的是,静美的安宁,不是水的发源地,却备受水的滋润。大自然的恩惠使我们拥有区别于这个国家其他地方的这么壮观的植物和动物。

3. The following year, the Florida legislature authorized the Tropical Everglades National Park Commission to take over the responsibilities of the Tropic Everglades National Park Association and with the power to acquire land by purchase, gift, beques or condemnation.

第二年,佛罗里达立法机关批准热带大沼泽地国家公园委员会权力接管热带大沼泽地国家公园协会,因此有了通过购买,赠送,遗赠和征用来获得土地的权力。

4. The Act provided for the classification of wildlife species as "endangered" or "threatened", and mandated legal protection for species so listed. In justification for such protection, the Act also recognized that the various species of fish, wildlife and plants have aesthetic, educational, historical and scientific value.

法案规定了濒危和受到危害的野生动物的分类级别,强制推出列出物种的合法保护政策。在这个保护法案中,还陈述了不同种类的鱼,野生动物和植物有审美、教育、历史和科学价值。

Questions

1. Why was the protection of the Everglades National Park so special?

2. What did President Harry S. Truman say on 6 December 1947 in a ceremony held at

Everglades City?

3. Who is Ernest F. Coe and what did he do to influence the Everglades National Park?
4. What are the special plants and animals in the Everglades National Park?
5. What measures were taken to preserve the Everglades National Park?

Chapter 15

Mount Rainier National Park

雷尼尔山国家公园

【简介】

雷尼尔山国家公园是一座以雷尼尔山为中心的公园,位于华盛顿西部,西雅图南面,1889 年为保护雷尼尔山自然景色而建立了这座国家公园。雷尼尔山是美国最高的火山,拥有除了阿拉斯加以外最大的单一冰河以及最大的冰河系统。由于太平洋吹来的东风湿度较高,地球上有史以来全年最大的降雪量就出现在这里。雷尼尔山是世界上最雄伟的山岭之一,从山顶向四周望去,可以看到 1 500 米以下的景色全被隐没在雾海之中,只有较高的山峰探出一角,仿佛海中的浮岛。山顶终年被冰雪覆盖,有 27 道冰河向四周喷射而出。另外,在山腹的草原地带,每到 7、8 月间,冰雪融化,花开满山,成了一片美丽的花海。

Mount Rainier established on 02 March 1899, contains vast expanses of pristine(原始的) old-growth forests, subalpine(亚高山带的) flower meadows, spectacular alpine(高山的) scenery, and great opportunity for stimulating outdoor activities. Mount Rainier National Park is the fifth oldest national park in the United States. The park has the greatest single-peak glacial system in the United States. Glaciers radiate(延展) from the summit and slopes of the 14,411 foot volcano. The park encompasses 378 square miles. Elevation ranges from 1,880 at the Carbon River rain forest to 14,411 feet at the summit. Annual visitation exceeds 2 million visitors, with the majority of visitors coming to the park during the months of May through October.

The first recorded view of Mount Rainier was made by Captain George Vancouver, a British explorer, while mapping Puget Sound in 1792. He named it after his friend Peter Rainier. The Indians called it Takhoma and had many legends about it.

In 1833, William Fraser Tolmie, became the first white man to enter what is now Mount Rainier after Tolmie's search for some medicinal herbs while working at Fort Nisqually. In 1857, a well documented attempt to climb Mount Rainier was made by an Army lieutenant, August Valentine Kautz. He was stationed at Fort Steilacoom and had the mountain practically(实际上) in his backyard. He developed a great desire to reach it's summit and with some companions hired a Nisqually Indian as their guide and headed for(前往) the foot of the Nisqually Glacier in July. After 6 days of traveling through forest and thicket(灌木丛), they finally began their climb up toward the summit. By the eighth day, Wahpowety the Nisqually guide, was suffering from snow blindness. His companions gave out(筋疲力尽) as well, but Kautz went forward and continued until he reached what was probably the 14,000 feet level. He was about 400 feet shy of(还不到) the summit. It was a great disappointment, however he proved that legends of Indians climbing Mount Rainier could be true. Mount Rainier could be climbed.

Settlements continued to spreadacross the seacoast, and so did the notoriety of the Northwest's greatest mountain. The US naval squadron(中队) mapping the area, used Mount Rainier as a landmark and giving it the altitude of 12,330 feet. Today it is considered to be 14,411 feet.

People continued to come and visit this beautiful, peaceful, challenging mountain. As the railroad, roads, and vehicle sightseeing became better and more efficient, and made it easier to visit Mount Rainier National Park, more people came. There are now over two million visitors annually.

Mount Rainier is an episodically(偶尔发生的) active volcano. The volcano began to grow between one half and one million years ago. The slopes of lava flows on opposite sides of the mountain probably projected more than 1,000 feet above the present summit. The upper portion of the cone was likely removed by explosions and landslides. The current summit, Columbia Crest, lies at 14,411 feet above sea level, on the rim of the recent lava cone(熔岩锥).

Mount Rainier, the highest (4392 m) volcano in the Cascade Range, towers over a population of more than 2.5 million in the Seattle Tacoma metropolitan area, and its drainage system via the Columbia River potentially impacts another 500,000 residents of southwestern Washington and northwestern Oregon. Mount Rainier is the most hazardous volcano in the Cascades in terms of its potential for magma(岩浆) water interaction and sector collapse, and major eruptions or debris flows even without eruption. It poses significant dangers and economic threats to the region but despite such hazards and risk, Mount Rainier has received little study.

In 1989 the International Association of Volcanology and Chemistry of the Earth's Interior (IAVCEI) established a Task Group for the International Decade for Natural Disaster Reduction.

The Task Group conceived the idea of selecting several volcanoes for focused study during the next decade as "Decade Volcano Demonstration Projects", established criteria for a "Decade Volcano" and solicited(请求) nominations. The IAVCEI accepted the nominations of seven volcanoes in developing countries and two in the U. S. Mount Rainier and Mauna Loa. Mount Rainier was chosen to be studied because it is representative of one or more volcanic hazards: it is geologically active as evidenced by surface manifestation(显示) of heat (geothermal activity), it has had recent volcanic events (last eruption was about 150 years. ago), and it is likely to erupt again, based on past history; its location poses significant hazards to a heavily populated area; it is a well known volcano (a number of research publications have been written on it); it is politically and physically accessible for study; and its volcanic geology is well exposed.

Glaciers are among the most conspicuous and dynamic geologic features on Mount Rainier in Washington state. They erode the volcanic cone and are important sources of stream flow(河川径流) for several rivers, including some that provide water for hydroelectric(水力发电) power and irrigation. Together with perennial(多年生的) snow patches(雪区), glaciers cover about 36 square miles of the mountain's surface, about nine percent of the total park area, and have a volume of about one cubic mile.

Glaciers may seem to be rigid and unchanging but in fact, they deform(变形) and flow continuously. Glaciers flow under the influence of gravity by the combined action of sliding over the rock on which they lie and by deformation, the gradual displacement between and within individual ice crystals(冰晶). Maximum speeds occur near the surface and along the centerline of the glacier. During May, 1970, Nisqually Glacier was measured moving as fast as 29 inches per day. Flow rates are generally greater in summer than in winter, probably due to the presence of large quantities of meltwater at the glacier base.

Climatic conditions in large part regulate the size of a glacier because they control the quantities of snowfall and melt. The position of the snout(吻状突起), or terminus, of a glacier may change as the relative quantities of snowfall and glacier melt change. If summer melt exceeds winter snowfall, the terminus retreats, whereas if snowfall exceeds summer melt, the terminus advances. These changes in terminus position do not occur instantaneously(即刻的), but typically take several years or more to become apparent. Glaciers are therefore sensitive indicators of climate changes.

Scientists measure winter snow accumulation and summer melt of snow and ice to analyze the response of glaciers to climate; however, it is very time-consuming and potentially a hazardous task. Consequently, alternative data, which are obtained by mapping of terminus positions and surveying of glacier surface elevations, are commonly used. At Mount Rainier, annual measurements of Nisqually Glacier's terminus position were begun in 1918 by National Park Service (NPS) personnel(全体人员) and are currently made by U. S. Geological Survey (USGS) personnel.

Changes in terminus position may actually be forecast by precise surveys of a glaciers surface elevation. For example, a rise in surface elevation, which reflects an increase in ice thickness, is typically followed within a few years or decades by terminus advance. The surface-elevation record at Nisquaffy Glacier is the lengthiest of any made in North America. The record, which was started in 1931, shows the glaciers dramatic responses to about half a century of small but significant climatic variations. These measurements of surface elevation were begun by personnel of the Tacoma City Light Co. because of their interest in water for hydroelectric power. Measurements were later done by USGS personnel and most recently by NPS personnel.

Retreat of the Little Ice Age glaciers was slow until about 1920 when retreat became more rapid. Between the height of the Little Ice Age and 1950, Mount Rainier's glaciers lost about one-quarter of their length. Beginning in 1950 and continuing through the early 1980's, however, many of the major glaciers advanced in response to relatively cooler temperatures of the mid-century. The Carbon, Cowlitz, Emmons, and Nisqually Glaciers advanced during the late 1970's and early 1980's as a result of high snowfalls during the 1960's and 1970's. Since the early 1980's and through 1992, however, many glaciers have been thinning and retreating and some advances have slowed, perhaps in response to drier conditions that have prevailed at Mount Rainier since 1977.

The seasons and elevation play a big role in determining where you may best find wildlife. In summer, chipmunks(花栗鼠), chickarees(赤栗鼠), ground squirrels(地松鼠), marmots(土拨鼠), and pika(鼠兔) are commonly seen mammals. Clark's nutcrackers(胡桃夹子), gray jays(灰松鸡), Steller's jays(暗冠蓝鸦), and ravens are commonly seen birds. Deer are frequently seen, but black bear, elk, and mountain goats are more elusive(隐蔽的). Look for elk on the east side of the park in September. The color of black bear may be brown, tan or blond, and to see one is a rare treat. Mountain goats stay close to the high country cliffs.

Notes

Mount Rainier 雷尼尔山　喀斯喀特山脉的大火山之一,有不少冒汽的岩洞和温泉。不仅是华盛顿州的地标,许多器物皆以此山为图案,而且对该州人而言,更带有几分神秘与传说的色彩,颇具神圣的地位。雷尼尔山为喀斯喀特山脉的高峰,是世界上最雄伟的山岭之一,海拔4 323米,比邻近高峰高出约2 500米。雷尼尔山同时也是一个包覆着超过91平方千米雪和冰的活火山。火山为圆锥形,基盘为花岗岩,火山体为安山岩。它拥有除了阿拉斯加以外最大的单一冰河与最大的冰河系统。雷尼尔山的山顶终年冰封积雪,有27道冰河向四方放射而出。

George Vancouver 乔治·温哥华(1757—1798年) 英国航海家,1791—1794年率考察队从英国出发,精确勘测了从旧金山至不列颠哥伦比亚省的太平洋海岸,并测量了温哥华岛海域的小海湾和航道。

Puget Sound 普吉特海湾 位于美国太平洋西北区,通过胡安·德富卡海峡与太平洋相连。整个海湾周边地区集中了华盛顿州九大城市中的六个,即西雅图、塔科马、埃弗里特、肯特、贝尔维尤和费德勒尔韦,人口约400万。

Nisqually Glacier 尼斯夸力冰河 美国华盛顿州雷尼尔山南侧最大的冰河之一。在山上最容易看到的就是尼斯夸力冰河,在雷尼尔山国家公园内通过天堂游客设施可以到达。现在冰川却在逐渐消退,在2 800米的海拔的测量显示,在1994—1997年间,冰山厚度就消退了17米。

Tacoma 塔科马 美国华盛顿州西部港口城市。位于普吉特湾南端。海陆空交通要地。农林产品集散地和加工中心,有"美国木材业首都"之称。主要工业有木材加工、机器制造、食品加工以及有色金属冶炼、造船等,是前往雷尼尔山和奥林匹克山的旅游基地。有普吉特湾大学(1888年建)、太平洋路德大学(1890年建),是美国华盛顿州普吉特海湾南端的一个港口城市,约在西雅图和州府奥林匹亚之间。2006年人口为201 600人,是该州第三大城市、皮尔斯县县治。

Columbia River 哥伦比亚河 北美洲西部大河之一。源出加拿大南部洛矶山脉,西南流经美国,注入太平洋。长2 044千米,流域面积415 211平方千米。最大支流斯内克河。河口平均流量7 400米/秒。春夏高山冰雪融解时水量最大,冬季最小。干、支流可通航约1 000千米,大洋海轮可直达河口以上150千米的波特兰。水力资源丰富。

International Association of Volcanology and Chemistry of the Earth's Interior (IAVCEI) 国际火山学与地球内部化学协会 协会由每四年变换一次成员的执行委员会所管理。委员会决定协会的政策,使得他们能够通过一系列的手续费和任务。协会的目标是更对外化,探索和其他国际科学组织的关系,并参与国际科学项目研究。同时要维持一个稳健的出版政策,主要是通过国际期刊火山公告来奖励高水平火山研究结果。

Mauna Loa 冒纳罗亚火山 夏威夷海岛上的一个活跃盾状活火山,山顶的大火山口叫莫卡维奥,意思为"火烧岛"。火山爆发带来周期性和毁灭性破坏,凡岩浆流经之处,森林焚毁,房屋倒塌,交通断绝。冒纳罗亚火山海拔4 205米,是夏威夷的最高峰。其山顶常有白云缭绕,忽隐忽现。随着太平洋板块的缓慢漂泊,冒纳罗亚火山最终被带离热点,并将在50~100万年后停止喷发。

U.S. Geological Survey 美国地质勘探局 一个美国政府内政部的下属机构。此机构的科学家主要研究美国的地形、自然资源和自然灾害及其应付方法(如地震、火山喷发、泥石流等)。该部门负责四大科学范畴:物理学、地理学、地质学和水文学。

Little Ice Age 小冰河期　顾名思义指的是相对而言较冷的时期,但是比主要的冰期还要暖和,维持的时间也比较短。一般而言,小冰河期泛指15世纪至19世纪中叶气温偏低的时期。大约在16世纪到19世纪之间,地球上广大地区出现了寒冷气候,人们称为小冰河期。它始于13世纪,在17世纪达到巅峰,并最终在大约200年前减缓了活动趋势。在它的巅峰时期,北欧一带饥荒肆虐。有人认为挪威和瑞典有一半的人口在饥荒中丧生。与此同时,全球各地冰雪蔓延:埃塞俄比亚的部分地区白雪皑皑,中国的农作物歉收,苏必利尔湖面开始结冰。

Translation of Selected Sentences

1. Mount Rainier, the highest (4,392 m) volcano in the Cascade Range, towers over a population of more than 2.5 million in the Seattle Tacoma metropolitan area, and its drainage system via the Columbia River potentially impacts another 500,000 residents of southwestern Washington and northwestern Oregon.

雷尼尔山是位于喀斯喀特山脉的最高的火山(4 392 米),坐落在西雅图塔科马大都市地区的人口远远超过250万,流经哥伦比亚河的排水系统潜在地影响着华盛顿州西南部和俄勒冈州西北部的另外50万人口。

2. Mount Rainier is the most hazardous volcano in the Cascades in terms of its potential for magma water interaction and sector collapse, and major eruptions or debris flows even without eruption.

雷尼尔山是喀斯喀特山脉最危险的火山,因为它潜在的岩浆水的相互作用和部分坍塌,还有大量的喷发,甚至可以在没有爆发的情况下涌出泥石流。

3. The Task Group conceived the idea of selecting several volcanoes for focused study during the next decade as "Decade Volcano Demonstration Projects", established criteria for a "Decade Volcano" and solicited nominations.

任务组已经构想出选择几个火山在下一个十年作为"十年火山演示工程"来集中研究,并且已经建立了"十年火山"的标准并申请提名。

4. Glaciers flow under the influence of gravity by the combined action of sliding over the rock on which they lie and by deformation, the gradual displacement between and within individual ice crystals.

在重力的影响下,冰川漂动,流经岩石时,因附着在岩石上,或是发生变形的相互作用,逐渐代替个体冰晶。

5. Scientists measure winter snow accumulation and summer melt of snow and ice to analyze the response of glaciers to climate; however, it is very time-consuming and potentially a hazardous task. Consequently, alternative data, which are obtained by mapping of terminus positions and surveying of glacier surface elevations, are commonly used.

科学家测量冬季积雪量和夏季冰雪融化量来分析气候对冰山的作用,但是这是个费时并且有潜在危险的工作。因此,通常使用由最终位置的绘图和冰川表面海拔的研究所获得的变动的数据。

Questions

1. Why does Mount Rainier contain great opportunity for stimulating outdoor activities?

2. How was the first recorded view of Mount Rainier made? And where did the name come from?

3. Why was Mount Rainier chosen to be studied? And what is the specialty?

4. May the changes in terminus position actually be forecast by precise surveys of a glaciers surface elevation?

5. Do the seasons and elevation play a big role in determining where you may best find wildlife?

Chapter 16

The National Park of American Samoa
美属萨摩亚国家公园

【简介】

美属萨摩亚位于南太平洋之上,位于赤道与南回归线之间,而美国美属萨摩亚国家公园便位于这里。其主要岛屿有图图伊拉岛、奥福岛和塔乌岛。该地于 1988 年成为美国的国家公园,1993 年 9 月 9 日由美国国家公园管理局向当地政府提出签下为期 50 年的租约。该国家公园内有珊瑚礁、热带雨林,非常适合远足和潜水。美属萨摩亚国家公园的设立初衷就是为了保护萨摩亚独一无二的自然资源,公园占地面积约 36 平方千米,其中 10 平方千米的面积是水域。

Located some 2,600 miles southwest of Hawaii, this is one of the most remote national parks in the United States. It is also one of the newest; it was authorized by the U. S. Congress in 1988 and officially established in 1993 when a 50 year lease was signed.

American Samoa, the only US Territory south of the equator, consists of five rugged, highly eroded; volcanic islands, and two coral atolls(珊瑚环礁). The land area of the territory is 76 square miles. The population is approximately 60,000, with most people living on the main island of Tutuila. Tuna fishing(金枪鱼) and canning(罐头) are major industries. Many Samoans practice subsistence farming.

The nation of Samoa lies 60 miles to the west of Tutuila. The main islands of Upolu and Savaii constitute a much larger land area than American Samoa. Between these two large islands are the small islands of Manono and Apolima. Several uninhibited islets(小岛) lie just off the eastern shore of Upolu.

The geology of the Samoan islands is surprisingly interesting. First, we are living on a volcano, which is resting quietly at the moment. Second, our volcano is on the move, it's traveling towards China with us on it. And, finally and most unfortunately, our volcano is doomed and it will eventually sink back into the dark ocean depths.

The most recent volcanic eruptions were a lot more recent than many people realize. In Western Samoa, major eruptions occurred in 1905 when lava flows destroyed a village. In the Manua islands, subsurface volcanic eruptions and earthquakes occurred in 1866, causing dense clouds of smoke and pumice(浮石) to erupt from the ocean surface for several months. One hundred years ago is just a blink of the eye to a volcano, which measures time in the millions of years. We humans tend to forget how briefly people have lived upon these shores. Human habitation on Tutuila, even considering the whole 3000-year period that Samoans have dwelt here, represents a mere 0.2% of the time since the sun first shone on this new land.

Finally, all oceanic volcanos must come to an end. As time passes, two things happen. Our volcano erodes continuously as ocean waves attack its shorelines and rivers gouge into its terrain. In addition, the weight of a newly formed volcano is so heavy that it causes the volcano to sink slowly back down into the sea. Rose Atoll(罗斯环礁) and Swains Island(斯温斯岛) are good examples of sunken volcanos. In the distant past, Rose and Swains may have been magnificent mountainous islands with beautiful rainforests and coral reefs. But that's ancient history now, because those islands, over a period of several million years, eventually sank out of sight. All that remains are tiny amounts of coral that grew up from the peaks of the mountains as they slipped below the sea surface.

American Samoa is much larger than you might think it is. The whole Territory covers 117,500 square miles, which is about the size of New Zealand or the state of Oregon. The Territory is big because we claim jurisdiction of the ocean that surrounds American Samoa, from the shoreline out to 200 miles offshore. That is standard procedure throughout the world (each country with marine coasts wants to protect its coastline and marine resources from others). To be more precise, American Samoa has jurisdiction(管辖权) over territorial waters out to 3 miles, while the US federal government maintains control of the zone from 3 to 200 miles from shore.

Most of the Territory is open ocean, of course. Only a minuscule(极小的) 0.1% of the area consists of dry land (all 7 islands total only 76.1 square miles). The other 99.9% marine portion consists of two main habitats, the shallow coastal waters adjacent to the islands and the deep waters offshore (vasa). Shallow coastal habitats, with their coral reefs and colorful fish, are quite limited in total area because our islands slope steeply down into deeper water and depths of 2 000 feet are reached within 0.5 – 2 miles from shore. So, most of our coral reefs are restricted to a narrow ring around each island. In total, we have about 114 square miles (296 km) of coral reefs in the Territory.

The volcanic island of Tutuila is the largest of the seven islands that comprise American Samoa. Pago Pago Harbor, a collapsed volcanic caldera is one of the largest natural harbors in the South Pacific. It cuts deeply into the south-central coast almost dividing the island in two. From east to west, a steep mountainous spine runs the 20 – mile length of the island, punctuated in places by notable summits including Matafao Peak, Tutuila's tallest mountain at 2 142 feet; North Pioa Mountain, popularly known as Rainmaker Mountain, 1718; and Mount Alava, the steep ridgeline looming to the north of Pago Pago Harbor, marking the south boundary of the park area. About one mile southeast of Tutuila's eastern tip is the volcanic island of Aunu'u the smallest of the inhabited islands; it can be reached by boat and explored by foot in a day. After you visit the National Park, visits to the east and west ends of the island will reveal more island and ocean scenery and insights into Samoan culture.

The Manua Island group includes the volcanic islands of Ofu and Olosega (joined by a bridge), and Tau. These islands lie 60 miles east (a 30 – minute flight on a small plane) from Tutuila; flights are available between Ofu and Tau. These islands are sparsely populated, and each village has only a few hundred people.

Farther east lies the Rose Atoll, and to the north is Swains Island. These coral atolls are difficult to get to and not possible to visit. Rose Atoll is a US National Wildlife Refuge.

American Samoa is home to many different birds and fish and a few very interesting mammals.

Bats are the only native mammals found in American Samoa. They are harmless to people. There are two species of flying foxes (fruit bats)(果蝠)and one insect-eating bat species here. The fruit bats, the Samoan flying fox and the white-collared flying fox, are as large as one pound with a wing span of 3 feet. They eat pollen(花粉)and fruits and serve an essential role in the ecology of the native rainforest by pollinating(授粉)plants and dispersing seeds.

On oceanic islands, birds are usually the most abundant and diverse group of animals. This park is home to more than 35 species, both resident and migratory, including seabirds, water birds, forest birds, and shore birds. The cliffs and sea stacks(海柱) are ideal nesting habitat for tropical seabirds, while the rainforest is home to collared kingfishers(翠鸟), cardinal and wattled honeyeaters, bluecrowned lories, Samoan starlings, purple-capped fruit doves, many-colored fruit doves, Pacific pigeons and banded rails.

Coral reefs abound in our coastal waters and are the homes of nearly 900 fish species, over 200 coral species, and countless other invertebrates. Many of these are harvested by villagers.

From the mountaintops down to the ocean's edge, the islands are covered with mixed species, paleotropical (Old World)(古热带的) rainforest. Samoa's rainforest is closely related to that of Asia and Africa, as opposed to neo-tropical (New World) rainforests of Central and South America. In mixed species forests, no single tree or plant species dominates. This is the only rainforest of its kind protected within a US national park.

Of the park's total area of 10,500 acres about 8,000 acres are rainforest. Depending on the elevation, the forest can be characterized as coastal, lowland, ridge, montane(山地森林) or cloud forest. In addition to many species of trees, plants found here include vines, ferns, lichens (地衣), eoiohytes (air plants)(气生植物), and mosses.

Each of the islands of American Samoa supports a coral reef ecosystem in coastal waters. Two islands, Rose Atoll and Swain's Island, are coral atolls with well developed reef systems. The remaining five islands are extinct volcanoes surrounded by narrow fringing reefs(岸礁). Offshore from the volcanic islands of Tuutila, Aunu'u, Ofu, Olosega, and Tau are drowned barrier reef.

Coral reefs support a wide variety of marine life representative of the tropical Indo-Pacific waters. To date, nearly 900 species of fish and nearly 200 species of coral have been documented. The fringing reef on the south side of the island of Ofu within the park is one of the best preserved reefs in this part of the South Pacific. It is an excellent place for snorkeling(潜水). Remember that coral is made up of living organisms(生物体) and can be easily damaged. It is unlawful to collect coral, or any other marine life.

Notes

American Samoa 美属萨摩亚　地处太平洋波利尼西亚岛群的中南部,位于南纬13°~15°,西经171°~168°。面积197平方千米。由萨摩亚群岛171°西经线以东的图图伊拉岛、塔乌岛、奥富岛、奥洛塞加岛等大小7个岛屿组成。总人口3.4万(1984)。大部分为波利尼西亚人的萨摩亚族,也有少数美国人、新西兰人和华人。居民多信奉基督教。官方语言为萨摩亚语。首府帕果帕果,位于图图伊拉岛中部帕果帕果湾,是主要的港口,人口1.1万。

Tutuila 图图伊拉　太平洋中南部东萨摩亚群岛的主岛,位于南纬14°17′、西经170°41′。陆地面积137平方千米。人口2.7万。首府帕果帕果在南岸。东、中部多山,森林茂密,最高峰马塔福峰海拔702米。西部有肥沃平原。西南角的塔富纳有大型飞机场。出口椰干、鲔鱼罐头及手工艺品。太平洋西南部美属萨摩亚的最大岛屿。长约30千米,最宽处10千米,面积135平方千米。有山岭贯穿断断续续的全岛,山区林木茂密,多深谷。海岸地带土地肥沃。出口金枪鱼罐头、椰仁乾和手工艺品。主要城镇帕果帕果(Pago Pago),为太平洋最好港口之一。人口45 043(1990)。

Upolu 乌波卢岛　萨摩亚独立国两主岛中面积较小的一个,也是人口较多、经济较发达的一个,首都阿皮亚即位于该岛北部。该岛西隔阿波利马海峡与萨瓦伊岛相望。乌波卢岛是一个位于西南太平洋的火山岛,面积1 125 平方千米,人口约11 万。英国作家罗伯·路易斯·史蒂文生曾在岛上买下一座种植园长期居住,1894 年在此逝世并埋葬于岛上。在波利尼西亚神话(特别是萨摩亚)中,乌波卢是在同名岛上的第一个女人。18 世纪末19 世纪初,该岛有时被叫做 Ojalava 或 Ojolava。一种极小的蜘蛛生活在乌波卢岛上。据2005 版吉尼斯世界纪录记载,该蜘蛛大小相当于该书的一个句点。

Savaii 萨瓦伊岛　太平洋中南部西萨摩亚最西端和最大的岛屿。位于南纬13°35′、西经172°25′。与乌波卢岛(Upolu)隔有阿波利马(Apolima)海峡。长约80 千米,最宽处40 千米,面积1 707 平方千米。人口约43 150(1981)。岛上多火山,中部的毛加西利西利山(Mount Silisili)海拔1 858 米,为西萨摩亚最高峰。西北部为平原。西北端的阿绍于1966 年建有深水码头。出产椰干、香蕉、可可、芋头、咖啡和牛。多野猪。1970 年西北部的阿绍(Asau)开始发展木材工业。主要居住区为东岸的图亚西维(Tuasivi)。

Manono 马诺诺　大洋洲萨摩亚的一个岛屿,位于萨瓦伊岛与乌波卢岛之间的阿波利马海峡中。距乌波卢岛最西端3.4 千米。它是艾加伊勒泰区的一部分。

Apolima 阿波利马岛　萨摩亚的一个岛屿,位于萨瓦伊岛与乌波卢岛之间的阿波利马海峡中。该岛为四个有人居住的岛屿中最小的一个。该岛面积略小于1 平方千米。它是艾加伊勒泰区的一部分。该岛人口为88 人(2001 年统计),皆居住在该岛唯一的村庄阿波利马-犹他(Apolima-Uta)中。该村位于该岛北端。

Pago Pago 帕果帕果　太平洋中南部美属萨摩亚首府和主要港口。位于图图伊拉岛南岸中部帕果帕果湾内。人口4 000 人(1990)。为天然良港。美国重要的海军基地。输出金枪鱼及椰干等。1872 年帕果帕果被选为美国海军的装煤站,1951 年以前一直是个活跃的海军基地,现为船只的定期停靠港。1964 年机场开放,促进了旅游交通和现代化。

Translation of Selected Sentences

1. In the Manua islands, subsurface volcanic eruptions and earthquakes occurred in 1866, causing dense clouds of smoke and pumice to erupt from the ocean surface for several months.

在魔奴岛上,于1866 年发生的地下火山喷发和地震导致了浓密的烟雾云和浮石从海洋表面喷发了数月之余。

2. All that remains are tiny amounts of coral that grew up from the peaks of the mountains as they slipped below the sea surface.

所剩的是细小珊瑚生长于山峰之巅,缓缓地滑落到海洋表面以下的。

3. To be more precise, American Samoa has jurisdiction over territorial waters out to 3 miles, while the US federal government maintains control of the zone from 3 to 200 miles from shore.

更准确地说,美属萨摩亚拥有领水周围三英里的管辖权,美国联邦政府拥有从海岸算起的 3 ~300 英里的管辖区域。

4. From east to west, a steep mountainous spine runs the 20-mile length of the island, punctuated in places by notable summits including Matafao Peak, Tutuila's tallest mountain at 2,142 feet; North Pioa Mountain, popularly known as Rainmaker Mountain, 1 718; and Mount Alava, the steep ridgeline looming to the north of Pago Pago Harbor, marking the south boundary of the park area.

从东到西,陡峭的山脊绵延海岛长达 20 英里,在一些地方被著名的包括穆塔夫峰在内的山峰所断开,图图伊拉最高的山有 2 142 英尺,北朴奥山,或者又叫做造雨山有 1 718 英尺,阿拉瓦山的陡峭的山脊线在北帕果帕果湾隐约可见,是公园地区的南部边界。

6. The cliffs and sea stacks are ideal nesting habitat for tropical seabirds, while the rainforest is home to collared kingfishers, cardinal and wattled honeyeaters, bluecrowned lories, Samoan starlings, purple-capped fruit doves, many-colored fruit doves, Pacific pigeons and banded rails.

悬崖和海柱是热带海鸟理想的栖居地,热带雨林是翠鸟、红色垂肉蜜雀、蓝冠鹦鹉、萨摩亚八哥、紫盖果鸽、彩色果鸽、太平洋鸽和带秧鸡的家园。

Questions

1. Why is the geology of the Samoan islands surprisingly interesting?

2. Finally all oceanic volcanos must come to an end and as time passes, which two things happened?

3. Which volcanic island is the largest of the seven islands that comprise American Samoa?

4. Which is the only native mammal found in American Samoa?

5. Depending on the elevation, what can the forest be characterized? In addition to many species of trees, what do plants found here include?

Chapter 17

Yosemite National Park
约塞米蒂国家公园

【简介】

约塞米蒂国家公园为中国大陆的译名，台湾译作"优胜美地国家公园"，其他译名还有优山美地国家公园、约瑟米提国家公园等。它是美国西部最美丽、参观人数最多的国家公园之一，与大峡谷国家公园、黄石国家公园齐名，位于加利福尼亚州东部内华达山脉（Sierra Nevada，西班牙语：下雪的山）上。

Yosemite National Park was established on 1st October 1890, and is the Nation's third oldest national park. The park was established for the purpose of preservation of the resources that contribute to its uniqueness and attractiveness. Congress recognized the importance of preserving this great park for future public enjoyment when it established Yosemite National Park. Yosemite National Park is a showcase of spectacular geological features, including the greatest concentration of granite domes（沙漠拱丘）in the world and the largest exposed granite monolith（庞然大物）in the world.

Yosemite National Park embraces almost 1,200 square miles of scenic wild lands set aside in 1890 to preserve a portion of the central Sierra Nevada that stretches along California's eastern flank（侧面）. The park ranges from 2,000 feet above sea level to more than 13,000 feet and has these major attractions; alpine wilderness, three groves of Giant Sequoias and the glacially carved Yosemite Valley with impressive waterfalls, cliffs and unusual rock formations.

Yosemite National Park is open 24 hours a day, year round with the highest visitation in June, July and August and the lowest in December, January and February.

The first application of a park concept originated in Yosemite with the grant of 1864 (Federal land given to California for preservation) signed by Abraham Lincoln and since that time the park has played an important role in pioneering park management concepts.

Yosemite National Park possesses outstanding recreational values and supreme scenic attractions, including alpine and subalpine wilderness, three groves(小树林)of giant sequoia(巨杉)trees and thundering(如雷鸣般的)waterfalls that are among the world's highest. Yosemite was the birthplace of the idea of the Sierra Club and plays an important role in wildlife preservation and preserving biological diversity.

Yosemite National Park is a world heritage site which has made a significant contribution to California's cultural heritage, to the national park movement, and to Yosemite's 4,000 years of cultural heritage by Native Americans. The park provides solitude and inspiration and serves as an outdoor classroom for environmental education.

While man has lived in Yosemite for thousands of years, the park's human history is far shorter than its geological history. At one time, this area was made up of gentle rolling hills, crisscrossed with a maze of stream systems. Millions of years ago, California's Sierra Nevada was formed by a gradual series of earth upheavals. As the mountains rose, the land tilted and the westward flowing Merced River accelerated, carving deep, v-shaped river canyons. Later, massive glaciers flowed down the canyons. Colder temperatures slowed melting and eventually glaciers formed and began to carve away at the v-shaped canyons, transforming them into u-shaped valleys. Tributary streams did not carve their canyons as deep as Merced Canyon. Glaciers sheared off(切断)these canyons leaving them as "hanging valleys". Tributary creeks, which had once joined the main stream at the same elevation, now plummeted(垂直落下)off of shear cliffs, giving birth to the park's famed waterfalls. Eventually, sediment washed down out of the high country, filled in Lake Yosemite to form the present valley floor.

The area's first residents were Native Americans who inhabited the region perhaps as long ago as 7,000 to 10,000 years. Various tribes lived in the area over the years, the most recent of which was a Miwok tribe that called Yosemite Valley Ahwahnee which is believed to mean, "place of the gaping mouth". They referred to themselves as the Ahwahneechee.

The Ahwahneechee lived off the land, harvesting acorns, hunting and fishing. The discovery of gold in the foothills of California ended this idyllic lifestyle when some of the tribe, angered by the encroachment of the western miners, attacked a trading post in the Merced River Canyon. In retaliation, the miners organized state-sanctioned Mariposa Battalion, which entered Yosemite Valley on 27 March 1851 in pursuit of the Yosemite Indians. Tenaya, the Yosemite chief, had been leading his tribe in raids on white settlers in the foothills of the Sierra Nevada. The Battalion captured Tenaya and his tribe. They marched them to reservations in the foothills and eventually let the Indians return to the valley, which was named after them.

By 1855, the first party of tourists arrived and nine years later, encouraged by a group of influential Californians, Abraham Lincoln signed the Yosemite Grant which set aside Yosemite Valley and the Mariposa Grove of Giant Sequoias as a state supervised public reserve.

In 1890, Robert Underwood Johnson, editor of Century Magazine, and John Muir, were concerned that the high country and watershed for Yosemite Valley were being destroyed by grazing and timber interests. The two launched a successful campaign to persuade Congress to set aside the high country as a national park.

In 1906, Yosemite Valley and the Mariposa Grove were returned to federal jurisdiction. In 1932, the Wawona Basin, including the Wawona Hotel and golf course were purchased and included in the National Park.

Wawona was once an Indian encampment and, later, was the site of a wayside hotel built in 1856 by Galen Clark. Known as Clark's Station, it served as a stop for visitors in the transit between Yosemite Valley and Mariposa. In 1864, when Yosemite Valley and the Mariposa Groves were set aside for protection, Clark became the first guardian of the area. In 1875, the year the original Wawona road opened, the Washburn brothers purchased the area and build the Wawona Hotel that is still in operation today. Wawona focuses on Yosemite's human history. It is the setting of the Pioneer Yosemite History Center, a collection of relocated historic buildings and horse-drawn coaches.

Open since 1927, the Ahwahnee is one of America's most distinctive hotels, unparalleled in magnificence and charm. The hotel is a great American castle, massive and warm with huge cathedral ceilings, enormous stone hearths and richly colored Native American and Oriental rugs. The hotel was designated a National Historic Landmark on 02 June 1987.

The story of Yosemite began about 500 million years ago when the Sierra Nevada region lay beneath an ancient sea. Thick layers of sediment lay on the sea bed, which eventually was folded and twisted and thrust above sea level. Simultaneously molten rock(熔岩)welled up(涌出) from deep within the earth and cooled slowly beneath the layers of sediment to form granite. Erosion gradually wore away almost all the overlying rock(覆盖岩石)and exposed the granite. And even as uplifts continued to form the Sierra, water and then glaciers went to work to carve the face of Yosemite. Weathering and erosion continue to shape it today.

Yosemite National Park is home to 300 – 500 American black bears, Ursus americanus(美洲黑熊). Although usually referred to as the black bear, very few are black, and they are more likely to be found in a variety of colors ranging from black to brown, blond, or cinnamon.

Black bears are omnivores(杂食动物) and will eat almost anything. They spend most of their days foraging for(搜查) grasses, seeds, berries, acorns(橡树子), and insects and occasionally feed on carrion(腐肉). Bears tear open rotten logs or old stumps in search of insect larvae. Meadows also furnish(提供) a wide variety of food, such as grass, clover(三叶草), lily, wild onion, and brodiaea bulbs. Research in Yosemite shows that plants, including acorns, comprise 75% of the diet of Yosemite bears. Bears are also fond of fruit, particularly manzanita (石兰科常绿灌木), service berry, elderberry(莲灰), and wild cherry. Chipmunks(花鼠), ground squirrels, marmot, pocket gophers, and mice are also a part of their diet. In the fall, bears are often seen beneath oak trees searching for acorns. Unfortunately, many Yosemite bears have also perfected the skill of obtaining food from humans.

Bears are also opportunists, which means they can easily adapt to new foraging habitats, from meadows to manzanita bushes, from cars to picnic coolers. Opportunism is often seen as an indication of intelligence in animals and most researchers agree that bears are highly intelligent. Anyone who has lost food to a bear learns that a bear's strength and intelligence should never be underestimated. This enterprising nature of the black bear can be linked to the fact that cubs stay with their mothers for well over a year after birth. This allows mother bears time to teach cubs survival techniques that directly relate to opportunism. It is this characteristic which allows cubs to learn from their mothers how to break into cars for food.

The biggest threat to the survival of the black bear in Yosemite is the availability of human food; in cars, campgrounds, picnic areas, and the wilderness. Oncea bear is rewarded by obtaining human food, it will often continue to seek it out and some may even resort to intimidating(吓人) humans in order to get more. As their natural fear of people fades, they may become more aggressive. When they become too aggressive and human safety is threatened, bears are sometimes killed by park rangers.

Obtaining human food alters the natural foraging habits, population dynamics(种群动态), biology, and behavior of bears. It is the ultimate goal of wildlife managers is to have all bears in Yosemite eating their natural diet, avoiding humans and our food altogether. It is your responsibility to store food properly when visiting Yosemite. Your actions can affect the lives of bears!

When pioneers came to this scenic area more than 100 years ago, many hundreds of bighorns (大角羊) roamed these mountains. Soon after, however, the California bighorn sheep were forced to extinction in Yosemite National Park by hunting, diseases, and the competition for food by domestic sheep. Today, through the efforts of the National Park Service, U. S. Forest Service, California Department of Fish and Game and the Yosemite Association, the bighorns are once again part of Yosemite's wildlife population. Found along the eastern edge of the park, the bighorns are sometimes seen beside Highway 120 beyond Tuolumne Meadows (泰伦恩草地).

The population of Golden Eagles, Aquila chrysaetos (金雕), within the boundaries of Yosemite is doing well. Golden eagles are not commonly seen in Yosemite Valley. They are superb fliers and hunters. Look for their golden crown of feathers. Young birds have white areas on wings and tails.

Yosemite is home of the rare and endangered Great Gray Owl, Strix nebulosa. A current study is underway documenting the status, distribution, numbers, habits and health of the park population of this large, noble species. In some areas of the park you may hear their distinctive, deeply-toned hoot.

This silver-grey member of the canine(犬科的) family is seen year round. At night you may hear them singing in a chorus of howls, barks and yodels. Coyotes, Canis latrans, are primarily predators(捕食者) of field mice and squirrels, though they have learned to beg from people. Please do not feed them, as human food is harmful to them, they may bite and conditioning them to seek food from people makes them vulnerable to being hit by passing cars.

All the deer in Yosemite are mule deer. Often seen in or near meadows browsing(吃草) or grazing(放牧), the naturally timid mule deer have grown accustomed to seeing people. More attacks on humans occur by deer than by bears. Though they appear to be tame and may even approach you, the California mule deer is a wild animal and will charge if cornered or threatened. Its hooves and antlers(鹿角) are sharp. Always leave the deer a wide area to walk away and, like all other animals in the park, never tempt them with food.

A variety of tree and ground squirrels can be found throughout Yosemite. One of the traditional favorites is the western gray squirrel, Sciurus griseus. During mating season (twice a year) they become animated, chasing one another, fighting and making noise. Look for the impressive gray, bushy(多毛的) tail.

Of the 27 varieties of trees in the park, these four are easy to identify, due to their impressive size and distinctive characteristics:

California black oak is abundant in Yosemite Valley. These large deciduous trees(落叶树), with yellow-green leaves and dark trunks, produce acorns which the Miwok Indians ground into nutritious flour.

The incense-cedar(翠柏) has a feathery reddish bark which is often confused with the giant sequoia(巨松). Incense-cedars grow abundantly throughout the Sierra, while sequoias grow only in limited numbers of groves.

The giant sequoias are the largest trees on earth. Three groves of them are located in the park: the Mariposa Grove near the southern entrance (Highway 41); the Tuolumne Grove, near Crane Flat on the Tioga Road; and the Merced Grove, off the Big Oak Flat Road between Crane Flat and the Big Oak Flat Entrance. The giant sequoia often lives from 1,000 to 3,000 years. A few of these big trees were planted by settlers in Yosemite Valley, though they are not native to the Valley.

The ponderosa pine(杰克松) can be identified by its bark made up of irregularly shaped plates separated by dark furrows. Mature trees are considerably wide at the base with a straight trunk rising many feet before reaching the branches.

Notes

Sierra Nevada 内华达山脉　美国西南部花岗岩断块山,科迪勒拉山系西缘山地的组成部分。北起拉森峰,南至蒂哈查皮山口,长 640 千米,宽 80 千米~130 千米,以大致南北走向绵亘于加利福尼亚州东部。山体形成于侏罗纪末至白垩纪初的内华达造山运动,一般海拔1 800~3 000 米,山势自东向西倾斜;东坡断崖陡峻,拔起于东邻大盆地之上,高差达1 500~3 000米,植被稀疏,多为灌木草类;西坡平缓倾向加利福尼亚中央谷地,多河流切割成深邃峡谷,森林茂密,迎太平洋湿润气流而雨雪丰沛,河流、湖泊众多,水力资源丰富,使中央谷地成为富饶的农业区。

Yosemite Valley 约塞米蒂谷　一译"尤塞米提谷"。美国加利福尼亚州中东部内华达山西坡的冰川槽谷,在圣弗朗西斯科以东约 251 千米处。印第安人称为"阿赫瓦尼",意为"深草谷地"。谷底宽平,谷壁陡峭,具冰蚀 U 形谷的典型特征。两侧多悬谷。圣华金河支流默塞德河上游流贯其中,形成一系列瀑布,其中约塞米蒂瀑布分三段降落 796 米,是北美洲最长的瀑布。1890 年连同附近地区的湖泊、草甸、丛林(其中有巨大的红杉树种)等辟为约塞米蒂国家公园。已修建三条汽车公路供旅游者登临游览。

Sierra Club 塞拉俱乐部　又译作山岳协会、山峦俱乐部和山脉社等,是美国的一个环境组织,著名的环保主义者约翰·缪尔(John Muir)于1892 年 5 月 28 日在加利福尼亚州旧金山创办了该组织,并成为其首任会长。塞拉俱乐部拥有百万会员,分会遍布美国,且与加拿大塞拉俱乐部(Sierra Club Canada)有着紧密的联系。

Merced River 莫瑟得河　位于美国加利福尼亚中心部分,是圣华金河的一个支流,全长 145 英里。莫瑟得河因贯穿于约塞米蒂国家公园南部,因迅猛和陡峭的流式而著名。河一旦流到山麓或者低洼的地面就会发生急剧的变化,形成缓慢的流式,贯穿于灌溉土地之中。

Miwok 米沃克人　加利福尼亚印第安人,操佩纽蒂语(Penutian),原分 7 个居民集团,在方言方面及地区方面互无联系:沿海米沃克人,居住在今旧金山市以北;湖滨米沃克人,居住在克利尔湖盆地(Clear Lake Basin);海湾米沃克人(或萨克兰人〔Saclan〕)居住在圣华金河(San Joaquin)及沙加缅度河的三角洲;平原米沃克人,居住在沙加缅度及圣华金河下游;再往东是山地米沃克人,分为北部、中部及南部三个部分,均居住在内华达山西麓。山地各支米沃克人是米沃克人中间的最大多数,与欧洲人接触初期,有100多个村庄。关于沿海一带的米沃克人——沿海米沃克人、湖滨米沃克人及海湾米沃克人——的原住民文化所知甚少。他们采集橡实、捕鱼、用弓箭猎鹿及其他猎物;住房是半地下式的木柱覆泥建筑。所编箩筐,不透水,饰以念珠或鸟羽。男人通常无衣著,女人穿短裙。内地米沃克文化包括山地及平原文化,我们知之较详。各居民集团居住在山麓或低地,只在夏季或狩猎时才移往高山地带。他们的主要住房是半地下式的泥土房屋,不过夏季的山地居所仅是树皮搭成的披屋。主食是橡实,储存于筐状粮仓之内。编织各种筐篮,但无陶器。社会由不同家系及半偶族组成,各管诸如继承、婚配等事务;大概也分等级,有首领及副首领,妇女可以根据男系获得此等称号。内地米沃克人信奉库克苏教(Kuksu)。20世纪晚期,沿海米沃克人和内地米沃克人各剩100多人。

the Ahwahnee 阿瓦尼酒店　位于优胜美地国家公园内,为来到优胜美地国家公园的商务和休闲旅客提供最便捷的位置。它是一家四星级酒店,饭店所有123间客房都装修一流,舒适现代。在优胜美地国家公园拥有优越的地理位置,饭店同时也提供旅客精致优雅的设备和贴心的服务。

Translation of Selected Sentences

1. Yosemite National Park is a showcase of spectacular geological features, including the greatest concentration of granite domes in the world and the largest exposed granite monolith in the world.

约塞米蒂国家公园汇集了各种壮观的地貌特征,包括世界上最集中的沙漠拱丘和最大的暴露着的花岗岩整块石料。

2. The first application of a park concept originated in Yosemite with the grant of 1864 (Federal land given to California for preservation) signed by Abraham Lincoln and since that time the park has played an important role in pioneering park management concepts.

第一个公园观念的申请起源于约塞米蒂国家公园,1864年被授予(联邦土地归加利福尼亚保护),并由亚伯拉罕·林肯签署,从那以后公园就在创园管理理念上起了很重要的作用。

3. Yosemite National Park is a world heritage site which has made a significant contribution to California's cultural heritage, to the national park movement, and to Yosemite's 4,000 years of cultural heritage by Native Americans.

约塞米蒂国家公园是世界遗产址,为加利福尼亚文化遗产、国家公园运动、约塞米蒂印第安人4 000年的文化遗产做了重要贡献。

4. Tributary creeks, which had once joined the main stream at the same elevation, now plummeted off of shear cliffs, giving birth to the park's famed waterfalls.

曾经在相同维度流入主流小溪的支流现在从陡峭的悬崖上垂直落下,形成了公园著名的瀑布。

5. Thick layers of sediment lay on the sea bed, which eventually was folded and twisted and thrust above sea level. Simultaneously molten rock welled up from deep within the earth and cooled slowly beneath the layers of sediment to form granite.

厚重的沉积物层位于海床上,最终褶皱扭曲并被推向海平面。同时熔岩从深深的地表涌出来,慢慢的在沉积物层冷却形成花岗岩。

6. Bears are also opportunists, which means they can easily adapt to new foraging habitats, from meadows to manzanita bushes, from cars to picnic coolers. Opportunism is often seen as an indication of intelligence in animals and most researchers agree that bears are highly intelligent.

熊也是机会主义者,他们可以轻松的适应新的觅食习惯,不管是草地、灌木丛、车里还是野餐冷饮。机会主义经常被看作是一种动物智力的象征,而且大部分研究者都同意熊是高度聪明动物的说法。

Questions

1. Why do we say that Yosemite National Park possesses outstanding recreational values and supreme scenic attractions?
2. Who were the area's first residents?
3. Is the Ahwahnee one of America's most distinctive hotels? And how could you describe it?
4. Why do we say that bears are opportunist?
5. What are the special animals in Yosemite National Park?

Chapter 18

Adirondack Park
阿迪朗达克公园

【简介】

这处荒野看来有些不可思议,它的面积比几个州还要大,甚至比有些国家还要大,几乎与比利时的面积相等。但是它又紧靠美国人口最密集的地方——离纽约市仅320千米。这个地方就是阿迪朗达克州立公园。整个荒野面积约12 950平方千米,其中包括湖泊、森林、河流和低矮浑圆的山丘,这个地区在19世纪30年代以前实际上还鲜为人知。法国人、英国人和印第安人为争夺其东北边界上的尚普兰湖,打了几乎整整两个世纪,因为这是加拿大和哈得孙河之间的一条良好的水路。1894年当纽约州宣布"这一森林保护地作为野生林地永久保存下来"之后,该地区便成了美国首批州立公园之一。

The Adirondack Park is a publicly protected area in northeast New York. It is the largest park and the largest state-level protected area in the contiguous United States, and the largest National Historic Landmark.

The park covers some 6.1 million acres (24,700 km^2), a land area greater than Vermont, Wales, or than Yellowstone, Yosemite, Grand Canyon, Glacier, and Great Smoky Mountains National Parks combined.

Much of the land is directly controlled by the state's Forest Preserve, but more than half the land within the Adirondack Park is privately owned, including several villages and hamlets(小村庄).

The Adirondack Park boundary, commonly referred to as the "Blue Line", contains the entire Adirondack Mountain range, as well as some surrounding areas, all within the state of New

York. The park includes all of Hamilton and Essex counties, as well as considerable portions of Clinton, Franklin, Fulton, Herkimer, St. Lawrence, and Warren counties and small portions of Lewis, Oneida, Saratoga, and Washington counties as well. The Clinton County towns of Altona and Dannemora, despite being entirely within the park boundary, are specifically excluded from the park by statute, due to the large prison facilities in both towns.

Not all of the land within the park is owned by the state, although new sections are frequently purchased or donated. State land comprises 2.7 million acres (11,000 km^2), about 45% of the park's area, including the highest peaks in New York State, as well as Mount Marcy, the highest elevation in the state. About one million acres (4,000 km^2) of this is classified as wilderness, with most of the remainder managed under the somewhat less stringent wild forest classification. Villages and hamlets comprise less than 1% of the area of the park; the remaining area of more than 3 million acres (12,000 km^2) is privately held but is generally sparsely developed. There is often no clear demarcation(划分) between state, private, and wilderness lands in the park. Signs marking the Adirondack Park boundary can be found on most of the major roads in the region, but there are no entrance gates and no admission fee.

The vast majority of the Adirondack Mountains are within the bounds of the traditional territory of the Mohawk First Nation(摩和克族国) until at least 1720. A sedentary agrarian democratic society before Contact, they controlled the eastern part of the Haudenosaunee (Iroquois) Confederacy or Six Nations. The Mohawk population had already been decimated by European diseases when they were largely driven out of their homeland by foreign European settlers. Although the Confederacy was divided, most of the Iroquois sided with the British during the American Revolution. After the war, most of the Iroquois land which fell within the American border was forcefully signed over through treaties or seized outright. During the 19th century, the U. S. Government forced most of the Iroquois off their traditional territory into reservations in the Midwest. Many escaped to Canada, where they were granted land in the British colony of Upper Canada, modern day Ontario. Citizens of the Mohawk Nation, numbering around 125,000 today, now live mostly north of the Saint Lawrence River in Canada along with the other member nations of the Haudenosaunee Confederacy. The Iroquois are now in the midst of land claims proceedings over parts of the state of New York.

The thinking that was to lead to the creation of the park first appeared in George Perkins Marsh's Man and Nature, published in 1864. Marsh argued that deforestation(森林采伐) could lead to desertification(沙漠化): referring to the clearing of once lush lands surrounding the Mediterranean, he asserted "the operation of causes set in action by man has brought the face of the earth to a desolation(荒芜) almost as complete as that of the moon".

The idea for the park itself first occurred to surveyor Verplanck Colvin in 1870, while taking in the view from atop Seward Mountain. He wrote to the state government that action was necessary to protect the forests or it would be wasted, which would lead to the drying up(枯竭) of the water needed to keep the Erie Canal in operation. Three years later he was appointed to a committee formed to consider how to do this.

While his term "Adirondack Park" led to some derision(嘲笑) and fears from longtime residents of the area that they might be bought out and evicted(驱逐), proponents of the idea began to use "Adirondack Forest Preserve" instead. Both terms continue in use to this day, with the former referring to the land inside the Blue Line and the latter to that portion owned by the state.

In 1878, Seneca Ray Stoddard produced a topographical survey(地形测量) of the Adirondacks that was influential in the creation of the Park.

Serious efforts to protect this land began in 1882, when businessmen in New York began to be concerned about the effects of widespread logging. Without trees, the many steep slopes(陡坡) on the mountains in the region were likely to erode, and the silt(淤泥) from the slopes could conceivably have silted up(淤塞) the Erie Canal and the Hudson itself, choking off New York State's economic backbone.

In 1885, legislation declared that the land in the Adirondack Park and the Catskill Park was to be conserved and never put up for sale or lease. The park was established in 1892, due to the activities of Colvin and other conservationists. The park was given state constitutional protection in 1894, so that the state-owned lands within its bounds would be protected forever ("forever wild"). The part of the Adirondack Park under government control is referred to as the Adirondack Forest Preserve. Further, this became a National Historic Landmark in 1963.

On September 18, 2008, the Nature Conservancy purchased Follensby Pond—about 14,600 acres (59 km^2) of private land inside the park boundary—for $16 million. The group plans on selling the land to the state which will add it to the forest preserve once the remaining leases for recreational hunting and fishing on the property expire(期满).

Ongoing efforts have been made to reintroduce native fauna that had been lost in the park during earlier exploitation. Animals in various stages of reintroduction include the American beaver, the fisher, the American marten, the moose, the Canadian lynx, and the osprey. Not all of these restoration efforts have been successful yet. There are 53 species of mammals that live in the park.

The park has a year-round population of about 130,000 people in dozens of villages andhamlets. Seasonal residents number about 200,000, while an estimated 7 – 10 million tourists visit the park annually. It is estimated that 84 million people live within a day's drive of the park.

There are more than 3,000 lakes and 30,000 miles (48,000 km) of streams and rivers. Many areas within the park are devoid of(缺乏)settlements and distant from usable roads. The park includes over 2,000 miles (3,200 km) of hiking trails; these trails comprise the largest trail system in the nation. With its combination of private and public lands, its large scale and its long history as a place people have tried to coexist with nature, many see the Adirondacks as a model for the ways natural areas with human populations can be protected into the future. There are parks in India and other countries that use the Adirondacks concept.

The New York State Department of Environmental Conservation (DEC) is responsible for the care, custody(监护), and management of the 2,700,000 acres ($10,900\ km^2$) of public (state) land in the Adirondack Forest Preserve. The Adirondack Park Agency (APA, created 1971) is a governmental agency that performs long-range planning for the future of the Adirondack State Park. It oversees development plans of private land-owners, as well as activities within the Forest Preserve. Development by private owners must be reviewed to determine if their plan is compatible(与和谐共处的)with the park.

This system of management is distinctly different from New York's state park system, which is managed by different agencies, primarily the state's Office of Parks, Recreation and Historic Preservation. While it is frequently referred to as a state park, the Adirondack Park has much more in common with a national forest: it mixes private and public land and has year-round residents within its boundaries in long-established settlements. "Adirondack Park" was Verplanck Colvin's original term for the area; it and the park itself predate by several decades the formal establishment of state parks in New York.

The southern side of the park is closer to major population centers, and lies just north of the New York State Thruway (Interstate 90). Interstate 87 (the Adirondack Northway) traverses the eastern side of the park between the Capital District of New York and Montreal, Canada. The northern and western portions of the park are somewhat more remote, but can be reached from Interstate 81, NY 3, NY 28, and US 11. The park is also served by the Adirondack Regional Airport, Amtrak's Adirondack Route along the shores of Lake Champlain and the Adirondack Scenic Railroad.

Notes

Great Smoky Mountains National Park 大烟山国家公园　位于美国东部,跨越田纳西州和北卡罗来纳州。由小罗斯福总统于 1940 年 9 月 2 日在公园内的纽芳隘口宣布成立。大烟山国家公园号称是美国东部最后一片广大的原始林。占地2 100平方千米。大烟山国家公园以其丰富的物种而闻名。由于大烟山国家公园内的海拔高度变化剧烈,加之降雨量充沛,因而大烟山公园极富生物多样性。大烟山国家公园也是切罗基人的世居地。由于441号公路横贯大烟山公园,使得大烟山公园成为美国最多人游览的国家公园之一。但另一方面,大烟山国家公园也面临严重的空气污染(来自观光者的汽车尾气及周边的发电厂)、酸雨等问题。

Adirondack Mountain 阿迪朗达克山脉　美国纽约州东北部的一处山地,位于圣罗伦斯河(北)、莫华克河谷(南)、安大略湖(西)和尚普兰湖－乔治湖(东)之间。最高点为马西山,是纽约州的第一高峰。一般被视为阿巴拉契亚山脉的一部分,但地质上与劳伦琴山更相似。

Mount Marcy 马西山　美国纽约州最高点(1 629米)。位于该州东北部阿第伦达克(Adirondack)山区。哈得逊河的主要源流奥帕莱森特(Opalescent)河源出该山。

Six Nations 易洛魁联盟　一群居住于今日美国纽约上州、宾州、俄亥俄州和加拿大魁北克的印第安人。16世纪末由五大部族摩和克人、奥奈达人、奥农达加人、瑟内萨人和卡尤加人在纽约地区北部组成联盟。后来休伦人加入成为第六个部族。易洛魁联盟逐渐成为美国东北部和加拿大东部最强大的原住民势力。该联盟在17世纪末达到巅峰,辖地东西自乞沙比克湾(Chesapeake Bay)到密西西比河及伊利诺河,南北自肯塔基到安大略。他们缔造了很多成就,包括建立彼此制衡的政治体系,这也可能就是美国建国者民主思想的渊源。易落魁联盟的中央思想是,彼此永不互相争战。而在对抗敌人时,他们的联合政治及军事力量往往所向披靡。

Iroquois 易洛魁人　1570年组成易洛魁联盟,易洛魁人生活在美国纽约州、威斯康星州、宾夕法尼亚州、俄亥俄州,以及加拿大的安大略省和魁北克省。易洛魁人长期实行母系氏族制,亲属关系按母系确定,经济活动也以母亲为中心,妇女地位比较高。易洛魁人以农耕闻名,也从事渔猎、畜牧和制陶、编筐等手工业。除了"六部落"之外,伊利人、切罗基人是易洛魁人的旁支。加上他们,总计共10余个部落。

Erie Canal 伊利运河　全长为584千米(363英里),整条运河12米(39英尺)宽、1.2米(4英尺)深,共83个水闸,每个水闸为27米乘4.5米(90英尺乘15英尺),最高可以行驶排水量68公吨(75吨)的平底驳船。伊利运河是第一条提供美国东海岸与西部内陆的快速运输工具,这比当时最常用的以动物拉动的拖车还快许多。伊利运河不只加快运输的速度,也将沿岸地区与内陆地区的运输成本减少了95%。快捷的运河交通使得纽约州西部更便于到达,因此也造成中西部的人口快速成长。

Catskill Park 卡兹奇公园　旅游和郊游胜地,有包括州属卡兹奇林苑和较多的私人山地。山中有人工湖多处,是纽约都会区的重要水源,其中包括阿肖肯(Ashokan)、唐斯维尔(Downsville)、内弗辛克(Neversink)、朗道特(Rondout)及斯科哈里(Schoharie)水库。

The Nature Conservancy 大自然保护协会　从事生态环境保护的国际民间组织，致力于在全球保护具有重要生态价值的陆地和水域，以维护自然环境、提升人类福祉。由于坚持采取合作而非对抗性的策略，以及用科学的原理和方法来指导保护行动，经过50余年的不懈努力，协会已跻身美国十大慈善机构行列，位居全球生态环境保护非营利民间组织前茅。

Translation of Selected Sentences

1. The park includes all of Hamilton and Essex counties, as well as considerable portions of Clinton, Franklin, Fulton, Herkimer, St. Lawrence, and Warren counties and small portions of Lewis, Oneida, Saratoga, and Washington counties as well.

公园包括所有汉密尔顿和艾塞克斯县、以及克林顿、富兰克林富尔顿、赫尔基蒙、沃伦县的大部分地区和劳伦斯·刘易斯、奥奈达、华盛顿、萨拉托加各地县的小部分地区。

2. Marsh argued that deforestation could lead to desertification: referring to the clearing of once lush lands surrounding the Mediterranean, he asserted "the operation of causes set in action by man has brought the face of the earth to a desolation almost as complete as that of the moon".

马什认为砍伐森林会导致沙漠化，具体是指：清理周围一次就会导致地中海郁郁葱葱的土地消失，他说"人类的行为致使地球表面荒芜，面积近乎月球的大小"。

3. Without trees, the many steep slopes on the mountains in the region were likely to erode, and the silt from the slopes could conceivably have silted up the Erie Canal and the Hudson itself, choking off New York State's economic backbone.

没有树木，山上许多陡峭边坡区域可能被侵蚀，山坡上滑落的泥沙已经淤塞了伊利运河和哈德逊，阻止了作为经济支柱的纽约州的发展。

4. With its combination of private and public lands, its large scale and its long history as a place people have tried to coexist with nature, many see the Adirondacks as a model for the ways natural areas with human populations can be protected into the future.

私人和公共土地的混合，其规模之大，历史之悠久，使得此处成为人们尝试与自然共存的地方。许多人认为阿迪朗达克山脉是人口与自然区域保护共存于未来的典范。

5. While it is frequently referred to as a state park, the Adirondack Park has much more in common with a national forest: it mixes private and public land and has year-round residents within its boundaries in long-established settlements.

虽然经常被称为州立公园，阿迪朗达克公园却具备很多国家森林特点——私人土地和公共土地共存，境内有常住人口长期定居。

Questions

1. Who hascontrolled much of the land in the Adirondack Park?

2. Whose traditional territory was the vast majority of the Adirondack Mountains until at least 1720?

3. In 1878, who produced a topographical survey of the Adirondacks that was influential in the creation of the Park?

4. In what aspectis the system of management distinctly different from New York's state park system?

5. Which department alsoserves the park?

Chapter 19

Dinosaur National Monument
恐龙国家纪念公园

【简介】

恐龙国家纪念公园位于美国科罗拉多州和犹他州之间。为保护当地的恐龙化石层，于1915年建成的陈列和展览史前爬行动物化石的保护区。面积达20万英亩。区内有国家恐龙纪念中心。中心有一面墙壁系石山陡壁的断面，它是世界上最奇特的恐龙遗骨贮藏所。古生物学家在此清理出恐龙、鳄鱼、龟等爬虫类动物遗骨和热带植物化石。其中，有生活在大约1.5亿年前的侏罗纪时期的圆顶龙骨化石。在保护区为游人可以观赏到经过专业人员复原的这些动植物原形。这里还有风景点多处。格林河和杨帕河流经的峡谷有著名的75英里长的“白水”奇观，游客可乘橡皮筏游览。山岩间还有保存完整的古印第安人的生活遗迹，还有博物馆及供游人停息的宿营地等。

Dinosaur National Monument is a National Monument located on the southeast flank(侧面) of the Uinta Mountains on the border between Colorado and Utah at the confluence(汇合) of the Green and Yampa Rivers. Although most of the monument area is in Moffat County, Colorado, the Dinosaur Quarry is located in Utah just to the north of the town of Jensen, Utah. The nearest communities are Vernal, Utah and Dinosaur, Colorado. This park has fossils of dinosaurs including Allosaurus, Abydosaurus (a nearly complete skull, lower jaws and first four neck vertebrae of the specimen DINO 16488 found here at the base of the Mussentuchit Member of the Cedar Mountain Formation is the holotype for the description) and various long-neck, long-tail sauropods. It was declared a National Monument on October 4, 1915.

The rock layer enclosing the fossils is asandstone and conglomerate(砾岩) bed of alluvial(冲

积的)or river bed origin known as the Morrison Formation from the Jurassic Period some 150 million years old. The dinosaurs and other ancient animals were washed into the area and buried presumably(大概)during flooding events. The pile of sediments were later buried and lithified (岩化的)into solid rock. The layers of rock were later uplifted and tilted to their present angle by the mountain building forces that formed the Uintas. The relentless(残酷的)forces of erosion exposed the layers at the surface to be found by paleontologists(古生物学者).

The Uinta Basin is a large depression(洼地)in remote northeast Utah, encircled by several mountain ranges and bisected(一分为二的)by the Green River—the basin extends approximately from Vernal in the north to the town of Green River in the south and is largely inaccessible by road. Habitation in this area is restricted to the Uintah/Ouray Indian Reservation, reached only by several long dirt tracks. North of the basin, the Green River crosses from Utah into Colorado and is joined by the Yampa River, and for many miles both rivers flow through deep twisting canyons and gorges cut through the sandstone mountains.

The canyons around the river junction(连接)form the centerpiece(核心部分)of Dinosaur National Monument, although almost all visitors come for the other main attraction—a steep cliff face, now enclosed within a large building, covered by hundreds of large fossilized dinosaur bones. This dinosaur quarry is located close to the Green River, 6 miles from US 40 near Vernal, however it has been closed since 2006 because of subsidence(下沉); instead a few dinosaur bones can be viewed in situ(在原地)on a nearby trail. For 50 miles north and east, the land becomes mountainous, and a large area of colorful canyons and ridges is protected within the monument boundary, viewable from various overlooks, maintained paths, or extended off trail hiking; the main access route is the 30 mile Harpers Corner Road.

The dinosaur fossil beds (bone beds) were discovered in 1909 by Earl Douglass, a paleontologist working and collecting for the Carnegie Museum. He and his crews excavated(发掘)thousands of fossils and shipped them back to the museum in Pittsburgh, Pennsylvania for study and display. President Woodrow Wilson proclaimed the dinosaur beds as Dinosaur National Monument in 1915. The monument boundaries were expanded in 1938 from the original 80-acre (320 000 m^2) tract surrounding the dinosaur quarry in Utah, to its present extent of over 200,000 acres (800 km^2) in Utah and Colorado, encompassing the spectacular river canyons of the Green and Yampa.

Though lesser-known than the fossil beds, the petroglyphs(岩石雕刻)in Dinosaur National Monument are another treasure the monument holds. Due to problems with vandals(破坏文化遗产的人),many of the sites are not listed on area maps.

The Dinosaur National Monument sits on a vast area of desert land in Northwestern Colorado and Northeastern Utah. Typical of "high deserts", summer temperatures can be exceedingly hot, while winter temperatures can be very cold. Snowfall is common, but the snow melts rapidly in the arid(干旱的) and sunny climates of these states. Rainfall is very low, and the evaporation (蒸发) rate classifies the area as desert, even though the rainfall just barely exceeds 10 inches.

The Dinosaur wall located within the Dinosaur Quarry building in the park consists of a steeply tilted (67° from horizontal) rock layer which contains hundreds of dinosaur fossils. The enclosing rock has been chipped away to reveal the fossil bones intact for public viewing. In July 2006, the Quarry Visitor Center was closed indefinitely due to structural problems that have plagued the building since 1957 as it was built on unstable clay. The current preferred plan is to rehabilitate the existing Quarry building, destroying the serpentine ramp, administrative, and museum facilities. A new facility to house the visitor center and administrative functions would be constructed elsewhere in the monument. This plan would provide a safe structure while still retaining a portion of the historic Mission 66 era exhibit hall. It was announced in April 2009 that Dinosaur National Monument would receive $13.1 million to refurbish(刷新) and reopen the gallery as part of President Obama's administration $750 billion stimulus plan.

The approach to Dinosaur National Monument from the southwest is dominated by Split Mountain, a rocky sandstone ridge that really does look as if it has been recently cleaved(裂开) by some great force, exposing many twisted strata(地层). At Jensen, a turn off (country route 149) from US 40 leads across a belt of farmland into the west section of the monument, by far the most visited area. The famous quarry building is located a quarter of a mile from the entrance road at the edge of the Split Mountain foothills; this protects the source of the largest single collection of bones from the Jurassic period ever found. The July 2006 closure was a result of many years of gradual damage to the building caused by movement of the underlying soft soils. A reconstruction project plan was completed late 2008 which calls for relocation of the fossil museum and installation of a brand new quarry building, a process likely to be completed in 2011. Until then, the visitor center is located in a temporary building along the main road; this has various exhibits, examples of native plants, a bookstore, and one fiberglass(玻璃纤维) dinosaur. The Fossil Discovery Trail starts outside and leads to a number of unexcavated fossil dinosaur bones in the surrounding hills. There are no other fossils on display elsewhere in the monument, and although they may possibly be found by casual prospectors, it is illegal to remover specimens, and discoveries should instead be reported to the authorities.

Scenic Drive—Tour of the Tilted Rocks: Past the temporary visitor center, road 149 continues for ten miles, crossing the Green River, and passing two nature trails, many panels of ancient Indian petroglyphs, two campsites, a historic log cabin and two shady box canyons. The first group of petroglyphs is reached after one mile—at the base of the reddish cliffs lining the road, though most of the pictures are rather faded. Not far beyond, route 149 passes the trailhead (小道的起点) for the Sound of Silence nature trail, and then a side road that descends to the shores of the Green River just as it emerges from Split Mountain Canyon, and leads to a picnic area, boat launch ramp and campsite, plus another trailhead, for the Desert Voices loop path (see QTVR). There is no trail into the canyon though walking upstream is quite easy, along beaches and sand banks on the west side of the river. The largest park campground (Green River) is a mile south, offering 88 tent or RV sites, with drinking water but no hook-ups(冷却). The main road then crosses the Green River on a slender bridge and enters private land for a few miles (Chew Ranch), passing fields and crops, before re-entering the monument where it becomes narrow and unpaved, and meets a turn-off for a rougher track to the Blue Mountain/Cliff Ridge area. The scenic drive continues a further 1.5 miles, running along the shallow canyon of Cub Creek, through red rocks bearing many more petroglyphs, most in good, crisp condition. Some are signposted but the majority is unmarked. The road ends at a clearing in a patch of woodland beside a log cabin built by pioneer Josie Morris, who lived here from 1914 to 1963. The roof is a modern replacement but most of the rest of the structure is genuine, and the site includes various other relics(遗骸) such as livestock enclosures, drainage ditches(排水沟) and corrals(围栏). Two short trails start here, to Box Canyon and Hog Canyon.

East—the Canyons: Most of the 200,000 acres of Dinosaur National Monument lie in Colorado and contain colorful ridges, plateaus and canyons surrounding the Green and Yampa rivers, which form an approximately inverted T-shape. This rocky and hostile area was not part of the original monument established in 1915, but was added in 1932 once the need to protect the land was realized. The canyons are different in character to others of the Colorado river system such as Canyonlands, Glen Canyon or the Grand Canyon, with their eroded flat layers, since here many of the strata have been twisted by geological forces and the rocks are usually white and grey, not the dominant red found further south. Apart from road 149 to the dinosaur quarry and Split Mountain, there are only four other paved routes leading into the wilderness area around the rivers:

U.S. Bureau of Reclamation plans for a ten-dam, billion dollar Colorado River Storage Project began to arouse opposition in the early 1950s when it was announced that one of the proposed dams would be at Echo Park, in the middle of Dinosaur National Monument. The controversy assumed major proportions, dominating conservation politics for years. David Brower, executive director of the Sierra Club, and Howard Zahniser of The Wilderness Society led an unprecedented nationwide campaign to preserve the free-flowing rivers and scenic canyons of the

Green and Yampa Rivers. They argued that, if a national monument was not safe from development, how could any wildland be kept intact?

On the other side of the argument were powerful members of Congress from western states, who were committed to the project in order to secure water rights, obtain cheap hydroelectric power and develop reservoirs as tourist destinations. After much debate, Congress settled on a compromise that eliminated Echo Park Dam and authorized the rest of the project. The Colorado River Storage Project Act became law on April 11, 1956. It stated, "that no dam or reservoir constructed under the authorization of the Act shall be within any National Park or Monument."

Historians view the Echo Park Dam controversy as signaling the start of an era that includes major conservationist political successes such as the Wilderness Act and the Wild and Scenic Rivers Act.

Notes

Uinta Mountains 尤因塔山脉　美国犹他州东北部山脉,一小部分伸入怀俄明州东南部,绵亘160千米以上。为落矶山脉中南段的一部分。多4 000米以上山峰,包括犹他州最高点金斯峰(Kings Peak),海拔4 126米。为普罗沃、迪谢纳诸河源头。境内有自然保护区、国家森林和印第安人保护区。

Allosaurus 异特龙　又称跃龙或异龙,是兽脚亚目肉食龙下目恐龙的一属。异特龙是种大型的二足、掠食性恐龙,平均身长为10米,最长可达12到13米。它们生存于晚侏罗纪,于1.55亿年前到1.45亿年前。自从在1877年被奥塞内尔・查利斯・马什(Othniel Charles Marsh)命名以来,已有许多种类被归类于异特龙属,但只有少数被认为是有效种。

Abydosaurus 阿比杜斯龙　蜥脚下目腕龙科下的一属恐龙,生存于白垩纪的美国犹他州。阿比杜斯龙是少数发现完整头颅骨的蜥脚类恐龙之一,也是第一个被发现完整头颅骨的北美洲蜥脚类恐龙。阿比杜斯龙的牙齿狭窄,其他早期腕龙科的牙齿较宽。阿比杜斯龙是美国古生物学家发现的一种恐龙新物种,它是一种体型庞大、长脖颈的素食类恐龙,它们独特之处在于并不会咀嚼食物,只能直接吞咽。

sauropods 蜥脚类　古爬行动物蜥臀目恐龙的一亚目。体躯庞大,是所有陆生脊椎动物中最巨大的。头很小,颈和尾都长,四肢粗壮,以四足行走。以植物为食。主要生活于沼泽地带。侏罗纪最盛,至白垩纪末灭绝。著名的迷惑龙、梁龙、腕龙以及我国的天山龙、马门溪龙、盘足龙皆属此类。

Morrison Formation 摩里逊岩层　从美国西部蒙大拿州,向南延伸至新墨西哥州的一条薄沉积地层,属于侏罗纪的上层沉积层,以科罗拉多州的摩里逊镇命名。整条结构层中蕴含丰富的恐龙化石,包括鸟脚龙、异特龙、梁龙和腕龙。摩里逊岩层非常接近犹他州的维纳尔镇，当地亦是国家恐龙纪念碑所在地，今日游人可在此一窥被埋在地底的恐龙骸骨。

Green River 格林河　美国西部河流，源出怀俄明州中西部的温德河山脉（Wind River Range），大致向南流经怀俄明州西南部。在拉巴奇（La Barge）以下筑有拦河坝，形成丰特奈尔（Fontenelle）水库。流过弗拉明（Flaming）峡谷国家游乐区时再次形成巨大水库。后穿越犹他州东部进入科罗拉多州西北，经恐龙化石国家保护区又进入犹他州。再向西南流，过格林河市折向东南，在莫阿布（Moab）以南汇入科罗拉多河。全长 1 175 千米，流域面积 11.7 万平方千米。河流大部在崎岖山地穿行，只在涨水时可行小船。

Colorado river 科罗拉多河　美国西南方、墨西哥西北方的河流。长度大约有 2 333 千米，有一部分在美国洛矶山脉区域西边消失。整个科罗拉多河河系大部份流入了加利福尼亚湾，是加州淡水的主要来源，另一部份则往南流向墨西哥，但没有出海。1540 年西班牙探险家梅尔基奥·迪亚兹（Melchior Diaz）是第一个探索并记录了科罗拉多河的探险家。科罗拉多河是北美洲主要河流之一。发源于美国西部科罗拉多州中北部落矶山脉，由大面积厚雪融化提供水源，向西南流，开辟一条 2 330 千米曲折蜿蜒水道，注入加利福尼亚湾。

U. S. Bureau of Reclamation 美国垦务局　创建于 1902 年，下属于美国内政部。它在美国西部地区 17 个州的大坝、水电站和渠道的建设中是久负盛名的。这些水利工程促进了人们在美国西部建立家园并推动了当地经济的大发展。垦务局已经建设了 600 座大坝和水库，包括著名的科罗拉多河上的胡佛坝和哥伦比亚河上的大古力坝。

Sierra Club 塞拉俱乐部　或译作山岳协会、山峦俱乐部和山脉社等，是美国的一个环境组织，著名的环保主义者约翰·缪尔（John Muir）于 1892 年 5 月 28 日在加利福尼亚州旧金山创办了该组织，并成为其首任会长。塞拉俱乐部拥有百万会员，分会遍布美国，且与加拿大塞拉俱乐部（Sierra Club Canada）有着紧密的联系。

The Wilderness Society 荒野协会　一个全国性的环保社团组织，其主要目的和宗旨是保护和改善生态环境，恢复地球生命演变的自然进程和原始面貌。

Translation of Selected Sentences

1. In July 2006, the Quarry Visitor Center was closed indefinitely due to structural problems that have plagued the building since 1957 as it was built on unstable clay.

在 2006 年 7 月，砂石场参观中心可能是因为结构问题被关闭，因为建在不稳定的泥土上，从 1957 年就已经出现问题。

2. The famous quarry building is located a quarter of a mile from the entrance road at the edge of the Split Mountain foothills; this protects the source of the largest single collection of bones from the Jurassic period ever found.

著名的砂石场建筑坐落于离断山山麓丘陵边缘的入口四分之一英里处，这样可以保护在侏罗纪时期发现的最大的骨骼收藏来源。

3. Scenic Drive—Tour of the Tilted Rocks: Past the temporary visitor center, road 149 continues for ten miles, crossing the Green River, and passing two nature trails, many panels of ancient Indian petroglyphs, two campsites, a historic log cabin and two shady box canyons.

风景路——倾斜岩石旅游:通过临时旅游中心,149 号公路继续向前 10 英里,穿过格林河,经过两个自然小径,有许多古印第安人岩石雕刻的镶板,两个露营地,一个历史著名的小木屋和两个成荫的箱形峡谷。

4. The main road then crosses the Green River on a slender bridge and enters private land for a few miles (Chew Ranch), passing fields and crops, before reentering the monument where it becomes narrow and unpaved, and meets a turn-off for a rougher track to the Blue Mountain/Cliff Ridge area.

主干道则穿过格林河朝向一座窄桥,进入私人用地几英里后,再穿过场地和农作物,最后重新进入现在变得狭窄和无铺面的历史遗迹,和一条通往蓝山或悬崖山脊的粗野小路的岔路口。

5. U. S. Bureau of Reclamation plans for a ten-dam, billion dollar Colorado River Storage Project began to arouse opposition in the early 1950s when it was announced that one of the proposed dams would be at Echo Park, in the middle of Dinosaur National Monument.

美国垦务局计划开始一个 10 水坝组,10 亿美元的科罗拉多河存储计划。20 世纪 50 年代,在其宣布该水坝之一将建在恐龙国家纪念碑中间的回音公园后招致民众的反对。

Questions

1. Where is Dinosaur National Monument located?
2. What are the functions of the canyons around the river?
3. How did the July 2006 closure happen?
4. Where does most of the 200,000 acres of Dinosaur National Monument lie?
5. How do historians view the Echo Park Dam controversy?

Chapter 20

Acadia National Park
阿卡迪亚国家公园

【简介】

阿卡迪亚国家公园在美国缅因州大西洋沿岸，弗伦奇曼湾两侧，面积 168 平方千米。1916 年初建。1919 年成为拉斐特国家公园，1929 年改为现名。主要部分为芒特迪瑟特岛森林地带，以凯迪拉克山为主体，有阿内蒙洞和西厄尔德芒茨泉；其他地区还包括峭壁耸立的半个欧岛和斯库迪克半岛。该园处在北方和温带交接处，寒冷的浅水海湾中栖息着大量海生动物。

Acadia National Park is a National Park located in the U. S. state of Maine. It reserves much of Mount Desert Island, and associated smaller islands, off the Atlantic coast. Originally created as Lafayette National Park in 1919, the first National Park East of the Mississippi, it was renamed Acadia in 1929.

Acadia National Park is one of the most beautiful and enchanting natural places on the North American continent. Mountains, islands and ocean combine in a unique sensory splendor that has become the crown jewel(皇冠之珠)of Maine and all of America. Each year, millions of people visit for a memorable(难忘的) vacation and many more hope to. Towns, such as Bar Harbor, are intertwined with Acadia National Park land creating a unique blend of pristine landscapes with a Downeast flavor and character. Once visited, you want to return. Once felt, the experience stays with you forever.

The area first was inhabited by the Wabanaki people. In the fall of 1604, Samuel de Champlain observed a high-notched island composed of seven or eight mountains rising to bare-rock summits from slopes of birch(桦树), fir, and pine. In spite of many changes over nearly 400 years, the area remains essentially(本质上) the same.

The landscape architect(景观设计师) Charles Elliot is credited with the idea for the park. It first attained(取得) federal status when President Woodrow Wilson, established it as Sieur de Monts National Monument on July 8, 1916, administered by the National Park Service. On February 26, 1919, it became a national park, with the name Lafayette National Park in honor of the Marquis de Lafayette, an influential French supporter of the American Revolution. The park's name was changed to Acadia National Park on January 19, 1929.

From 1915 to 1933, the wealthy philanthropist(慈善家) John D. Rockefeller, Jr. financed, designed, and directed the construction of a network of carriage trails throughout the park. He sponsored the landscape architect Beatrix Farrand, with the nearby family summer home Reef Point Estate, to design the planting plans for the subtle carriage roads at the Park. The network encompassed over 50 miles (80 km) of gravel carriage trails, 17 granite bridges, and two gate lodges, almost all of which are still maintained and in use today. Cut granite stones placed along the edges of the carriage roads act as guard rails(防护轨) of sort and are locally known as "coping stones" to help visitors cope with the steep edges. They are also fondly called "Rockefeller's teeth".

One of the most unique aspects of how Acadia National Park came to be is that it is due to the vision and donations of private citizens. George B. Dorr and Charles W. Eliot anticipated the dangers that over-development would bring to the area and acted quickly to prevent it. John D. Rockefeller, Jr., played a critical role by building the carriage roads (1915—1933) and donating over 11,000 acres of land towards what became Acadia National Park. There have been countless others since who have donated their time and resources to help realize this dream so that we may all experience its raw beauty and power.

Most of the park is onMount Desert Island. To the Northeast is Schoodic Peninsula and to the Southwest (accessible only by boat) is Isle Au Haut. Baker Island (Southeast coast) and Bar Island (near Bar Harbor) also have national park land.

The park includes mountains, an ocean shoreline, woodlands, and lakes. In addition to Mount Desert Island, the park comprises much of the Isle au Haut, parts of Baker Island, and a portion of the Schoodic Peninsula on the mainland.

In total, Acadia National Park consists of more than 47,000 acres, including 30,300 acres on Mount Desert Island, 2,728 acres on Isle au Haut and 2,366 acres on the Schoodic Peninsula.

Cadillac Mountain, named after the French Explorer of the same name, is on the eastern side of the island. Its green, lichen-covered(长满地衣的), pink granite summit is, because of a combination of its eastern location and height, one of the first places in the United States to see the sunrise. Miles of carriage roads were originally built by Rockefeller, Jr. The mountains of Acadia National Park offer hikers and bicycle riders views of the ocean, island lakes, and pine forests.

The inletSomes Sound, often described as the "only fjord(海湾)on the East Coast",is now called a fjard by officials

On October 17, 1947, 10,000 acres of Acadia NationalPark were burned in a fire that began along the Crooked Road several miles west of Hulls Cove. The forest fire was one of a series of fires that consumed much of Maine's forest as a result of a dry year. The fire burned for days and was fought by the Coast Guard, Army, Navy, local residents, and National Park Service employees from around the country. Restoration of the park was supported, substantially(实质上), by the Rockefeller family, particularly John D. Rockefeller, Jr. Regrowth was mostly allowed to occur naturally and the fire has been suggested to have actually enhanced the beauty of the park, adding diversity to tree populations and depth to its scenery.

In order to preserve scenic values and define its permanent boundary, the park began purchasing small tracts of land and easements(地役权)in 1986. True to the spirit of the original vision, many landowners continue this tradition today by placing easements on their property to limit any potential future development.

Rocky coastlines, granite mountains, lakes and ponds, moss and evergreen, crashing waves and abundant wildlife mix with local character and seafood in this magical place. Borders of the park are accented by picturesque harbor villages such as Bar Harbor, Northeast Harbor, Southwest Harbor and more.

There is a National Park information center on Thompson Island just before crossing onto Mount Desert Island. The main island has two centers. The main Visitor Center is near Route 3 in Hulls Cove and the Park Headquarters & Winter Visitor Center is on Route 233 near Eagle Lake.

The National Park Service, as part of their Centennial Initiative celebrating its 100th anniversary in 2016, created a project to promote voluntary, multimodal(多峰的)park access for present and future generations. Going "car free" offers visitors the opportunity to explore Acadia by foot, bicycle, shuttle bus, commercial tour bus, private automobile, or private and commercial vessels(船舶). The project includes an inter-modal transportation center on state-owned land four miles north of the park, multiple-use trails to connect gateway communities with the park, and rehabilitation of historic carriage roads surrounding Eagle Lake.

The park is home to some 40 different species of mammalian wildlife. Among these are red and graysquirrels, chipmunks, white-tailed deer(白尾鹿), moose(驼鹿), beaver (Castor canadensis), porcupine(豪猪), muskrats(麝鼠), foxes, coyote, bobcats(山猫) and black bears. Many other marine species have been observed in the surrounding area and waters.

Excavations(洞穴)of old Indian sites in the Mount Desert Island region have yielded remains of the native mammals. Bones of wolf, beaver, deer, elk, Gray seal (Halichoerus grypus), the Indian dog, and Sea Mink(海貂)(Neovison macrodon), as well as large numbers of raccoon, lynx, wolf, muskrat, and deer. Although beaver were trapped to extinction on the island, two pairs of beaver that were released in 1920 by George B. Dorr at the brook between Bubble Pond and Eagle Lake have repopulated it. The large fire in 1947 cleared the eastern half of the island of its coniferous trees(针叶树) and permitted the growth of aspen(山杨), birch, alder, maple and other deciduous trees(落叶树)which enabled the beaver to thrive.

Species that used to inhabit the island include themountain lion (美洲狮)(or puma) and the gray wolf. It is thought that these predators(捕食者) have been forced to leave the area due to the dramatic decrease in small prey and proximity to human activity.

Notes

Mount Desert Island 山沙岛　位于美国缅因州的汉考克县,是缅因沿着海岸线最大的岛屿。拥有108平方英里的土地,是美国连续地区内第六大岛屿。经常被称为美国东海岸第三大岛屿的山沙岛其实是位于长岛之后的第二大岛屿。

Bar Harbor 巴尔港　阿卡迪亚的核心城镇就是一个小小的旅游中心,叫做巴尔港。这个小市镇是整个阿卡迪亚国家公园的旅游经济的核心,国家公园是非赢利性质的,巴尔港就把整个地区的旅游业集中于一点。酒店、旅馆、餐馆、住宅都在那里,去阿卡迪亚,必去巴尔港。巴尔港是一个不大的城镇,居民人口也就几千人,不过这里酒店、饭馆、酒吧、咖啡店林立,人头涌涌,一派旅游壮观的景象。因为旅游而形成的市镇不少,但是大多数具有其他的功能,这个巴尔港则是纯粹的旅游开发结果,很有典型意义。

Wabanaki 瓦班纳基人　操阿尔冈昆语(Algonquin)的印第安部落联盟。居住在北美洲东北部。建立联盟是为了自保,抵抗易洛魁联盟(Iroquois League),特别是摩和克人(Mohawk)。最早的组织包括今纽约州以东及东北诸部落:马莱西特人(Malecite),在今新伯伦瑞克(New Brunswick);帕萨马科迪人(Passamaquoddy)和佩诺布斯科特人(Penobscot),在今缅因州;还有今佛蒙特和新罕布夏两州的各部落。后来又包括了一些东部部落,南部则包括德拉瓦州部落。主要以渔猎为生,沿岸地区各部落种植玉蜀黍。北部典型住房是桦皮覆盖的棚屋,几个家庭共住一处。桦皮小舟为日常交通工具。狩猎使用陷阱、罗网或弓箭。各部落都由若干较小的群组构成,群组有头目或文职首领,专管咨询但无强制性权威;另有一名作战首领。由包括全部落男女成员在内的全体大会决定战争问题;由首领及各家庭代表组成的最高会议决定其他重要问题。人与人之间终生保持互相尊重的同志之谊,相信有英雄人物能在民族危亡时回来拯救他们,至今不变。20 世纪晚期约 5000 人。

Samuel de Champlain 萨缪尔·德·尚普兰　法国探险家,地理学家,魁北克城的建立者。也是法国同北美贸易,特别是皮毛贸易的开拓者。

Marquis de Lafayette 拉法耶特侯爵(1757 年 9 月 6 日—1834 年 5 月 20 日)　又译拉法叶,法国将军、政治家。生于奥弗涅省(今上卢瓦尔省)沙瓦尼亚城堡一贵族家庭。1771 年在巴黎普莱西中学毕业后入伍。1775 年晋升骑兵上尉。1775 年美国独立战争开始。拉法耶特认为"美国的独立将是全世界热爱自由人士的福祉"。1777 年拉法耶特自备战舰,募集人员,参加美国独立战争,同美洲殖民地人民共同抗击英军。1780 年任华盛顿前卫部队司令。1779 年离开美国,回到法国,一年后,拉法叶说服法国王室让他带 6 000 名法军前往美国参战。五个月后,英军总司令投降。美国成为一个独立的国家。因为拉法叶侯爵赞成革命,引起许多被迫附庸国(如当时的波兰、爱尔兰等)革命运动持续不断,直到近代独立为止。为了纪念拉法叶侯爵,欧洲各国和美国首府华盛顿、纽约州等地都有一条"拉法叶"路,仅次于主干道。而法国政府为了纪念拉法叶侯爵的贡献,将新型先进巡防舰命名为"拉法叶级巡防舰",中华民国海军亦有数艘,由法国售予。

American Revolution 北美独立战争　也称"美国独立战争",指 1775 年波士顿人民在美国莱克星顿与英军交战拉开独立战序幕到 1783 年英军在法国签订《巴黎条约》投降的战争。美国独立战争既是一次民族独立战争,又是一次资产阶级革命。

Schoodic Peninsula 斯库德克半岛　绕过东边的一道海峡,就可以看到斯库德克半岛伸出大海,这里人迹罕至,巨大的花岗岩屹然挺立,迎接着海浪无休止地拍打。从斯库德克半岛放眼望去,海天相接,景色十分优美

Cadillac Mountain 卡迪拉克山　据说是美国境内太阳最早照耀到的几个地方之一。卡迪拉克山主峰是大西洋东海岸北起加拿大的加斯帕半岛(Gaspe Penisula)、南到巴西里约热内卢一带的最高点。阳光照耀山顶,衬着紫色、红色和蓝色的浪涛,蔚为壮观。

Somes Sound 萨姆斯·桑德峡湾　两边是连绵的卡迪拉克山,中间是海水奔流的幽幽深谷,萨姆斯·桑德峡湾是毗邻的 48 个州中唯一真正的峡湾。峡湾日落,摄人心魄。

Translation of Selected Sentences

1. Mountains, islands and ocean combine in a unique sensory splendor that has become the crown jewel of Maine and all of America.

山脉、群岛和海洋组成了奇特的视觉盛宴,成为缅因州和全美国的皇冠之珠。

2. Towns, such as Bar Harbor, are intertwined with Acadia National Park land creating a

unique blend of pristine landscapes with a Downeast flavor and character.

类似巴尔港的几个城镇和阿卡迪亚国家公园相互交织着，创造出一种奇特的原始风景和海岸风情特色的融合。

3. The landscape architect Charles Elliot is credited with the idea for the park.

景观设计师查尔斯·艾略特因公园的构想而受到赞誉。

4. Cut granite stones placed along the edges of the carriage roads act as guard railsof sort and are locally known as "coping stones" to help visitors cope with the steep edges. They are also fondly called "Rockefeller's teeth".

切割放置在运输公路边上的作为防护轨的花岗岩石在当地也被称为"基石"，来帮助参观者应对陡峭的边缘。他们被亲切的叫做"洛克菲勒的牙齿"。

5. Restoration of the park was supported, substantially, by the Rockefeller family, particularly John D. Rockefeller, Jr. Regrowth was mostly allowed to occur naturally and the fire has been suggested to have actually enhanced the beauty of the park, adding diversity to tree populations and depth to its scenery.

公园的修复实质上是由洛克菲勒家族，特别是由约翰·洛克菲勒二世支持的，修建主要是在自然和火灾方面的建设，事实上已经提升了公园的美感，为景色增添了树木的数量和种类多样性。

Questions

1. Before it was renamed Acadia in 1929, what was its original name? And where did the original name come from?

2. From 1915 to 1933, who financed, designed and directed the construction of a network of carriage trails throughout the park?

3. What is one of the most unique aspects of Acadia National Park?

4. In order to preserve scenic values and define its permanent boundary, what did the park do in 1986?

5. What kind of plants and animals does Acadia National Park have?

Chapter 21

Denali National Park and Preserve
迪纳利国家公园和保护区

【简介】

迪纳利国家公园原名麦金利山,是阿拉斯加州的第一个国家公园,建于 1917 年。公园于 1980 年更名为迪纳利国家公园。公园以巍峨的麦金利山为标志,麦金利山被阿拉斯加当地人称作迪纳利山,是“大家伙”的意思。它海拔 6 193.5 米,是北美最高峰,也是世界上最宏伟的山脉之一。迪纳利是美国地貌的一处奇观,这里地形开阔,一望无垠,摄人心魄。半年一次的驯鹿迁徙是迪纳利国家公园的一大奇观。

The word “Denali” means “the high one” in the native Athabaskan language and refers to the mountain itself. The mountain was named after president William McKinley of Ohio in 1897 by local prospector (采矿者) William A. Dickey, although McKinley had no connection with the region. The name is only used by those outside of Alaska. Charles Alexander Sheldon took an interest in the Dall sheep (野大白羊) native to the region, and became concerned that human encroachment (侵犯) might threaten the species. After his 1907—1908 visit, he petitioned (请求) the people of Alaska and Congress to create a preserve for the sheep. The park was established as Mount McKinley National Park on February 26, 1917. However, only a portion of Mount McKinley (not even including the summit) was within the original park boundary. The park was designated an international biosphere reserve in 1976. A separate Denali National Monument was proclaimed by Jimmy Carter on December 1, 1978.

Mount McKinley National Park, whose name had been subject to local criticism from the onset (有力的开始), and Denali National Monument were incorporated and established into

Denali National Park and Preserve by the Alaska National Interest Lands Conservation Act, December 2, 1980. At this time the Alaska Board of Geographic Names changed the name of the mountain back to "Denali", even though the U. S. Board of Geographic Names maintains "McKinley". Alaskans tend to use "Denali" and rely on context to distinguish between the park and the mountain. The size of the national park is over 6 million acres (24,500 km^2). On December 2, 1980, a 2,146,580 acre (8,687 km^2) Denali Wilderness was established within the park.

Denali habitat is a mix of forest at the lowest elevations, including deciduous (落叶的) taiga (针叶树林地带). The preserve is also home to tundra (苔原) at middle elevations, and glaciers, rock, and snow at the highest elevations. Today, the park hosts more than 400,000 visitors who enjoy wildlife viewing, mountaineering, and backpacking (背包旅行). Wintertime recreation includes dog-sledding (狗拉雪橇), cross-country skiing, and snowmobiling (雪上汽车) where allowed.

Denali is home to a variety of Alaskan birds and mammals, including a healthy population of grizzly bears(大灰熊) and black bears. Herds of caribou (驯鹿) roam throughout the park. Dall sheep are often seen on mountainsides, and moose (驼鹿) feed on the aquatic plants (水生植物) of the small lakes and swamps. Despite human impact on the area, Denali accommodates gray wolf dens, both historic and active. Smaller animals, such as hoary marmots, arctic ground squirrels, beavers, pikas (鼠兔), and snowshoe hares (美洲兔) are seen in abundance. Foxes, martens, lynx, wolverines also inhabit the park, but are more rarely seen due to their elusive natures.

The park is also well known for its bird population. Many migratory species reside in the park during late spring and summer. Bird watchers may find waxwings, arctic warblers(北极鸣鸟), pine grosbeaks, and wheatears, as well as ptarmigan and the majestic tundra swan. Predatory birds (捕食鸟) include a variety of hawks, owls, and the gyrfalcon, as well as the abundant but striking golden eagle.

Ten species of fish, including trout, salmon, and arctic grayling share the waters of the park. Because many of therivers and lakes of Denali are fed by glaciers, glacial silt and cold temperatures slow the metabolism of the fish, preventing them from reaching normal sizes. A single amphibious species, the wood frog, also lives among the lakes of the park.

Denali parkrangers maintain a constant effort to keep the wildlife wild by limiting the interaction between humans and park animals. However, the number of wild bears necessitates their wearing collars to track movements. Feeding animals is strictly forbidden, as it may cause adverse affects on the feeding habits of the creature. Visitors are encouraged to view animals from safe distances. Despite the large concentration of bears in the park, efforts by rangers to educate backpackers and visitors about preventative measures and BRFCs have greatly reduced the number of dangerous encounters. Certain areas of the park are often closed due to uncommon wildlife activity, such as denning areas (驻穴的地方) of wolves and bears or recent kill sites. These restricted areas may change throughout the year. Through the collective care of park staff and visitors, Denali has become a premier destination for wildlife viewing.

The Alaska Range, a mountainous expanse running through the entire park, provides interesting ecosystems (生态系统) in Denali. Because the fall line lies as low as 2,500 feet (760 m), wooded areas are rare inside the park, except in the flatter western sections surrounding Wonder Lake, most of the park is vast expanses of tundra and lowlands (低地) of the park where flowing waters melt the frozen ground. Spruces (云杉) and willows (柳树) make up the majority of these treed areas. Because of mineral content, ground temperature, and a general lack of soil, areas surrounding the bases of mountains are not suitable for sufficient tree growth, and most trees and shrubs do not reach full size.

Having a range of elevations, there is a variety of vegetation(植被) zones. From lowest to highest, there is low brush bog, bottomland spruce-poplar forest, upland spruce-hardwood forest, moist tundra, and finally the highest of elevations, alpine tundra.

Throughout Denali's history, there has been a patchwork pattern of different plants relying on fire. Because of this, the fire history is too complicated to explain. North of the Alaskan Range, fires are common, occurring when old forests need replacement.

Tundra is the predominate ground cover of the park. Layers of topsoil(表层土) collect on rotten, fragmented rock moved by thousands of years of glacial activity. Mosses, ferns (蕨类植物), grasses, and fungi (真菌类) quickly fill the topsoil, and in areas of muskeg (苔藓) "wet tundra", tussocks form and may collect algae (藻类). The term muskeg includes very spongy waterlogged (浸满水的) tussocks as well as deep pools of water covered by solid looking moss. Wild blueberries and soap berries thrive in this landscape, and provide the bears of Denali with the main part of their diets.

Over 450 species of flowering plants fill the park, and can be viewed in bloom throughout summer. Images of goldenrod (秋麒麟草属植物), fireweed, lupine, bluebell (野风信子), and gentian filling the valleys of Denali are often used on postcards and in artwork.

Long winters are followed by short growing seasons. Eighty percent of the bird population returns after cold months, raising their young. Most mammals and other wildlife in the park spend the brief summer months preparing for winter and raising their young.

Summers are usually cool and damp, but temperatures in the 70s are not rare. The weather is so unpredictable that there have even been instances of snow in August.

The north and south side of the Alaskan Range have a completely different climate. The Gulf of Alaska carries moisture to the south side, but the mountains block water to the north side. This brings a drier climate and huge temperature fluctuations（起伏）to the north. The south receives transitional maritime continental climates, with moister, cooler summers and warmer winters

The Denali Wilderness is a wilderness area in the Denali National Park and Preserve. It encompasses（包含）the high heart of the Alaska Range, including Denali, the centerpiece of the wilderness, which comprises about one-third of the national park.

Denali Wilderness covers the area formerly known as Mount McKinley National Park from 1917 until the park was expanded and renamed in 1980. It is 2,146,580 acres（8,687 km^2）in area; the entire park is larger than the state of Massachusetts.

An immense collection of cultural sites gives more and more clues as to what and who used to live there. Thousand of years ago, grassland was abundant, and mammoths（猛犸象）utilized the flat Mammoth Steppe to move and graze. Around 11,000 to 13,500 years ago, these grasslands shrunk and woody shrubs began to appear. Back then, the North-Alaskan Range was predominantly ice free.

187 cultural sites tell about Denali's past, eighty-four have prehistoric（史前的）items. Native Americans have lived in this environment for 11,000 years, using every resource the wild provided. The Koyukon, Denaina, Athna, Kolchan, Tanana, and Athabaskans are particularly known.

Denali is emerging as a site of interesting fossils, including footprints ichnites（足迹化石）that were credited with being the first evidence of prehistoric wading birds（涉禽）, probing in mudflats for food during the Late Cretaceous Period, when they were first publicly reported in May 2006. A footprint of a carnivorous theropod（兽脚亚目食肉恐龙）had previously been reported in the park.

Notes

Athabaskan language 阿萨帕斯卡语　居住于北美的土著人,分为北美的南部和北部两大族。他们使用的语言被称为阿萨帕斯卡语。在北美,阿萨帕斯卡语仅次于使用范围最广的阿兹特克语,是第二种使用人数最多的语言。

William McKinley, Jr. 威廉·麦克金利(1843 年 1 月 29 日—1901 年 9 月 14 日)　美国第 25 届总统(1897—1901)。他经历激烈的竞选最终胜出,并且因为力主推行金本位制和高关税政策而闻名于世。在他的领导下,美国用 100 天的时间,赢得了美西战争的胜利。

biosphere reserve 生物圈保护区　联合国教科文组织在其“人与生物圈计划”(MAB)中提出的一项概念,指受到保护的陆地、海岸带或海洋生态系统的代表性区域。截至2007年,该组织已经认定了109个国家的527个地区为生物圈保护区。按照联合国教科文组织的解释,生物圈保护区是被创立以用来“展示和推广人与自然界和谐相处”的地区。具体来说,一个生物圈保护区必须具有一块被立法保护的核心地区,周边必须有缓冲区域,缓冲地区之外还须有过渡区域。通过良好的管理,生物圈保护区的生态系统和生物多样性必须得到良好的保护。除了具有保护功能外,生物圈保护区更重要的是要具有促进资源可持续利用的发展功能和开展科学研究、监测、教育、培训、信息交流等后勤支持功能。另外,保护区还必须通过特殊的区域设计来发挥这些功能。

Koyukon 科育肯人　科育肯人是生活在北阿拉斯加的阿萨帕斯卡人的一个分支。他们主要居住于科育肯和育肯河沿岸,靠狩猎为生。即使在当今社会,科育肯人仍保持着几千年沿袭下来的生活习惯。

Denaina 德纳亚娜人 德纳亚娜人是生活在阿拉斯加的土著人,属于美州印第安人的一个分支,是阿拉斯加中南部地区最早的居民。

the Late Cretaceous Period 白垩纪后期　白垩纪距今1.37亿年~6700万年,是中生代第三个纪。这一时期形成的地层称白垩系,位于侏罗系之上、新生界之下。白垩纪是中生代地球表面受淹没程度最大的时期,在此期间北半球广泛沉积了白垩层,1822年比利时学者奥马·利达鲁瓦将其命名为白垩系。白垩层是一种极细而纯的粉状灰岩,是生物成因的海洋沉积,主要由一种叫做颗石藻的钙质超微化石和浮游有孔虫化石构成。

Translation of Selected Sentences

1. The park was designated an international biosphere reserve in 1976. A separate Denali National Monument was proclaimed by Jimmy Carter on December 1, 1978.

1976年公园被指定为国际生物圈保护区。1978年12月1日吉米·卡特总统宣布竖立迪纳利国家纪念碑。

2. At this time the Alaska Board of Geographic Names changed the name of the mountain back to “Denali”, even though the U. S. Board of Geographic Names maintains “McKinley”.

当时,虽然美国地名委员会仍然称这座山峰为“麦金利”,但阿拉斯加的地名委员会已决定恢复这座山原来的名字——迪纳利。

3. Despite human impact on the area, Denali accommodates gray wolf dens, both historic and active.

虽然人类对迪纳利的生物圈或多或少存在着影响,但迪纳利一直以来都是灰狼栖息的地方。

4. Predatory birds include a variety of hawks, owls, and the gyrfalcon, as well as the abundant but striking golden eagle.

捕食鸟包括各种鹰、猫头鹰、矛隼以及鹫。

5. Feeding animals is strictly forbidden, as it may cause adverse affects on the feeding habits of the creature.

在迪纳利公园喂食动物是被严格禁止的,因为人类的喂食行为会影响动物的摄食习性。

6. Over 450 species of flowering plants fill the park, and can be viewed in bloom throughout summer.

整个夏季,公园中有450多种鲜花竞相开放,争奇斗艳。

Questions

1. Who named the McKinley Mountain first?
2. How did Denali National Park get the name?
3. How did the park help to prevent human impact on the wild preserve of the Denali National Park?
4. Why feeding animals is strictly forbidden in Denali Wild Preserve?
5. Why is it rare for wooded area to be found in Denali National Park?

Chapter 22

Shenandoah National Park
山那都国家公园

【简介】

山那都国家公园位于美国维吉尼亚州的北部，首都华盛顿特区以西大约 112 千米，仅约两小时车程。山那都国家公园沿蓝岭山脉分布，南北长约 161 千米。山岭上有天际线公路贯穿公园。从山那都国家公园可以沿着山岭上长达 430 多千米的蓝岭公路一直开到大雾山国家公园。公园的主要地貌特征是山谷及森林地型。公园占地面积 793.21 平方千米，园区内有一个总统度假山庄，因为距华盛顿很近，交通也方便得多，是当时的总统们喜欢来这里休闲的地方，地以人贵，也就因此成名。公园景色秀丽，春天到处野花盛开，经常举办特别的野花导览活动。

Shenandoah National Park lies in the eastern portion of the state of Virginia along the spine (山脊) of the Blue Ridge Mountains. Located near the Washington, D. C. metropolitan (首都的) area, it is one of America's most popular national parks. It contains mountains, such as the 4,011 foot Stony Man and 4,049 foot Hawkshill, hollows (山谷), creeks, forests, streams, waterfalls, and vistas of the Shenandoah Valley and Piedmont areas of Virginia. Although the Skyline Drive road attracts most of the park's annual 2 million visitors, there are also 500 miles of hiking trails, including 45 miles of the East Coast's famous Appalachian Trail.

Shenandoah National Park encompasses(包围、环绕) part of the Blue Ridge Mountains in the U. S. state of Virginia. This national park is long and narrow, with the broad Shenandoah River and valley on the west side, and the rolling hills of the Virginia Piedmont on the east. Although likely the most prominent feature of the Park is the scenic Skyline Drive, almost 40% of the land area 79,579 acres (322.04 km^2) has been designated as wilderness and is protected as part of the National Wilderness Preservation System. The highest peak is Hawksbill Mountain at 4 051 feet (1,235 m).

Shenandoah National Park passes through parts of eight counties. It lies along the Blue Ridge Mountains in north-central Virginia. These mountains form a distinct highland (高地) rising to elevations above 1,200 meters, local topographic relief between the Blue Ridge Mountains and Shenandoah Valley exceeds 1,000 meters at some locations. The crest of the range divides the Shenandoah River drainage basin (流域盆地), part of the Potomac River (波拖马可河) drainage, on the west side, from the James and Rappahannock River drainage basins on the east side.

Rocks exposed in the park are among the oldest in Virginia dating to over one billion years in age. The park is located along the western part of the Blue Ridge anticlinorium (复背斜层), a regional-scale Paleozoic structure at the eastern margin of the Appalachian fold and thrust belt. Rocks within the park were folded, faulted, distorted, and metamorphosed during the late Paleozoic Alleghanian Orogeny (280 to 330 million years ago).

Shenandoah was authorized in 1926 and fully established on December 26, 1935. Prior to being a park, much of the area was farmland and thereare still remnants (残迹) of old farms in several places. The Commonwealth of Virginia slowly acquired the land through eminent domain (征用权) and then gave it to the U. S. Federal Government provided it would be designated a National Park.

In the creation of the park, a number of families and entire communities were required to vacate portions of the Blue Ridge Mountains. Many residents in the 500 homes in eight affected counties of Virginia were vehemently (激烈地) opposed to losing their homes and communities.

Nearly 90% of the inhabitants worked the land for a living. Many worked in the apple orchards (果园) in the valley and in areas near the eastern slopes. The work to create the National Park and Skyline Drive began following a terrible drought in 1930 which destroyed the crops of many families in the area who farmed in the mountainous terrain (地形), as well as many of the apple orchards where they worked picking crops. Nevertheless, it remains a fact that they were displaced, often against their will, and even for a very few who managed to stay, their communities were lost. A little-known fact is that, while some families were removed by force, a few others (who mostly had also become difficult to deal with) were allowed to stay after their properties were acquired, living in the park until nature took its course and they gradually died. The policy allowed the elderly and disabled who so wished to remain with life tenancy. The lost communities and homes were a price paid for one of the country's most beautiful National Parks and scenic roadways.

In the early 1930s, the National Park Service began planning the park facilities and envisioned(预想) separate provisions for "colored guests", as African Americans were described in contemporaneous (同时代的) government documents. At that time, Virginia initially attempted to ban African Americans entirely from the park, but settled for enforcing its segregation laws in the park's facilities.

By the Thirties, there were several concessions operated by private firms within the area that would become the park, some going back to the late 19th Century. These early private facilities were operated only for whites. By 1937, the Park Service accepted a bid from Virginia Sky-Line Company to take over the existing facilities and add new lodges, cabins, and other amenities (令人愉快之物), including Big Meadows Lodge. Under their plan, all the sites in the parks, save one, were for "Whites Only". Their plan included a separate facility for African Americans at Lewis Mountain—a picnic ground, a smaller lodge, cabins and a campground. The site opened in 1939, and it was substantially inferior to the other park facilities. By then, however, the Interior Department was increasingly anxious to eliminate segregation from all parks. Pinnacles picnic ground was selected to be the initial integrated site in the Shenandoah, but Virginia Sky-Line Company continued to balk (妨碍), and distributed maps showing Lewis Mountain as the only site for African Americans. During World War II, concessions closed and park usage plunged. But once the War ended, in December 1945, the NPS mandated that all concessions in all national parks were to be desegregated(废止种族隔离). In October 1947 the dining rooms of Lewis Mountain and Panorama were integrated and by early 1950, the mandate was fully accomplished.

The climate of the park, and in turn also its flora(植物) and fauna (动物群), is largely typical for mountainous regions (山区) of the eastern Mid-Atlantic woodland, and indeed a large portion of common species are typical of ecosystems at lower altitudes as well. On southwestern faces of some of the southernmost (最南的) hillsides pine predominates and there is also the occasional prickly pear cactus (仙人掌) which grows naturally. In contrast, some of the northeastern aspects are most likely to have small but dense stands of moisture loving hemlocks (芹叶钩吻) and mosses in abundance. Other commonly found plants include oak, hickory (山胡桃树), chestnut, maple, tulip poplar, mountain laurel (月桂树), milkweed, daisies, and many species of ferns. The once predominant American chestnut tree was effectively brought to extinction by a fungus (菌类) known as the Chestnut blight (枯萎病) during the 1930s, though the tree continues to grow in the park, it does not reach maturity and dies back (枝叶枯萎) before it can reproduce. Various species of oaks superseded the Chestnuts and became the dominant tree species. Gypsy moth（舞毒蛾）infestations beginning in the early 1990s began to erode the dominance of the oak forests as the moths would primarily consume the leaves of oak trees. Though the Gypsy moths seem to have abated some, they continue to affect the forest and have destroyed almost 10 percent of the oak groves.

The Blue Ridge Mountains of Virginia had been occupied and used by Native Americans long before the arrival of Europeans. They hunted, gathered nuts and berries, and altered the land itself by burning vegetation to confine and kill game. Europeans later occupied the land, cutting most of the forest in the process of farming and cultivating the land, and engaged in some mining of copper, iron, and manganese（锰）.

When the search for national park land in the east began a number of Virginians promoted the Blue Ridge site. Ultimately, the state of Virginia, using state money and additional funds donated by some 24,000 Virginians, and through the untiring efforts of William E. Carson, acquired some 3,870 tracts of land and donated the land to the federal government for the purposes of creating a national park. Approximately 2000 residents, "mountain folk", had to move as a result of the creation of the park. The park itself was established on December 26, 1935. Among its supporters was Franklin Delano Roosevelt, who dedicated the park, and whose Depression Area Civilian Conservation Corps project was responsible for the construction of many of the park's existing recreational facilities.

Unlike most of America's national parks, Shenandoah was not founded to preserve wilderness. Much of the land contained within the present park boundaries had been in private ownership for two centuries, and much of the original forest land had been eliminated through logging（伐木业）activities which date back to 1700. The wilderness which exists in the part now has largely been recaptured by nature since the founding of the park. The park itself has served as a laboratory for the observation of the way in which nature heals the wounds caused by human activity. The forest is rapidly returning to its original state, representative of the hardwood（阔叶树）shield which once covered most of the northeastern United States. The forest has currently reclaimed approximately 95% of the park. In the winter and early spring, before the leaves have returned to the deciduous trees（落叶树）, it is possible to see for miles across the returning forests. The land truly appears as a wilderness, without any clue of the centuries of human activities. The current forest contains a wide variety of species of trees, including hemlock, grey birch and balsam fir. Sadly, one type of tree—the chestnut—which once represented a quarter of the trees in the forest—is no longer presented.

The Blue Ridge Mountains are named for the haze which often hangs over the mountains themselves and the valleys on either side of the ridge. Fog frequently blankets both mountains and valleys, creating interesting views and vistas.

Notes

the Blue Ridge Mountains 蓝岭　美国俄勒冈东北部、华盛顿东南部山脉,由哥伦比亚高原逐渐升高和被侵蚀的部分构成。在俄勒冈的哥维克·巴特岩部分高达27 773米(9 106英尺)。山上树木郁郁葱葱,远远望去,仿佛山岭之上笼罩在一层淡淡的蓝色薄雾之中,蓝岭便因此而得名。

Piedmont areas 皮埃蒙特山麓高原　美国东部一高原地区,从纽约州延伸至阿拉巴马州,位于阿巴拉契亚山脉和大西洋沿岸平原之间。

Paleozoic 古生代　地质时代中的一个时代,属于显生宙,上一个代是新元古代,下一个代是中生代。动物群以海生无脊椎动物中的三叶虫、软体动物和棘皮动物最繁盛。在奥陶纪、志留纪、泥盆纪、石炭纪相继出现低等鱼类、古两栖类和古爬行类动物。鱼类在泥盆纪达于全盛。石炭纪和二叠纪昆虫和两栖类繁盛。古植物以海生藻类为主。

Translation of Selected Sentences

1. This national park is long and narrow, with the broad Shenandoah River and valley on the west side, and the rolling hills of the Virginia Piedmont on the east.

西部是宽阔的山那都河及其河谷,东边是起伏延绵的佛吉尼亚皮的蒙特高原,而狭长的国家公园就位于这两者之间。

2. The Commonwealth of Virginia slowly acquired the land through eminent domain and then gave it to the U. S. Federal Government provided it would be designated a National Park.

佛吉尼亚联邦通过征用,逐渐将土地收回,并将之交给美国联邦政府。希冀联邦政府将此地建成国家公园。

3. The work to create the National Park and Skyline Drive began following a terrible drought in 1930 which destroyed the crops of many families in the area who farmed in the mountainous terrain, as well as many of the apple orchards where they worked picking crops.

建设国家公园和天际高速路的工程刚刚开始,正遇到1930年大旱。这场旱灾使当地许多家庭农庄的庄稼颗粒无收,而且苹果园也难逃厄运。

4. The lost communities and homes were a price paid for one of the country's most beautiful National Parks and scenic roadways.

当地人背井离乡,付出了巨大的代价。正是因为他们的付出,才有了日后全美最美丽的国家公园和沿途风景最秀丽的行车道。

5. The once predominant American chestnut tree was effectively brought to extinction by a fungus known as the Chestnut blight during the 1930s—though the tree continues to grow in the park, it does not reach maturity and dies back before it can reproduce.

20世纪30年代一种菌类引发美洲栗树患上了枯萎病。患病的栗树虽然枝繁叶茂,但无法长到成熟,并且无法结果。

6. The Blue Ridge Mountains of Virginia had been occupied and used by Native Americans long before the arrival of Europeans.

早在欧洲开拓者到达之前,美洲印第安土著人就祖祖辈辈生活在佛吉尼亚的蓝岭山区。

7. The forest has currently reclaimed approximately 95% of the park. In the winter and early spring, before the leaves have returned to the deciduous trees, it is possible to see for miles across the returning forests.

到目前为止,公园里的森林已经恢复了95%。冬天或初春时节,透过树叶凋零的枝桠,可以望见绵延数英里的森林。

8. The Blue Ridge Mountains are named for the haze which often hangs over the mountains themselves and the valleys on either side of the ridge.

因为山岭上和河谷两侧常常笼罩着一层薄薄的蓝色雾霭,蓝岭便由此得名。

Questions

1. Who named the McKinley Mountain first?
2. When was Shenandoah being fully established? What the area used to be?
3. How did the government manage to displace residents around the area of Blue Ridge Mountains?
4. Was Shenandoah initially planned for all people in the nation? When was it desegregated?
5. Who were the original residents along the Blue RidgeMountains? Who helped to raise funds for establishing the Shenandoah National Park?
6. Why is it believed that Shenandoah was not to preserve wilderness? How did the wilderness return to the area?
7. How has the Blue Ridge Mountains got its name?

Chapter 23

Bryce Canyon National Park
布莱斯峡谷国家公园

【简介】

布莱斯峡谷国家公园是位于美国犹他州西南部的国家公园。其名字虽有峡谷一词,但其并非真正的峡谷,而是沿着庞沙冈特高原东面,由侵蚀而成的巨大自然露天剧场。其独特的地理结构称为岩柱,由风、河流里的水与冰侵蚀和湖床的沉积岩组成。位于其内的红色、橙色与白色的岩石形成了奇特的自然景观,因此被誉为天然石俑的殿堂。峡谷地带由摩门教信徒首先于19世纪50年代开发,并在1875年埃本尼泽·布莱斯移居至此后命名。此附近的地带于1924年成为美国国家保护区,并于1928年被设计为国家公园。

Bryce Canyon National Park is a national park located in southwestern Utah in the United States. The major feature of the park is Bryce Canyon which, despite its name, is not a canyon but a giant natural amphitheater created by erosion along the eastern side of the Paunsaugunt Plateau. Bryce is distinctive due to geological structures called hoodoos（岩柱）, formed by wind, water, and ice erosion of the river and lake bed sedimentary rocks. The red, orange, and white colors of the rocks provide spectacular（壮观的）views for park visitors.

The Bryce area was settled by Mormon pioneers in the 1850s and was named after Ebenezer Bryce, who homesteaded in the area in 1874. The area around Bryce Canyon became a U. S. National Monument in 1923 and was designated as a national park in 1928. The park covers 56 square miles（145 km^2）and receives relatively few visitors compared to Zion National Park and the Grand Canyon, largely due to its remote location. The town of Kanab, Utah, is situated at a central point between these three parks.

Little is known about early human habitation in the Bryce Canyon area. Archaeological（考古学的）surveys of Bryce Canyon National Park and the Paunsaugunt Plateau show that people have been in the area for at least 10,000 years. Basketmaker-period Anasazi artifacts（史前古器物）several thousand years old have been found south of the park. Other artifacts from the Pueblo-period（普韦布洛时期）Anasazi and the Fremont culture（up to the mid-12th century）have also been found.

The Paiute Indians moved into the surrounding valleys and plateaus in the area around the same time that the other cultures left. These Native Americans hunted and gathered for most of their food, but also supplemented their diet with some cultivated products. The Paiute in the area developed a mythology surrounding the hoodoos（pinnacles）in Bryce Canyon. They believed that hoodoos were the Legend People whom the trickster Coyote turned to stone. At least one older Paiute said his culture called the hoodoos Anka-ku-was-a-wits, which is Paiute for "red painted faces".

It was not until the late 18th and the early 19th century that the first European Americans explored the remote and hard-to-reach area. Mormon scouts visited the area in the 1850s to gauge（测量）its potential for agricultural development, use for grazing（放牧）, and settlement.

The first major scientific expedition to the area was led by U. S. Army Major（陆军少校）John Wesley Powell in 1872. Powell, along with a team of mapmakers and geologists, surveyed the Sevier and Virgin River area as part of a larger survey of the Colorado Plateaus. His mapmakers kept many of the Paiute place names.

Small groups of Mormon pioneers followed and attempted to settle east of Bryce Canyon along the Paria River. In 1873, the Kanarra Cattle Company started to use the area for cattle grazing.

The Church of Jesus Christ of Latter-day Saints sent Scottish immigrant Ebenezer Bryce and his wife Mary to settle land in the Paria Valley because they thought his carpentry skills would be useful in the area. The Bryce family chose to live right below Bryce Canyon Amphitheater. Bryce grazed his cattle inside what are now park borders, and reputedly（据说）thought that the amphitheaters were a "helluva place to lose a cow". He also built a road to the plateau to retrieve firewood and timber, and a canal to irrigate his crops and water his animals. Other settlers soon started to call the unusual place "Bryce's canyon", which was later formalized into Bryce Canyon.

A combination of drought, overgrazing and flooding eventually drove the remaining Paiutes from the area and prompted the settlers to attempt construction of a water diversion channel from the Sevier River drainage. When that effort failed, most of the settlers, including the Bryce family, left the area. Bryce moved his family to Arizona in 1880. The remaining settlers dug a 10 miles (16 km) ditch from the Sevier's east fork into Tropic Valley.

Bryce Canyon Lodge was built between 1924 and 1925 from local materials.

These scenic areas were first described for the public in magazine articles published by Union Pacific and Santa Fe railroads in 1916. People like Forest Supervisor J. W. Humphrey promoted the scenic wonders of Bryce Canyon's amphitheaters, and by 1918 nationally distributed articles also helped to spark interest. However, poor access to the remote area and the lack of accommodations kept visitation to a bare minimum.

Visitation steadily increased, and by the early 1920s the Union Pacific Railroad became interested in expanding rail service into southwestern Utah to accommodate more tourists.

In 1928 the canyon became a National Park. It now has this visitors' center.

At the same time, conservationists (天然资源的保护管理论者) became alarmed by the damage overgrazing and logging on the plateau, along with unregulated visitation, were having on the fragile features of Bryce Canyon. A movement to have the area protected was soon started, and National Park Service Director Stephen Mather responded by proposing that Bryce Canyon be made into a state park. The governor of Utah and the Utah Legislature, however, lobbied for national protection of the area. Mather relented and sent his recommendation to President Warren G. Harding, who on June 8, 1923 declared Bryce Canyon National Monument into existence.

A road was built the same year on the plateau to provide easy access to outlooks over the amphitheaters. From 1924 to 1925, Bryce Canyon Lodge was built from local timber and stone.

Members of U. S. Congress started work in 1924 on upgrading Bryce Canyon's protection status from a U. S. National Monument to a National Park in order to establish Utah National Park. Aprocess led by the Utah Parks Company for transferring ownership of private and state-held land in the monument to the federal government started in 1923. The last of the land in the proposed park's borders was sold to the federal government four years later, and on February 25, 1928, the renamed Bryce Canyon National Park was established.

In 1931, President Herbert Hoover annexed an adjoining area south of the park, and in 1942 an additional 635 acres (2.57 km^2) was added. This brought the park's total area to the current figure of 35,835 acres (145.02 km^2). Rim Road, the scenic drive that is still used today, was completed in 1934 by the Civilian Conservation Corps. Administration of the park was conducted from neighboring Zion Canyon National Park until 1956, when Bryce Canyon's first superintendent started work.

The USS(美国军舰) Bryce Canyon was named for the park and served as a supply and repair ship in the U. S. Pacific Fleet from September 15, 1950 to June 30, 1981.

Bryce Canyon Natural History Association (BCNHA) was established in 1961. It runs the bookstore inside the park visitor center and is a non-profit organization created to aid the interpretive（解释的）, educational and scientific activities of the National Park Service at Bryce Canyon National Park. A portion of the profits from all bookstore sales are donated to public land units.

Responding to increased visitation and traffic congestion, the National Park Service implemented a voluntary, summer-only, in-park shuttle system in June 2000. In 2004, reconstruction began on the aging and inadequate road system in the park.

Mule deer（长耳鹿）are the most common large animals found in the park.

More than 400 native plant species live in the park. There are three life zones in the park based on elevation: The lowest areas of the park are dominated by dwarf forests of pinyon pine and juniper（刺柏属丛木或树木）with manzanita（石兰科常绿灌木）, serviceberry（花楸树）, and antelope bitterbrush in between. Aspen（白杨）, cottonwood（三叶杨）, Water Birch, and Willow grow along streams. Ponderosa Pine forests cover the mid-elevations with Blue Spruce（蓝叶云杉）and Douglas-fir（花旗松）in water-rich areas and manzanita and bitterbrush as underbrush. Douglas-fir and White Fir, along with Aspen and Engelmann Spruce（英国针枞）, make up the forests on the Paunsaugunt Plateau. The harshest areas have Limber Pine and ancient Great Basin Bristlecone Pine—some more than 1,600 years old—holding on.

The forests and meadows of Bryce Canyon provide the habitat to support diverse animal life, from birds and smallmammals to foxes and occasional bobcats, mountain lions, and black bears. Mule deer are the most common large mammals（美洲野猫）in the park. Elk and pronghorn antelope（叉角羚）, which have been reintroduced nearby, sometimes venture into the park.

Bryce Canyon National Park forms part of the habitat of three wildlife species that are listed under the Endangered Species Act: the Utah Prairie Dog（犹他草原土拨鼠）, the California Condor（加利福尼亚秃鹰）, and the Southwestern Willow Flycatcher（霸鹟）. The Utah Prairie Dog is a threatened species that was reintroduced to the park for conservation, and the largest protected population is found within the park's boundaries.

About 170 species of birds visit the park each year, including swifts（雨燕）and swallows（燕子）. Most species migrate to warmer regions in winter, although jays, ravens, nuthatches, eagles, and owls stay. In winter, the mule deer, mountain lion（美洲狮）, and coyotes migrate to lower elevations. Ground squirrels and marmots（旱獭）pass the winter in hibernation（冬眠）.

Eleven species of reptiles（爬虫动物）and four species of amphibians（两栖动物）have been found at in the park. Reptiles include the Great Basin Rattlesnake, Short-horned Lizard, Side-blotched Lizard, Striped Whipsnake, and the Tiger Salamander.

While humans have greatly reduced the amount of habitat that is available to wildlife in most parts of the United States, the relative scarcity of water in southern Utah restricts human development and helps account for the region's greatly enhanced diversity of wildlife.

Notes

the Paunsaugunt Plateau 庞苏刚特高原　位于美国犹他州西南部,海拔2 100 ~2 800 米。东西宽度大约16 千米,南北长度约40 千米。

Mormon 摩门教徒　约瑟·史密斯(Joseph Smith)最开始在1830 年所成立的教会并无正式的名称,所代表的是耶稣基督的使徒在第一世纪时所建立的最原始的教会,故一般称“基督教会”,1838 年4 月26 日才因启示而正式称为“耶稣基督后期圣徒教会”(The Church of Jesus Christ of Latter-Day Saints),为的是要与耶稣基督的其他教会有所区别。此正式名称已延用至今日。刚成立的后期圣徒教会,被当时美国的主流宗教基督教(新教)所逼迫,因为其教宣传一夫多妻制度,其后约瑟·史密斯被杀害,由杨·百翰带领教徒到美国中部犹他州大盐湖山谷(Utah, Salt Lake City),并于1847 年在该处安定发展起来,也因此教会的总会位于盐湖城。事实上,盐湖城是此教会早期的教友凭借对神的信心拓荒所建成的一座城市。这在全世界的城市发展史上极为罕见且特殊。该教会的教友目前已遍布全球各国。据教会最新发表数字,信徒人数达至1 200 多万,超过半数居于北美洲以外。

Basketmaker 编篮者　编篮者文化(公元前800 年—公元700 年)以盛行使用编织物而得名,经济以农业和狩猎为主,并行采集。种植玉米、南瓜、豆类、向日葵,猎取鹿、羚羊、兔子和小动物。早期聚落多建在山脊顶部,房屋圆形,村外有卵石围墙。工具有鹿角尖状器、石刀和磨石,使用编织的筐篮、皮囊。人死后用皮毛、毯子包裹埋葬,并随葬工具、用具和武器。晚期出现圆形、椭圆形和方形半地穴居址,面积9 ~25 平方米。中间有火塘,火塘旁有一个被称为锡帕普的小洞,印第安人认为这是祖先现身进出的地方。房屋排成半月形,门向东。已产生宗教建筑基瓦,制造红褐和灰黑色的粗糙陶器,有钵、细颈圜底罐,饰以曲线纹。

Anasazi 阿那萨吉文化　美西南地区的新石器时代文化。“阿那萨吉”在纳瓦霍印第安语中意为“古代的人们”。主要分布在美国亚利桑那州北部高原的半沙漠半灌木林地区。年代为公元前800—公元1550 年。包括编篮者文化和普韦布洛文化两个阶段。1550 年西班牙人入侵后逐渐衰落。

Paiute 派尤特人　两个独特的美国印第安族群,分为南派尤特人和北派尤特人。南派尤特人原居住在犹他州南部、亚利桑那州西北部等地。传统上都以采集食物为生,主要采集种籽、松果,兼猎捕小猎物,不过许多南派尤特人也种植小型园圃。他们居住临时用柴草搭成的住所,不穿衣服,只披兔皮毯子,编织各种筐篮,用以采集食物或调制食物。派尤特人的古老风俗多已消失,现在只有巫医。20世纪晚期,北派尤特人及南派尤特人大都居住在居留地或其附近,共约4 000人。

Warren G. Harding 沃伦·哈丁　美国历史上第29届总统(1921—1923年)。他是来自俄亥俄州的共和党人。他曾是有着深远影响力的报刊发行人。曾担任俄亥俄州的副州长及美国会参议员。他是美国历史上第一位当选总统的参议员,也是第一位当选总统的报刊发行人。

Translation of Selected Sentences

1. Bryce is distinctive due to geological structures called hoodoos, formed by wind, water, and ice erosion of the river and lake bed sedimentary rocks.

由于风、水和冰对于河床和湖底的腐蚀,水成岩成为岩柱。布莱斯峡谷国家公园即因这样的岩柱而独成一景。

2. These Native Americans hunted and gathered for most of their food, but also supplemented their diet with some cultivated products.

这些美洲土著人靠打猎和采集获得大部分食物,同时也种植一些粮食作为补充。

3. The Paiute in the area developed a mythology surrounding the hoodoos (pinnacles) in Bryce Canyon. They believed that hoodoos were the Legend People whom the trickster Coyote turned to stone.

生活在这一地区的派尤特人中流传着关于布莱斯峡谷岩柱的神话。他们认为岩柱是误中歹人奸计而被咒语变成石头的伟人。

4. The Church of Jesus Christ of Latter-day Saints sent Scottish immigrant Ebenezer Bryce and his wife Mary to settle land in the Paria Valley because they thought his carpentry skills would be useful in the area.

耶稣基督后期圣徒教会派遣来自苏格兰的移民埃比尼泽·布莱斯和他的妻子来帕利亚峡谷开荒,因为他们认为布莱斯的木工手艺在这片土地上能够发挥作用。

5. He also built a road to the plateau to retrieve firewood and timber, and a canal to irrigate his crops and water his animals. Other settlers soon started to call the unusual place "Bryce's canyon", which was later formalized into Bryce Canyon.

他还修了一条通往高原的路,以便运输柴火和木材;同时修了一条运河,来灌溉他的庄稼和供牲畜饮水。其他的开拓者不久就称这片土地为"布莱斯利的峡谷"。之后正式命名为"布莱斯利峡谷"。

6. These scenic areas were first described for the public in magazine articles published by *Union Pacific and Santa Fe* railroads in 1916.

《太平洋联盟》这本杂志于1916年发表文章，第一次向公众描述这一地区的迷人景色。

7. At the same time, conservationists became alarmed by the damage overgrazing and logging on the plateau, along with unregulated visitation, were having on the fragile features of Bryce Canyon.

同时，令天然资源保护管理论的支持者感到震惊的是：过度放牧、毫无节制的树木砍伐以及管理无序的旅游业，已经对布莱斯峡谷脆弱的生态环境造成了严重影响。

8. The USS Bryce Canyon was named for the park and served as a supply and repair ship in the U. S. Pacific Fleet from September 15, 1950 to June 30, 1981.

美国海军一艘补给舰被命名为布莱斯峡谷号。这艘军舰从1950年9月15日开始一直在美国太平洋舰队服役，直到1981年6月30日退役。

9. The Utah Prairie Dog is a threatened species that was reintroduced to the park for conservation, and the largest protected population is found within the park's boundaries.

犹他草原的土拨鼠是濒临灭绝的物种。在将这种土拨鼠引进国家公园并进行保护后，其数量已经大幅上涨。

10. While humans have greatly reduced the amount of habitat that is available to wildlife in most parts of the United States, the relative scarcity of water in southern Utah restricts human development and helps account for the region's greatly enhanced diversity of wildlife.

由于人类的开发和居住，虽然在美国大部分地区，适合野生物生存的区域不断缩小，但是由于犹他州南部相对缺水，野生物的多样性反而得到明显的提高。

Questions

1. Who settled in Bryce Canyon initially?
2. How did "Bryce Canyon" get its name?
3. Why did most of thesettlers move out the area in 1880s?
4. How did the journalists help to promote the scenic wonders of Bryce Canyon in 1920s?
5. Why did the visitors prefer to visit other areas instead of beautiful Bryce Canyon in the early 20th century?
6. What did President Herbert Hoover do to help enlarge Bryce Canyon?

Chapter 24

Grand Teton National Park
大提顿国家公园

【简介】

大提顿国家公园位于美国怀俄明州西北部壮观的冰川山区,1929 年建立,占地 126 平方千米。公园内最高的山峰是大提顿峰,海拔 4 198 米,有存留至今的冰川。分布在该地的冰湖以珍尼湖为最著名。斯内克河上用水坝拦堵形成的杰克森湖为当地最大的水域。高耸入云的山巅,覆盖着千年的冰河,山连山,峰连峰,宛如进入人间仙境。公园内有成群的美洲野牛、麋鹿和羚羊,还有其他许多种哺乳动物。

Grand Teton National Park is a United States National Park located in northwestern Wyoming, U. S. Grand Teton National Park consists of approximately 310,000 acres and includes the major peaks of the 64 km long Teton Range (特顿山脉) as well as most of the northern sections of the valley known as Jackson Hole (杰克逊潭). Only 16 km south of Yellowstone National Park, the two parks are connected by the National Park Service managed John D. Rockefeller, Jr. Memorial Parkway. These three protected areas in conjunction with (与…协力) surrounding National Forests constitute the Greater Yellowstone Ecosystem (大黄石生态系统), which at almost 18,000,000 acres, is one of the largest intact (完整无缺的) mid-latitude temperate ecosystems in the world.

Human history of the Grand Teton region dates back at least 11,000 years, when the first nomadic (游牧的) hunter-gatherer Paleo-Indians would migrate into the region during warmer months in pursuit of food and supplies. In the early 19th century, the first caucasian (白种人的) explorers encountered the eastern Shoshone natives. Between 1810 and 1840, the region attracted various fur trading companies which vied（竞争）for control of the lucrative（有利的）beaver（海狸）fur. Organized U.S. Government explorations to the region commenced（开始）in the 1870s as an offshoot（分支）of exploration in Yellowstone. The first permanent settlers in Jackson Hole arrived in the 1880s. Efforts to preserve the region as a national park commenced in the late 19th century and by 1929, Grand Teton National Park was established, protecting the major peaks of the Teton Range. The valley of Jackson Hole remained in private ownership until conservationists led by John D. Rockefeller, Jr. in the 1930s began purchasing land in Jackson Hole to be added to the existing national park. Against public opinion and with repeated congressional efforts to repeal（废止）it, much of Jackson Hole was also set aside for protection as Jackson Hole National Monument in 1943. In 1950 the monument was abolished and most of the monument acreage（面积）was added to Grand Teton National Park.

Grand Teton National Park is named for Grand Teton which is the tallest mountain in the Teton Range. At 13,775 feet, Grand Teton rises abruptly more than 7,000 feet above Jackson Hole and is almost 850 feet higher than Mount Owen, the second highest summit in the range.

To the north of Jackson Hole, Yellowstone National Park had been established in 1872. By the late 19th century, conservationists worked to provide further protection to surrounding regions, leading U.S. President Grover Cleveland to create the Teton Forest Reserve, which included a portion of northern Jackson Hole. By 1902, the reserve was combined into the Yellowstone Forest Reserve, then divided again in 1908 by President Theodore Roosevelt, establishing the Teton National Forest, protecting most of the Teton Range. By 1907, the U.S. Bureau of Reclamation had constructed a temporary dam at the Snake River outlet of Jackson Lake. This dam failed in 1910 and a new concrete Jackson Lake Dam replaced it by 1911. The dam was further enlarged in 1916, raising lake waters 39 feet（12 m）as part of the Minidoka Project, designed to provide irrigation for agriculture in the state of Idaho. Though efforts to protect the Teton Range and Jackson Hole as part of an expanded Yellowstone National Park dated back to the late 19th century, proposals to construct more dams on some of the other lakes in Jackson Hole led Yellowstone National Park superintendent Horace Albright to block such efforts. Albright was originally an advocate of the expanded Yellowstone plan which was very unpopular with local residents. By the mid 1920's however, local sentiment had changed once the proposals for a new national park carved（开创）out of land currently protected as part of the Teton National Forest and including only the Teton Range and six lakes at the base of the range were put forward. With the general agreement of prominent Jackson Hole residents on this plan, President Calvin Coolidge signed the executive order establishing the 96,000 acres（39,000 ha）Grand Teton National Park on February 26, 1929.

The valley of Jackson Hole remained primarily in private ownership when John D. Rockefeller, Jr. and his wife visited the region in the late 1920s. Horace Albright had hoped to protect the valley of Jackson Hole north of the town of Jackson from commercial exploitation. Rockefeller agreed and through a private enterprise known as the Snake River Land Company was by 1927 buying land in Jackson Hole to be later turned over to the National Park Service. In 1930, this plan was revealed to the residents of the region and was met with strong disapproval. By 1942, John D. Rockefeller, Jr. became increasingly impatient that his land holdings in Jackson Hole might never be part of Grand Teton National Park, so he wrote the Secretary of the Interior（内政部长）Harold L. Ickes and informed him that he may sell the land to another party. Secretary Ickes recommended to President Franklin Roosevelt that the Antiquities Act (which permitted Presidents to set aside land for protection without the approval of Congress)（古物保护法）be used to establish a National Monument in Jackson Hole. Roosevelt agreed and the property belonging to the Snake River Land Company along with additional land carved from Teton National Forest became the 221,000 acres Jackson Hole National Monument in 1943. The monument and park were adjacent to each other and administered by the National Park Service, but the monument designation ensured no funding allotment（分配）, nor provided a level of resource protection equal to the park. Members of Congress repeatedly attempted to have the new National Monument abolished.

After the end of World War II, public sentiment shifted in favor of adding the monument to the park. Though there was still much local disagreement, the monument and park were combined in 1950. In recognition for John D. Rockefeller, Jr.'s efforts to establish and then expand Grand Teton National Park, a 24,000 acres parcel of land between Grand Teton and Yellowstone National Parks was added to the National Park service in 1972. This land along with the road from the southern boundary of Grand Teton National Park to West Thumb in Yellowstone National Park was designated as the John D. Rockefeller, Jr. Memorial Parkway. In 2001, the Rockefeller family donated the remnants（剩余的）of its JY Ranch for the establishment of the Laurance S. Rockefeller Preserve, dedicated on June 21,2008.

Grand Teton National Park is administered by the National Park Service, an agency of the Department of the Interior. Grand Teton National Park oversees the John D. Rockefeller, Jr. Memorial Parkway acreage and the associated scenic highway south of Yellowstone National Park. The park has an average of 100 permanent and 180 seasonal employees on staff annually. The park also manages 27 concession contracts which provide services such as lodging, restaurants, mountaineering guides, dude ranching（休养农场）, fishing and a boat shuttle on Jenny Lake. Grand Teton National Park works closely with other federal agencies such as the U. S. Forest

Service, the U. S. Fish and Wildlife Service and the Bureau of Reclamation. The park also works closely with the Federal Aviation Administration, a consequence of Jackson Hole Airport's presence in the park.

The Teton Range is the youngest mountain range in the Rocky Mountains (洛矶山脉), and began forming between 6 to 9 million years ago. The Teton Range runs roughly north to south and rises from the floor of Jackson Hole without any foothills (丘陵地带) along a 40 miles (64 km) long by 7 to 9 miles (11 to 14 km) wide active fault-block (断层石块) mountain front. The range tilts (使倾斜) westward, rising abruptly above Jackson Hole valley which lies to the east but more gradually into Teton Valley to the west. A series of earthquakes along the Teton Fault slowly displaced the western side of the fault upward and the eastern side of the fault downward at an average of 1 foot (0.30 m) of displacement every 300 – 400 years.

In addition to 13,775 ft (4,199 m) high Grand Teton, another nine peaks are over 12,000 ft (3,658 m) above sea level. Eight of these peaks between Avalanche and Cascade Canyons make up the often-photographed Cathedral Group. The most prominent peak north of Cascade Canyon is the monolithic Mount Moran (12,605 feet (3,842 m)) which rises 5,728 feet (1,746 m) above Jackson Lake. To the north of Mount Moran, the range eventually merges into the high altitude Yellowstone Plateau. South of the central Cathedral Group the Teton Range tapers off near Teton Pass and blends into the Snake River Range.

West to east trending canyons provide easier access by foot into the heart of the range as no vehicular roads traverse the range except at Teton Pass, which is south of the park. Carved by a combination of glacier activity as well as by numerous streams, the canyons are at their lowest point along the eastern margin of the range at Jackson Hole. Flowing from higher to lower elevations, the glaciers created more than a dozen U-shaped valleys throughout the range. Cascade Canyon is sandwiched between Mount Owen and Teewinot Mountain to the south and Symmetry Spire to the north and is situated immediately west of Jenny Lake. North to south, Webb, Moran, Paintbrush, Cascade, Death and Granite Canyons slice through Teton Range.

Most of the lakes in Grand Teton National Park were formed by glaciers and the majority of these lakes are located at the base of the Teton Range. In the northern section of the park lies Jackson Lake, the largest lake in the park at 15 miles (24 km) long 5 miles (8.0 km) wide and 438 feet (134 m) deep. Though Jackson Lake is natural, the Jackson Lake Dam was constructed at its outlet before the creation of the park (since modified) and the lake level was raised almost 40 feet (12 m) consequently. East of the Jackson Lake Lodge lies Emma Matilda and Two Ocean Lakes. South of Jackson Lake, Leigh, Jenny, Bradley, Taggart and Phelps Lakes can be found near the entrances to the canyons which lead into the Teton Range. Within the Teton Range, smaller lakes are sometimes found in high altitude cirques (盆地谷), and there are more than 100 alpine (高山的) lakes scattered throughout the high country. Lake Solitude, located at an elevation of 9,035 feet (2,754 m), is in a cirque at the head of the North Fork of Cascade Canyon. Other high altitude lakes can be found at over 10,000 feet (3,000 m) in elevation and a few, such as Icefloe Lake, remain ice clogged for much of the year.

The Snake River flows north to south through the park, entering Jackson Lake near the boundary of Grand Teton National Park and John D. Rockefeller, Jr. Memorial Parkway. The Snake River then flows through the spillways (泄洪道) of the Jackson Lake Dam and from there southward through Jackson Hole, exiting the park just west of the Jackson Hole Airport. The largest lakes in the park all drain either directly or by tributary streams into the Snake River. Major tributaries which flow into the Snake River include Pacific Creek and Buffalo Fork near Moran and the Gros Ventre River at the southern border of the park.

The diversity of wildlife communities in Grand Teton National Park and the John D. Rockefeller, Jr. Memorial Parkway complements the spectacular scenery. Part of the Greater Yellowstone Ecosystem, the two National Park Service areas offer wildlife a variety of habitats. Each habitat must supply the basic needs of wildlife: food, water, cover and living space.

Wind and snow limit life above treeline (about 10,000 feet). Some plants and animals have adapted to the harsh conditions. Plants are mat-like, animals are few. There are yellow-bellied marmots, pikas and bighorn sheep around the area.

Sagebrush(山艾树) is the most widespread habitat type in the park, sagebrush flats occur on dry, porous soils (多孔隙土). More than 100 species of grasses and wildflowers grow along with abundant sagebrush. Lack of cover makes large animals conspicuous, such as pronghorns, coyotes, bison, badgers, elk and Uinta ground squirrels.

Fromtree-line to valley floor, forests provide cover and food for many mammal species. Lodgepole pines (黑松) dominate, but forests also contain firs, aspens and spruces. Elk (麋鹿), mule deer, martens (貂鼠), red squirrels, black bears (黑熊) and snowshoe hares can been seen in the forests. Aquatic habitats and adjacent forests, marshes and meadows fulfill the needs of many forms of wildlife. Diverse and abundant vegetation offers excellent food and cover. Water is plentiful. Thus moose, river otters (水獭), beavers, muskrats (麝鼠), coyotes, bison (美洲野牛) and mule deer home in the forests of Grand Teton National Park.

Notes

John D. Rockefeller, Jr. 小约翰·洛克菲勒　老约翰·D·洛克菲勒唯一的儿子和继承人。洛克菲勒财团是美国十大财团之一,是以洛克菲勒家族的石油垄断为基础,通过不断控制金融机构,把势力范围伸向国民经济各部门的美国最大的垄断集团。创始人约翰·洛克菲勒以石油起家,1863 年在克利夫兰开办炼油厂,1870 年以该厂为基础,扩大组成俄亥俄标准(原译美孚)石油公司,很快垄断了美国的石油工业,并用其获得的巨额利润投资于金融业和制造业,经济实力发展迅猛。资产总额在 1935 年仅 66 亿美元,至 1960 年增至 826 亿美元,25 年中增长了 11.5 倍。其后又继续获得巨大发展,1974 年资产总额增长达 3 305 亿美元,超过了摩根财团,跃居美国十大财团的首位。美国最大的石油公司有 16 家,其中有 8 家属于洛克菲勒财团。

Paleo-Indians 古印度人　据说是在后冰期从亚洲的西伯利亚移居到美洲阿拉斯加的史前的一个民族。其中一些人一直向南至南美洲,而另一部分人向东直至后来的佛吉尼亚和西佛吉尼亚州才定居下来。考古学家认为,古印度人的移居路线是跟随着马群、驯鹿群或猛犸象群的迁徙方向。当时的古印度人以狩猎维持生计,也以采集居住地周围的食物果腹。

Shoshone 肖松尼族人　一支由三个分支组成的美国印第安人土著,他们分别是:住在爱达荷部分地区,犹他州北部,俄勒冈州东部,蒙大拿西部,现在大多数住在爱达荷州东南部的一支,也称作北肖松尼族人;居住在爱达荷州,犹他州和内华达州的大盆地区向南直到加利福尼亚的死谷,现在大多住在内华达州的一支,也称作西肖松尼族人;居住在怀俄明西部的温德河谷的一支,也称作东肖松尼族人。

Grover Cleveland 格罗弗·克利夫兰　美国南北战争后第一位当选总统的民主党人,也是美国历史上唯一一位于任期满离开白宫,又在四年后第二次当选总统并重新入主白宫的人。克利夫兰于 1884—1888 年第一次出任美国总统,他得到了民主党和共和党改革派的支持。他积极支持通过政策,限制给予任何一个利益集团特殊优惠的政策。克利夫兰于 1892 年第二次当选为美国总统。

Theodore Roosevelt 西奥多·罗斯福　人称老罗斯福,昵称泰迪。生于 1858 年 10 月 27 日,卒于 1919 年 1 月 6 日,是军事家、政治家。曾任美国第 26 任总统(1901—1909 年)。罗斯福曾任海军副部长,1900 年当选副总统。1901 年总统威廉·麦金莱(William McKinley)被无政府主义者刺杀身亡后,他继任成为美国总统,时年 42 岁。他曾成功调停日俄战争,并获得 1906 年度诺贝尔和平奖。他独特的个性和改革主义政策,使他成为美国历史上最伟大的总统之一。

the U. S. Bureau of Reclamation 美国垦务局　创建于 1902 年,下属于美国内政部。在美国西部地区 17 个州的大坝、水电站和渠道的建设中久负盛名。这些水利工程促进了人们在美国西部建立家园并推动了当地经济的大发展。今天,垦务局已经成为美国最大的水批发商,还是美国西部第二大的水力发电生产商。现在,垦务局已经成为一个水管理机构。

Calvin Coolidge 约翰·柯立芝　全名小约翰·卡尔文·柯立芝(John Calvin Coolidge,

Jr.),生于1872年7月4日,卒于1933年1月5日。美国第30任总统,共和党籍。佛蒙特州律师出身,在马萨诸塞州政界奋斗多年后成为州长。1920年大选时作为沃伦·哈定的竞选伙伴成功当选第29任美国副总统。1923年哈定在任内病逝,柯立芝随即递补为总统。1924年大选连任成功。政治上主张小政府,以古典自由派保守主义闻名。

Translation of Selected Sentences

1. These three protected areas in conjunction with surrounding National Forests constitute the Greater Yellowstone Ecosystem, which at almost 18,000,000 acres, is one of the largest intact mid-latitude temperate ecosystems in the world.

这三个自然保护区与临近的国家森林公园统称为大黄石生态系统,总面积达18,000,000英亩,是世界上保存最完整的中纬度温带生态系统。

2. Organized U. S. Government explorations to the region commenced in the 1870s as an offshoot of exploration in Yellowstone.

19世纪70年代,作为开发黄石国家公园项目的子项目,美国政府开始有组织地开发这一地区。

3. Against public opinion and with repeated congressional efforts to repeal it, much of Jackson Hole was also set aside for protection as Jackson Hole National Monument in 1943.

虽然与公众意见相悖,而且也多次遭到国会反对,1943年杰克逊潭的大部分地区作为杰克逊潭国家纪念园被保护起来(而免于被人为开发的厄运)。

4. The dam was further enlarged in 1916, raising lake waters 39 feet (12 m) as part of the Minidoka Project, designed to provide irrigation for agriculture in the state of Idaho.

1916年作为迷你多卡项目的一部分,大坝进一步扩建,将湖里的水位提升了39英尺(12米),此项目旨在为爱达荷州提供农业灌溉。

5. Though efforts to protect the Teton Range and Jackson Hole as part of an expanded Yellowstone National Park dated back to the late 19th century, proposals to construct more dams on some of the other lakes in Jackson Hole led Yellowstone National Park superintendent Horace Albright to block such efforts.

虽然早在19世纪保护特顿山脉以及杰克逊潭就已经是建设黄石国家公园项目的一部分,但是仍有人提议要在杰克逊潭建设更多大坝。这些提议遭到了黄石国家公园项目负责人贺瑞斯·奥尔布赖特的反对,并搁置下来。

6. Horace Albright had hoped to protect the valley of Jackson Hole north of the town of Jackson from commercial exploitation. Rockefeller agreed and through a private enterprise known as the Snake River Land Company was by 1927 buying land in Jackson Hole to be later turned

over to the National Park Service.

贺瑞斯·奥尔布赖特希望杰克逊镇北部的杰克逊潭峡谷能够得到保护,不要被用于商业开发。洛克菲勒同意奥尔布赖特的看法,并于1927年通过一家私人公司——斯内克河流土地公司购得了杰克逊潭的土地,这一片土地后来成为国家公园的一部分。

7. Secretary Ickes recommended to President Franklin Roosevelt that the Antiquities Act (which permitted Presidents to set aside land for protection without the approval of Congress) be used to establish a National Monument in Jackson Hole.

内务部长伊科斯建议富兰克林·罗斯福总统使用古物保护法,在杰克逊潭建立国家纪念公园。古物保护法允许总统可不经国会同意行使权力,将某些土地列为受保护对象。

8. In recognition for John D. Rockefeller, Jr.'s efforts to establish and then expand Grand Teton National Park, a 24,000 acres parcel of land between Grand Teton and Yellowstone National Parks was added to the National Park service in 1972.

为了纪念约翰·D·洛克菲勒为建立大提顿国家公园做出的不懈努力,政府于1972年将大提顿和黄石国家公园之间24 000英亩的土地也划归国家公园。

9. Grand Teton National Park is administered by the National Park Service, an agency of the Department of the Interior.

大提顿国家公园由内政部的一家代理机构——国家公园服务中心负责管理。

10. The range tilts westward, rising abruptly above Jackson Hole valley which lies to the east but more gradually into Teton Valley to the west.

山脉向西突起,下面就是杰克逊潭峡谷。潭水沿峡谷自东向西流淌,逐渐融入特顿峡谷。

11. West to east trending canyons provide easier access by foot into the heart of the range as no vehicular roads traverse the range except at Teton Pass, which is south of the park.

峡谷是东西向的,步行即可到达山脉腹地。在公园南部有一条山里唯一的车道——特顿路,游客可以驾车沿途游览。

12. Though Jackson Lake is natural, the Jackson Lake Dam was constructed at its outlet before the creation of the park (since modified) and the lake level was raised almost 40 feet (12 m) consequently.

虽然杰克逊湖是自然湖,但在国家公园建成之前,在湖水出口处曾修建了杰克逊湖大坝,使湖水的水位上升了大约40英尺(12米)。

13. The diversity of wildlife communities in Grand Teton National Park and the John D. Rockefeller, Jr. Memorial Parkway complements the spectacular scenery.

大提顿国家公园和约翰·D·洛克菲勒纪念大道两边多种多样的野生动植物,为公园的景色增添了亮丽的色彩。

14. Aquatic habitats and adjacent forests, marshes and meadows fulfill the needs of many forms of wildlife.

水生生境以及周围的森林、沼泽和草地,所有这些为野生动植物提供了生长栖息的良好环境。

Questions

1. When was Grand Teton national Park established?

2. Who helped to add acreage to theexisting national park in 1930s? How much do you know about Rockefeller Group?

3. What is Grand Teton National Park named for?

4. What was the purpose for constructing a dam at the Snake River outlet of Jackson Lake?

5. How did John D. Rockefeller, Jr. manage to add his land holdings in Jackson Hole into Grand Teton National Park?

6. When were the monument and the National Park combined?

7. How is the Grand Teton National Park managed?

8. Please give examples to illustrate how abundant the diversity of wildlife in Grand Teton National Park.

Chapter 25

Congaree National Park
坎格瑞国家公园

【简介】

美国坎格瑞国家公园位于美国南卡罗来纳州，面积为89平方千米。公园为一"冲蚀平原"，保留着美国最大面积生长于滩地的古老阔叶树林。这片森林的一部分是美国东部树林高度最高的，而且也是世界上最高的天然森林之一。1969年塞拉俱乐部发动了名为"草原根基"(grass roots)的活动，呼吁保护这片广袤而古老的阔叶树林。经过各环保人士的多方努力，其保护面积不断增大，最终于2003年11月10日被升级为国家公园。公园内设有木板环路，一方面是为了保护该地区沼泽环境下生长的植物，另一方面是为方便游客近距离地去欣赏公园内的美景。每个周日，游客还可以搭乘免费的游船在沼泽湿地中游览。夏天是游玩坎格瑞沼泽国家公园的大好时机，能从事的活动比较多，虽然蚊虫也多。另外，和其他赫赫有名的国家公园相比，这里的游客很少，但是这里的自然景观却很有特色。

Just 20 miles southeast of South Carolina's capital city lies Congaree National Park—the youngest national park ever established in this country. The Congaree River flows through the park where minor changes in elevation (海拔) cause regular flooding and make the land unsuitable for farming or development. The lush trees growing in this floodplain forest are some of the tallest in the Eastern U. S., forming one of the highest temperate deciduous (每年落叶的) forest canopies remaining in the world.

Congaree National Park protects the largest intact tract of old-growth bottomland hardwood forest in the United States. It features many state and national champion-sized trees. And, it provides suitable habitat for future champions. Because of minimal human contact, the monument is an ideal location for benchmark research. It is an ecosystem with high biodiversity.

In history, the area was claimed by Congaree Indians who unfortunately were wiped out by a smallpox epidemic (传染性疾病) introduced with the arrival of European settlers around 1700. Attempts were made through 1860 to make the land suitable for planting and grazing, not an easy task considering the swamp-like conditions.

By 1905, the Santee River Cypress Lumber Company had acquired much of the land. In the early 20th century, logging operations were attempted in an effort to make the land commercially profitable (有利可图的). But the land was too swampy to sustain the heavy equipment and only trees close to the waterways could be cut so that logs could be floated down river. It was soon discovered that in the constantly damp conditions, many of the cut trees remained too green to float. After a few years, operations were abandoned as unproductive, leaving the floodplain basically untouched.

Advances in timber technology and high lumber prices enticed loggers back to the area in the 1960s, but a grassroots effort to protect the ancient and biologically diverse ecosystem prevailed. The Congaree forest was designated a National Landmark in 1974.

The land was authorized(授权) as a National Monument on October 18, 1976, "in order to preserve and protect for the education, inspiration and enjoyment of present and future generations an outstanding example of a near-virgin southern hardwood forest." The monument was established as the result of a grass roots movement by South Carolinians united in the effort to protect the "greatest unprotected forest in the continental United States". Many partners were involved with its preservation (保留). Some of the key players were the Sierra Club, Audubon Society, South Carolina Environmental Coalition, the Congaree Swamp National Preserve Association, and local garden clubs. To ensure the continuation of the monument's preservation, another wave of partners has materialized. This new wave includes The River Alliance, Friends of Congaree Swamp, the National Guard, the South Carolina Department of Education, and the Richland County Delegation. Also among the new wave are those who hold the future of the monument in their hands, the children of South Carolina.

In 1983, this land wasdesignated a Biosphere Reserve, serving as a genetic library, protecting all of its species. It also acted as a barometer (晴雨表) of environmental change, indicating the relative health of our planet.

In 1988, it was designated as a wilderness for 98.4% of the monument's lands were in wilderness. It had attracted certain number of native visitors during those years for its natural beauty. But to say the word "swamp", and the first image that probably came to mind was of a wet, sticky, mosquito-infested mire that few people would want to visit. Such an image certainly might have kept some visitors away from this 22,000-acre forest. That's the reason why it was finally designated a National Park on November 10, 2003, being the youngest national park in the United States.

Technically speaking, Congaree is not a swamp, because it does not contain standing water throughout most of the year. One of the newest national parks is actually a floodplain forest that floods about ten times a year. Spreading northeast from the meandering Congaree River, the land is the largest contiguous (相邻的) tract of old-growth bottomland hardwoods in the United States. Bald cypress is a common tree in the park. The Congaree River, Wateree River, and several tributaries (支流) periodically flood the park, bringing rich silt that encourages tree growth. And do the trees of Congaree ever grow! Approximately 90 species attain heights and girths (周长) found nowhere else in the state. You'll find bald cypress with circumferences of 27.5 feet and extensive root systems that rise above the water in "knees" that can reach as high as 7.5 feet. Loblolly pines, rarely associated with hardwood floodplains, have stood here for about 300 years, soaring to record-breaking altitudes of 169 feet.

Sweet gums are the most ubiquitous trees, along with swamp chestnut oaks, laurel oaks, green ash, hickories, elms, sugarberries and sycamores. Poison ivy and wild grapevines climb the tree trunks to disappear into the high forest canopy. Most of these floodplain trees have shallow roots, and their great cathedral heights often cause them to topple, sometimes leaving a hole in the canopy up to a half acre wide. In these deadfall openings, brambles quickly grow, and forest succession begins again. Standing dead trees in the floodplain offer homes to seven species of Southeastern woodpeckers, three species of owls, and nine species of bats, and countless invertebrates.

Through this floodplain forest flows slow-moving Cedar Creek, a blackwater creek that drops only 20 feet in 16 linear miles. A marked 20-mile canoe trail on the Outstanding National Resource Waters of Cedar Creek provides a unique way to experience and enjoy Congaree National Park. The Congaree River forms the southern boundary of the Wilderness. Over 25 miles of this lazy, looping river provide another canoe route, now part of the Congaree River Blue Trail, a 50-mile designated recreational paddling trail extending from the state capital of Columbia downstream to Congaree National Park. More than 20 miles of hiking trails provide access to the northwestern portion of the floodplain forests within Congaree National Park, including a 2.4 mile boardwalk. All trails originate near the park Visitor Center. Due to the dynamic floodplain landscape, it is a good idea to check current trail conditions at the Visitor Center before setting out.

Large animals possibly seen in the park includebobcats, deer, feral pigs, feral dogs, coyotes, and turkeys. Its waters contain interesting creatures like amphibians, turtles, snakes, alligators, and many types of fish, including bowfin, largemouth bass, panfish, and catfish. Push back the ghostly Spanish moss that drips from the bald cypresses, and you enter a lush backcountry inhabited by bobcats, deer, and playful river otters. Yellow bellied sapsuckers drill holes into trees one day and return the next to feast on the sap that has filled the holes. The rapid-fire series of knocks you hear is from one of the many woodpeckers found in the park, also hard at work boring holes into trees.

At night in the fall and spring, rangers lead visitors on an "owl prowl", so they can hear the eerie calls of barred owls and see the glowing fungi that grow on the cypresses. According to local legends, the cypress tree's trademark "knees"—small, knobby wood growths that rise around the trunk's base—are really wood elves that come to life at night to dance through the forest.

Congaree National Park features primitive campsites and offershiking, canoeing, kayaking, and bird watching. Primitive and backcountry camping is available. Some of the hiking trails include the Bluff Trail (0.7 mi), Weston Lake Loop Trail (4.6 mi), Oakridge Trail (7.5 mi), and King Snake Trail (11.1 mi) where hikers may spot deer, raccoon, opossum, and even bobcat tracks. The National Park Service rangers have current trail conditions which can be found in the Harry Hampton Visitor's Center. Along with hiking trails, the park also has a 20-mile (32 km) marked canoe trail on Cedar Creek.

Most visitors to the park walk along the Boardwalk Loop, an elevated 2.4-mile (3.9 km) walkway through the swampy environment that protects delicate fungi and plant life at ground level. Canoe rides through the swamp are conducted for free every Saturday and Sunday. Congaree boasts both the tallest (169ft, 51.4m) and largest (42 cubic meters) Loblolly Pines (Pinus taeda) alive today.

In 2008, South Carolina Educational Television (SCETV) began shooting a documentary (纪录片) on the history of the Congaree National Park titled Roots in the River: The Story of Congaree National Park. The story features interviews with people involved in the movement that eventually led to the area's U.S. National Monument status, and observes the role the park plays in the surrounding community of the Lower Richland County area of South Carolina. The program was scheduled to air on the SCETV network in September 2009.

The park is open year-round but spring and fall remain the most pleasant seasons to visit. Not only is the landscape lush and vibrant (充满朝气的), but during these seasons, ranger-led walks take visitors on hikes to hear the calls of barred owls. Though mosquitoes are a kind of torture, the natural beauty of this national park is people's first choice to enjoy themselves in a hot summer.

Notes

South Carolina 南卡罗来纳州　位于美国东南部,北接北卡罗来纳州,东南临大西洋,海岸线总长4 200千米,十分曲折,西南接佐治亚州。全州轮廓呈三角形。面积80 432平方千米,在50州内列第40位。首府为哥伦比亚城。南卡罗来纳州不仅是风景卓越的“扇棕之州”、美国高尔夫球俱乐部的发源地,也拥有雄厚工业、科技实力和高效劳动力资源。

Smallpox 天花　由天花病毒引起的一种烈性传染病,也是到目前为止,在世界范围被人类消灭的第一种传染病。天花是感染痘病毒引起的,无药可治,患者在痊愈后脸上会留有麻子,“天花”由此得名。天花随西班牙人越过大西洋进入美洲大陆,及至16世纪末,美洲生存下来的人口估计刚刚超过100万。16—18世纪,每年死于天花的人数,欧洲约为50万人,亚洲约为80万人,而整个18世纪欧洲人死于天花的总数,则约在1.5亿人以上。18世纪,天花到达澳大利亚,杀死了50%的澳大利亚原住民。1798年詹纳发明了给人种牛痘预防天花,经过近200年坚持不懈的疫苗接种,20世纪70年代末,人类在地球上彻底控制了天花。1979年10月26日世界卫生组织宣布了这一结果,并在全世界停止了普遍种痘。停止种痘20多年来,再没有发现天花病例。

National Guard 美国国民警卫队　常简称为国民兵,是美国军队的一支,属美国预备役军事部队,由各州国民警卫队单位组成,不论是否现役,均受联邦政府认可。士兵皆为民兵,平常从事各行各业,旗下一分为二,有陆军国民警卫队和空军国民警卫队,分别是陆军和空军的预备队。国民警卫队散驻各州和其他国土,听命于州长或民兵指挥官。遇有紧急状态及灾害,如风灾、洪水或地震等,州长或民兵指挥官可下令召集国民警卫队。

bald cyprees 落羽松　杉科,水生植物,原产北美州湿地,落叶大乔木,花期在4月,叶形似羽毛,故得名。叶尖锐,冬季将落的叶片会转成橙褐色。

sweet gum 枫香树　虎耳草目枫香科下的一属,主要生长于气候温暖地带。大型落叶乔木,树干挺直,外形类似枫树,高度可达25~40米。叶掌状开裂,螺旋着生,叶子在秋天变红。全树含有芳香的挥发油,因而得名。

chestnut oaks 栗栎　山毛榉科栎属植物,叶形似栗,长于北美,树皮发黑色,叶披针形。

laurel oak 山月桂　又称美洲月桂。杜鹃花科山月桂属的常绿灌木,产于北美洲东部的大多数山区,1~6米高,叶呈卵形。玫瑰色、粉红色或白色的花朵在枝顶大批成簇开放。作为观景植物而被广泛种植。

green ash 绿白蜡树　落叶乔木,树冠卵圆形,树皮黄褐色,小枝光滑无毛,夏季开花,花萼钟状,无花瓣。白蜡树是白蜡虫的最适寄主,因而得名。白蜡树木材坚韧,可用于制作家俱、农具、车辆、胶合板等;枝条可编筐;树皮称“秦皮”,中医学上用于清热药。白蜡树白蜡是一种名贵天然药物原料。

hickory 美国山核桃　又名薄壳山核桃、长山核桃，为胡桃科山核桃属落叶乔木，产于美国南部和墨西哥北部，树高可达 20 ~ 70 米，其海拔要求在 800 ~ 1 600 米；树形高大，树姿优美，树冠开展，根系发达，少病虫害，生长速度快，是良好的行道树和庭园绿化树种。木材坚韧，纹理致密，不易伸缩和翘裂，可作雕刻、家具等用材。树体寿命长，可达百年以上。

bobcat 美洲野猫　指那些土生土长在北美洲的野猫，喜欢生活在有茂密覆盖物或者破碎的地形区，从针叶林到灌木丛、沼泽沙漠到湿地均可见。分布广泛，几乎遍布整个北美地区。由于人类的大肆捕杀，曾经高度繁殖的中西部、东部海岸的野猫已几近灭绝。

opossum 负鼠　有袋目负鼠科的通称，一种比较原始的有袋类动物，主要产自拉丁美洲。负鼠为中、小型兽类。大多数具有能缠绕的长尾，因此母负鼠能随身携带幼鼠到处奔跑。尾毛稀疏并覆以鳞片。少数种类尾短而具厚毛。四肢短。负鼠性情温顺，常常夜间外出，捕食昆虫、蜗牛等小型无脊椎动物，也吃一些植物性食物。

National Park Service 美国国家公园管理局　根据美国国会的相关法案，成立于 1916 年 8 月 25 日，隶属于美国内政部，主要负责美国境内的国家公园、国家历史遗迹、历史公园等自然及历史保护遗产的保护。1995 年开始，在华盛顿总部之下，设立 7 个地区局，并以州界划分管理范围。地区局下设 16 个支持系统。一般是将生态环境和资源类似的公园组成一个公园组，以便按其资源类型和特色开展相应的管理工作。

Translations of Selected Sentences

1. Congaree National Park protects the largest intact tract of old-growth bottomland hardwood forest in the United States. It features many state and national champion-sized trees. And, it provides suitable habitat for future champions.

坎格瑞国家公园保护着美国境内面积最大、保存最完整的滩地硬木原始森林。与本州或国内其他树木相比，园内很多树的高度处于冠军级别。此外，它也是未来树木之王生长的理想家园。

2. Advances in timber technology and high lumber prices enticed loggers back to the area in the 1960s, but a grassroots effort to protect the ancient and biologically diverse ecosystem prevailed.

20 世纪 60 年代，伐木业的技术发展和伐木价格之高，使得很多伐木工人重新回到了这个地区，但是，当时也有民间环保组织积极致力于对这个历史悠久、生态多样的生态系统的保护。

3. But to say the word "swamp", and the first image that probably came to mind was of a wet, sticky, mosquito-infested mire that few people would want to visit. Such an image certainly might have kept some visitors away from this 22,000-acre forest. That's the reason why it was

finally designated a National Park on November 10, 2003, being the youngest national park in the United States.

但是提到"沼泽"这两个字,或许给人的第一印象很糟,人们会把它想象成潮湿、黏乎乎、蚊虫肆虐的泥沼之地,因此鲜有人愿来此游玩。这样的形象当然使这个22公顷之大的森林拒人于千里之外。这就是为什么它最终在2003年改名为国家公园,从而成为美国最年轻的国家公园。

4. Congaree National Park features primitive campsites and offershiking, canoeing, kayaking, and bird watching. Primitive and backcountry camping is available.

坎格瑞国家公园有原始野营地,提供远足、划独木船、划皮划艇及观鸟等活动。公园内可以原始野营。

5. The story features interviews with people involved in the movement that eventually led to the area's U. S. National Monument status, and observes the role the park plays in the surrounding community of the Lower Richland County area of South Carolina.

这个纪录片含有对多人的采访记录,他们参与了这项活动,最终使得这个森林成为美国国家级公园,同时,纪录片也探讨了这个公园在南卡罗来纳州罗·里奇兰地区周边社区所扮演的角色。

Questions

1. Why was the Congaree swamp abandoned by the loggers in the early 20th century?

2. What was the purpose stated then the land was authorized as a National Monument on October 18, 1976?

3. Who were some of the key players in the preservation movement of this park?

4. Why is it said in the article that"technically speaking, Congaree is not a swamp"?

5. What is the image created when the area was named as Congaree Swamp National Park? What are the results of this image?

PART Ⅱ

加拿大篇

Chapter 1

Aulavik National Park
奥拉弗克国家公园

【简介】

奥拉佛克国家公园位于班克斯岛。由于地处加拿大地区北极最西端，壮观又富有野生动物的低地苔原是这个位于西北地极点的国家公园的特征。该公园保护着超过280个考古遗址和号称世界最北端的可航行水道。

Aulavik National Park (Inuvialuktun: place where people travel) is a national park located onBanks Island in the Northwest Territories of Canada. It is known for its access to the Thomsen River, one of the most northerly navigable rivers in North America. The park is a fly-in park, and protects approximately 12,274 km of Arctic Lowlands at the northern end of the island. The most practical way to visit the park is to charter a plane and currently the park has four landing sites. Aulavik is considered a polar desert and often experiences high winds. Precipitation (降水量) for the park is approximately 300 mm per year. In the southern regions of the park a sparsely (稀疏地) vegetated upland plateau reaches a height of 450 m above sea level.

The park has the highest concentration of muskoxen (麝牛) on earth, with estimates of 68,000 to 80,000 animals on the island, approximately 20% of which are thought to reside in the park. It is also home to the endangered Peary caribou (驯鹿) as well as the more common barren-ground caribou. Ptarmigan (雷鸟) and ravens (乌鸦) are considered the only year-round birds in the park, although 43 different species make seasonal use of the area. The park is completely treeless, and arctic foxes, brown and northern collared lemmings (旅鼠), arctic hares and wolves roam the rugged terrain. Birds of prey in the park include snowy owls (雪鸮),

rough-legged hawks, gyrfalcons, and Peregrine falcons, who feed on the lemmings.

Prior to the arrival of the Thule culture in southern Banks Island there were some Pre-Dorset culture people living in what is now the Aulavik National Park. Due to the cooling climate brought on by the Little Ice Age Banks Island was most likely deserted until the arrival of the Inuvialuit in the 17th century.

Aulavik National Park has two major bays, Castel Bay and Mercy Bay, and lies south of the McClure Strait. Captain Robert McClure, spent two winters in Mercy Bay on HMS Investigator, while searching for the missing John Franklin's lost expedition between 1850 and 1853. McClure's team abandoned their ship in Mercy Bay and hiked across the sea-ice of the strait to board another ship, HMS Resolute.

The Mercy Bay area was visited by the Copper Inuit of Victoria Island to salvage (抢救) materials left by McClure's party. They also hunted the caribou and muskox in the area as evidenced by the large number of food caches. In the 20th century the area was popular with Inuvialuit due to the large numbers of foxes. Until the fur trade went into decline, fox trapping provided a source of income for people from as far away as the Mackenzie Delta and the North Slope of Alaska. This influx (流入) of people led to the establishment of Sachs Harbor, the only community on the island.

Though people have lived in this part of the Arctic for almost 4,000 years, Banks Island has always been a land on the edge. Human use of the area that is now Aulavik National Park has been sporadic(零星的) and greatly influenced by climate change. During warmer periods, often lasting hundreds of years, parts of the island were occupied. During colder cycles, the inhabitants moved to areas that were more hospitable, occasionally returning to hunt or fish on the island.

The earliest archaeological sites found within Aulavik National Park are near Shoran Lake. These Pre-Dorset cultural sites date back to almost 1500 B. C and are characterized by oval dwellings (椭圆形的居室), usually outlined by a ring of stones that held down the edges of a tent. Site excavations have revealed flint scrapers, bone harpoon (鱼叉) heads and needles. The bones of hundreds of muskoxen indicate that, even then, this species was a vital source of food and provisions.

During the period of 1000 – 1450 A. D. Thule peoples occupied several sites along the south coast of Banks Island and made some use of what is now Aulavik National Park. At the coastal sites, the Thule peoples practiced an economy based on harvesting sea mammals, particularly bowhead whales (北极露脊鲸) and ringed seals. It also appears that they ventured into the northern reaches of the island to harvest muskoxen.

A major climatic cooling, known as the Little Ice Age, probably ended most occupation on

Banks Island for several hundred years. The deteriorating(恶化的) climatic conditions forced the Thule people to seek refuge in a smaller region where they developed very effective and specialized hunting skills. As the climate warmed, between 1650 and the 1820's, they occupied sites along the southern coast and hunted muskoxen in the area that is now the park.

Exploration of the area by Europeans began in the early 1800's. In 1820, Beechy, a member of the first Parry expedition, saw land to the southwest of Melville Island and christened (命名为) it Banks Island to honor a past president of the Royal Society.

Europeans did not actually visit Banks Island until thirty years later. In 1850, the H. M. S. Investigator, under the command of Captain Robert MClure, came in search of the Northwest Passage and the lost Franklin Expedition. In September1851, his ship became trapped in the ice at Mercy Bay at the northern end of Banks Island. In 1853, after almost two years locked in the pack ice, the ship was abandoned and the crew walked over the ice to Melville Island where they were rescued by the HMS Resolute. Later, the Investigator either sunk or drifted away. Today, all that is left of the site on the shores of Mercy Bay are wooden barrel staves (木桶板), a coal pile, and a large piece of metal.

From the late 1920's until the decline of the fur tradein the1970's, lucrative arctic fox trapping drew Inuvialuit to Banks Island from the Mackenzie Delta, Victoria Island, the Tuktoyaktuk Peninsula, and Alaska's North Slope. Families converged at the southwestern tip of Banks Island and set up the permanent community of Sachs Harbour.

Today, Sachs Harbour is home to approximately 120 people. In keeping with their cultural traditions, the Inuvialuit residents have been guaranteed the right of continued subsistence trapping, hunting and fishing within the national park boundaries. While residents occasionally use the north end of the island, Aulavik is seen as a reservoir for sustaining regional wildlife populations.

Aulavik National Park is a pristine(质朴的) arctic wilderness area. The park protects an area of the Western Arctic Lowlands, a representative natural region of national significance.

The climate of Aulavik National Park is typically arctic. The tundra(苔原) is frozen and snow-covered from September until June. Summers are brief and cool. The sun does not set between mid May and late July, and there is never true darkness from late April until late August.

During the summer (mid-June to mid August), temperatures in the centre of the park range from 1°C to 20°C with a daily average of 10°C. Closer to the coast, the temperatures tend to be cooler and range from −2°C to 12°C with a daily average of 8°C. Snow and freezing temperatures can occur at any time of the year.

Aulavik is considered a polar desert. The total annual precipitation for the park is approximately 300 mm, one third of which falls as rain during the summer. On average, there are only 20 mm of precipitation in each of May, June and July and 30 mm of precipitation in August. The wind is an almost constant presence and severe winds are common.

Banks Island's most abundant large mammal is, by far, the muskox. This species has experienced a dramatic increase in numbers over the last fifty years. Recent estimates (2001) place the islands muskox population at more than 68,000 and it is thought that 20% of these reside in the park. Considerable effort by Government of the Northwest Territories, Inuvialuit, and Parks Canada biologists is being directed towards understanding the growth of the muskox population and the impacts it may be having on the ecosystem.

Aulavik is also home to Peary caribou. The Banks Island population of Peary caribou was classified as endangered in 1991 due to a dramatic decline in numbers. Recent findings suggest that while the population is low, it is considered stable and potentially (潜在地) recovering. Current research, led by the Government of the Northwest Territories, is looking at population levels, seasonal habitat use, range conditions, and predation (掠夺行为), in order to better understand the population decline.

A variety of other mammals are found in Aulavik including arctic foxes, arctic wolves, ermines (貂), arctic hares, and both brown and collared lemmings. Marine mammals (海洋哺乳动物) along the north coast include polar bears, ringed seals (环斑海豹), bearded seals, beluga whales (白鲸) and bowhead whales.

Over forty-three species of birds make seasonal use of Aulavik, though only ptarmigan and ravens are considered permanent residents. Thirty-two bird species have been recorded as breeding within the park. Breeding bird densities are generally low, with the most abundant species being: lapland longspurs, horned larks, black-bellied plovers, long-tailed jaegers and sandhill cranes.

Birds of prey(猛禽) in Aulavik include snowy owls, rough-legged hawks, gyrfalcons, and peregrine falcons. The abundance of these species depends greatly on the abundance of lemmings. When the multi-year lemming cycle is low, so too are the numbers of those bird and mammal species which rely upon the lemming as a primary food source.

There are two Migratory Bird Sanctuaries (迁徙鸟类保护区) on Banks Island, one in the northern part of Aulavik National Park and the other covering the southwest corner of Banks Island. These two sanctuaries protect the seasonal nesting and moulting (脱毛) grounds of the Western Arctics 500,000 lesser snow geese. On most flights into the park you will be able to see thousands of these majestic birds across the island.

It is possible that the Thomsen River watershed is the most northerly example of a multi-species freshwater fish community. Farther north, there are only arctic char and nine spine stickleback（棘鱼）, or single species communities of arctic char（北极嘉鱼）. The Thomsen watershed is home to six freshwater species: lake trout, arctic char, least cisco, lake cisco, nine spine stickleback and fourhorn sculpin. Only lake trout is an exclusively freshwater species; the other five can spend a portion of their lives in salt water.

Overall, very little is known about the ecology of char on Banks Island. The Thomsen River is noted as important spawning（产卵）and nursery habitat for arctic char, and landlocked arctic char are found in several of the larger lakes in the park such as Nangmagvik.

Aulavik National Park contains a wide diversity of landscapes. The Thomsen River is characterized by a broad, lush（青葱的）valley flanked by gently undulating（起伏的）hills. Along the northwest coast of Aulavik, sheer（峻峭的）cliffs rise several hundred meters from the waters of MClure Strait. The central portion of the northern coastline is dominated by Castel Bay and Mercy Bay, which cut inland as much as 15 kilometres. To the southeast of Mercy Bay, steep canyons dominate the landscape. Further south, a sparsely vegetated（植被覆盖的）upland plateau rises to a height of 450 meters above sea level and forms a height of land between the Thomsen River and Prince of Wales Strait. West of Castel Bay, arid（干旱的）desert-like badlands（荒地）predominate. The main tributary of the Thomsen River is the Muskox River. The Muskox River flows into the Thomsen from the west and is characterized by poorly drained wetlands and frost polygon features.

Notes

Banks Island 班克斯岛　加拿大西北地区北极群岛中最西端的岛屿。长 400 千米，宽 177 - 290 千米，面积 70028 平方千米。地形以丘陵山地为主。栖息有北极狐、狼、北美驯鹿、北极熊和许多鸟类。西南岸的萨克斯港（Sachs）与大陆上的伊努维克（Inuvik）通飞机，为捕猎和探矿基地。班克斯岛位于北纬 73 度，全岛只有 87 个居民。他们祖祖辈辈生活在这里已有上千年的历史。他们均以捕猎为生，相貌长得跟中国人相似，应该是蒙古人的后裔。

Thomsen River 汤姆森河　位于加拿大西北部，是加拿大适合航运的最北端的河流。以划独木舟著名于世。汤姆森河大约有一半的长度穿越奥拉佛克国家公园。

Arctic Lowlands 北极低地　北极低地和哈德逊海岸低地都是地形的一种。北极低地位于加拿大地盾区。那里气候严寒无法耕种，全年大部分时间，地面被冰雪覆盖。北极低地富含褐煤、石油和天然气等自然资源。只有少数人定居那里，也是北极熊等生物的栖息地。

Thule culture 图勒文化　图勒人是所有现代因纽特人的祖先,大约于公元 1000 年生活在阿拉斯加海岸。后来他们向东部发展,到 13 世纪时曾到达加拿大的格陵兰地区。图勒之名源于他们在格陵兰西北部、面向加拿大曾经居住的地方的名称。图勒人与因纽特人有着血亲关系,他们在文化语言等方面都有联系。

Little Ice Age 小冰期　大约从 15 世纪初开始,全球气候进入一个寒冷时期,通称为"小冰期",此阶段在中国也称为"明清小冰期"。小冰期具有全球普遍降温的特点,通常认为小冰期始于 13 世纪,止于 19 世纪,然后出现一个相对温暖的气候环境时段,在 16 世纪中期和 19 世纪中期之间达到其最冷的顶点。

Translation of Selected Sentences

1. Aulavik National Park (Inuvialuktun: place where people travel) is a national park located on Banks Island in the Northwest Territories of Canada.

奥拉佛克国家公园(西加拿大因纽特语,意为:旅游的地方)坐落于加拿大西北部班克斯岛。

2. Ptarmigan and ravens are considered the only year-round birds in the park, although 43 different species make seasonal use of the area.

国家公园中各个季节会有 43 种不同的鸟类栖息,但常年栖息于此的只有雷鸟和乌鸦两种鸟类。

3. Captain Robert McClure, spent two winters in Mercy Bay on HMS Investigator, while searching for the missing John Franklin's lost expedition between 1850 and 1853.

罗伯特·麦克鲁尔船长花费两年时间,在摩尔西海湾搜索于 1850—1853 年间失踪的约翰·富兰克林探险队的踪迹。

4. Until the fur trade went into decline, fox trapping provided a source of income for people from as far away as the Mackenzie Delta and the North Slope of Alaska.

居住在麦肯齐三角洲和阿拉斯加北坡人们的生财之道一直是诱捕狐狸并贩卖其皮毛。直到皮毛生意被禁止后,这种捕猎才停止下来。

5. During colder cycles, the inhabitants moved to areas that were more hospitable, occasionally returning to hunt or fish on the island.

当气候变冷后,人们便迁往温暖的地区居住,只是偶尔回到岛上打猎或钓鱼。

6. The bones of hundreds of muskoxen indicate that, even then, this species was a vital source of food and provisions.

(考古挖掘出的)数以百计的麝牛骨证明,即使是在那个时期,麝牛也是人们的主要食物。

7. In 1820, Beechy, a member of the first Parry expedition, saw land to the southwest of Melville Island and christened it Banks Island to honor a past president of the Royal Society.

1820 年第一批帕里探险队里有一位名叫比奇队员,他看到了梅尔维尔岛西南方向的陆地,为纪念皇家学会已故主席班克斯先生,比奇将这片陆地命名为班克斯岛。

8. In keeping with their cultural traditions, the Inuvialuit residents have been guaranteed the right of continued subsistence trapping, hunting and fishing within the national park boundaries.

政府允许西因纽特人保持他们的文化传统,因此他们仍然可以在国家公园内捕猎和打渔。

9. The climate of Aulavik National Park is typically arctic. The tundra is frozen and snow-covered from September until June.

奥拉佛克国家公园的气候是典型的北极气候。属于苔原地带的公园,土地长期冰冻,每年 9 月一直到转年 6 月都被积雪覆盖。

10. Banks Island's most abundant large mammal is, by far, the muskox. This species has experienced a dramatic increase in numbers over the last fifty years.

至今为止,班克斯岛数量最多的哺乳动物是麝牛。在过去的 50 年中,麝牛的数量增长很快。

12. There are two Migratory Bird Sanctuaries on Banks Island, one in the northern part of Aulavik National Park and the other covering the southwest corner of Banks Island.

在班克斯岛上有两处迁徙鸟类保护区,一处坐落于奥拉佛克国家公园的北部,另一处是在班克斯岛西南角。

13. The Thomsen River is characterized by a broad, lush valley flanked by gently undulating hills.

汤姆森河两岸起伏延绵的山脉青葱翠绿,宽阔的山谷中河流缓缓流淌。

14. Further south, a sparsely vegetated upland plateau rises to a height of 450 meters above sea level and forms a height of land between the Thomsen River and Prince of Wales Strait. West of Castel Bay, arid desert-like badlands predominate.

在汤姆森河和威尔士王子海峡之间是海拔 450 米的、荒芜的高原。卡斯特尔海湾西边主要是干旱的、沙漠状的荒地。

Questions

1. Where isAulavik National Park located?
2. Who first settled on the Banks Island?
3. How did Banks Island get its name?

4. When did Europeans fist visit Banks Island? What happened to their ship?

5. How many habitants are living on Banks Island today? What do they do to make a living there?

6. What are the typical features of the climate for Aulavik National Park?

7. Why is Aulavik considered a polar desert?

8. How do the Sanctuaries help to protect migratory birds in Aulavik?

Chapter 2

Banff National Park
班夫国家公园

【简介】

班夫国家公园建于1885年,是加拿大历史最悠久的国家公园,也是加拿大第一个国家公园、避暑胜地。公园位于阿尔伯塔省西南部,与不列颠哥伦比亚省交界的洛矶山东麓。1885年建立,面积6666平方千米。班夫国家公园内有一系列冰峰、冰河、冰原、冰川湖和高山草原、温泉等景观,其奇峰秀水居北美大陆之冠。冰原公路从路易斯湖开始,一直连接到北部的贾斯珀国家公园。西面是省级森林和优鹤国家公园,南面与库特奈国家公园毗邻,卡纳纳斯基斯镇位于其东南方。公园内主要的商业区是位于弓河山谷的班夫镇。

Banff National Park is Canada's oldest national park, established in 1885 in the Rocky Mountains. The park, located 110 – 180 km west of Calgary in the province of Alberta, encompasses 6,641 km^2 of mountainous terrain (山岭地区), with numerous glaciers and ice fields, dense coniferous forest (针叶树林), and alpine landscapes. The Icefields Parkway extends from Lake Louise, connecting to Jasper National Park (贾斯珀国家公园) in the north. Provincial forests and Yoho National Park are neighbors to the west, while Kootenay National Park is located to the south and Kananaskis County (卡那那斯基斯郡) to the southeast. The main commercial centre of the park is the town of Banff, in the Bow River valley.

The Canadian Pacific Railwaywas instrumental in Banff's early years, building the Banff Springs Hotel and Chateau Lake Louise, and attracting tourists through extensive advertising. In the early 20th century, roads were built in Banff, at times by war internees (战俘), and through Great Depression-era public works projects. Since the 1960s, park accommodations have been open all year, with annual tourism visits to Banff increasing to over 5 million in the 1990s. Millions more pass through the park on the Trans-Canada Highway (横贯加拿大公路). As Banff is one of the world's most visited national parks, the health of its ecosystem has been threatened. In the mid-1990s, Parks Canada responded by initiating a two-year study, which resulted in management recommendations, and new policies that aim to preserve ecological integrity.

Throughout its history, Banff National Park has been shaped by tension between conservation and development interests. The park was established in 1885, in response to conflicting claims over who discovered hot springs there, and who had the right to develop the hot springs for commercial interests. Instead, Prime Minister John A. Macdonald set aside the hot springs as a small, protected reserve, which was later expanded to include Lake Louise and other areas extending north to the Columbia Ice-field.

Archaeological evidence found at Vermilion Lakes radiocarbon (放射性碳) dates the first human activity in Banff to 10,300 B. P. Prior to European contact, aboriginals (土著居民), including the Stoneys, Kootenay (库特奈人), Tsuu T'ina, Kainai, Peigans, and Siksika, were common in the region where they hunted bison and other game.

With the admission of British Columbia to Canada on 20 July 1871, Canada agreed to build a transcontinental (横贯大陆的) railroad. Construction of the railroad began in 1875, with Kicking Horse Pass chosen, over the more northerly (北方的) Yellow-head Pass, as the route through the Canadian Rockies. Ten years later, the last spike was driven in Craigellachie, British Columbia.

With conflicting claims over discovery of hot springs in Banff, Prime Minister John A. Macdonald decided to set aside a small reserve of 26 km^2 around the hot springs at Cave and Basin as a public park known as the Banff Hot Springs Reserve in 1885. Under the Rocky Mountains Park Act, enacted on 23 June 1887, the park was expanded to 674 Km^2 and named Rocky Mountains Park. This was Canada's first national park, and the second established in North America, after Yellowstone National Park. The Canadian Pacific Railway built the Banff Springs Hotel and Chateau Lake Louise to attract tourists and increase the number of rail passengers.

Early on, Banff was popular with wealthy European tourists, who arrived in Canada via trans-Atlantic luxury liner and continued westward on the railroad, as well as upper-class American and English tourists. Some visitors participated in mountaineering activities, often hiring local guides. Tom Wilson, along with Jim and Bill Brewster, was among the first outfitters (旅行用品商) in Banff. The Alpine Club of Canada, established in 1906 by Arthur Oliver Wheeler and Elizabeth Parker, organized climbs and camps in the backcountry (旅行用品商). By 1911, Banff was accessible by automobile from Calgary. In 1920, access to Lake Louise by road was available, and the Banff-Windermere Road opened in 1923 to connect Banff with British Columbia.

In 1902, the park was expanded to cover 11 400 km^2, encompassing areas around Lake Louise, and the Bow, Red Deer, Kananaskis and Spray Rivers. Bowing to (向…低头) pressure from grazing and logging interests, the size of the park was reduced in 1911 to 4,663 km^2, eliminating many foothills areas from the park. Park boundaries changed several more times up until 1930, when the size of Banff was fixed at 6 697 km^2, with the passage of the National Parks Act (《国家公园法》). The Act also renamed the park as Banff National Park, named for the Canadian Pacific Railway station, which in turn was named after the Banffshire region in Scotland. With the construction of a new east gate in 1933, Alberta transferred 0.84 km^2 to the park. This, along with other minor changes in the park boundaries in 1949, set the area of the park at 6,641 km^2.

In 1887, local aboriginal tribes signed Treaty, which gave Canada rights to explore the land for resources. At the beginning of the twentieth century, coal was mined near Lake Minnewanka in Banff. For a brief period, a mine operated at Anthracite, but was shut down in 1904. The Bankhead mine, at Cascade Mountain, was operated by the Canadian Pacific Railway from 1903 to 1922. In 1926, the town was dismantled, with many buildings moved to the town of Banff and elsewhere.

During World War I, immigrants from Austria, Hungary, Germany and Ukraine were sent to Banff to work in internment camps (俘虏收容所). The main camp was located at Castle Mountain, and was moved to Cave and Basin during winter. Much early infrastructure (下部构造) and road construction was done by Slavic Canadian internees.

Winter tourism in Banff began in February 1917, with the first Banff Winter Carnival (狂欢节). The carnival featured a large ice palace, which in 1917 was built by internees. Carnival events included cross-country skiing, ski jumping, curling (冰壶), snowshoe (雪鞋跑), and skijoring (滑雪游戏). In the 1930s, the first downhill ski resort, Sunshine Village, was developed by the Brewsters. Mount Norquay ski area was also developed during the 1930s, with the first chair lift installed there in 1948.

Since 1968, when theBanff Springs Hotel was winterized (做过冬准备), Banff has been a year-round destination. In the 1960s, the Trans-Canada Highway was constructed, providing another transportation corridor through the Bow Valley, in addition to the Bow Valley Parkway, making the park more accessible. Also in the 1960s, Calgary International Airport was built.

Canada launched several bids to host the Winter Olympics in Banff, with the first bid for the 1964 Winter Olympics which were eventually awarded to Innsbruck, Austria. Canada narrowly lost a second bid, for the 1968 Winter Olympics, which were awarded to Grenoble, France. Once again, Banff launched a bid to host the 1972 Winter Olympics, with plans to hold the Olympics at Lake Louise. The 1972 bid was most controversial (争议的), as environmental lobby groups provided strong opposition to the bid, which had sponsorship from Imperial Oil. Bowing to pressure, Jean Chrétien, then the Minister of Environment, the government department responsible for Parks Canada, withdrew support for the bid, which was eventually lost to Sapporo, Japan. The cross-country ski events were held at the Canmore Nordic Centre Provincial Park at Canmore, Alberta, located just outside the eastern gates of Banff National Park on the Trans-Canada Highway, when nearby Calgary hosted the 1988 Winter Olympics.

Since the original Rocky Mountains Park Act, subsequent acts and policies placed greater emphasis on conservation. With public sentiment tending towards environmentalism, Parks Canada issued major new policy in 1979, which emphasized conservation. The National Parks Act was amended in 1988, which made preserving ecological integrity the first priority in all park management decisions. The act also required each park to produce a management plan, with greater public participation.

In 1984, Banff was declared aUNESCO World Heritage Site, together with the other national and provincial parks that form the Canadian Rocky Mountain Parks, for the mountain landscapes containing mountain peaks, glaciers, lakes, waterfalls, canyons and limestone caves as well as fossils found here. With this designation came added obligations for conservation.

During the 1980s, Parks Canada moved to privatize(私有化) many park services such as golf courses, and added user fees for use of other facilities and services to help deal with budget cuts. In 1990, the Town of Banff was incorporated, giving local residents more say regarding any proposed developments.

In the 1990s, development plans for the park, including expansion at Sunshine Village, were under fire (受攻击) with lawsuits filed by Canadian Parks and Wilderness Society (CPAWS). In the mid-1990s, the Banff-Bow Valley Study was initiated to find ways to better address environmental concerns, and issues relating to development in the park.

The park has 56 mammal species that have been recorded. Grizzly(灰熊) and black bears inhabit the forested regions. Cougar (美洲狮), lynx (猞猁), wolverine (狼獾), weasel (鼬鼠), northern river otter and wolves are the primary predatory (食肉的) mammals. Elk, mule deer, and white-tailed deer are common in the valleys of the park, including around (and sometimes in) the Banff townsite (城镇所在地), while moose tend to be more elusive, sticking primarily to wetland areas and near streams. In the alpine regions, mountain goats, bighorn sheep (加拿大盘羊), marmots and pika (鼠兔) are widespread. Other mammals such as beavers, porcupines, squirrels, chipmunks, and Columbian ground squirrels are the more commonly observed smaller mammals. In 2005, a total of five caribou (北美产驯鹿) were counted, making this species one of the rarest mammals found in the park.

Due to the harsh winters, the park has few reptiles and amphibians with only one species of toad (蟾蜍), three species of frog, one salamander species and two species of snakes that have been identified. At least 280 species of birds can be found in Banff including bald and golden eagles, red-tailed hawk, osprey (鱼鹰), falcon and merlin (灰背隼), all of which are predatory species. Additionally, commonly seen species such as the gray jay (松鸦), American three-toed woodpecker (啄木鸟), mountain bluebird, Clark's nutcracker (星鸦), mountain chickadee (山雀) and pipit (田云雀) are frequently found in the lower elevations. The white-tailed ptarmigan is a ground bird that is often seen in the alpine zones. Rivers and lakes are frequented by over a hundred different species including loons, herons (苍鹭) and mallards (野鸭) who spend their summers in the park.

Endangered species in Banff include the Banff Springs snail which is found in the hot springs of Banff. Woodland caribou, found in Banff, are listed as a threatened species, as are grizzly bears.

Parks Canada management practices, notably fire suppression, since Banff National Park was established have impacted the park's ecosystem. Since the early 1980s, Parks Canada has adopted a strategy that employed prescribed burns, which helps to mimic effects of natural fires.

Notes

Kootenay 库特奈人　北美洛矶山一带的印第安人,曾聚居于英属哥伦比亚、爱达荷州和蒙大拿州。他们的语言被称为库特奈语。文中提及的 Stoneys, Tsuu T'ina, Kainai, Peigans 和 Siksika 均指曾经生活在班夫国家公园的北美土著人。

the Canadian Pacific Railway 加拿大太平洋铁路　简称 CPR 或加太铁路,在自治领建立之初的一个时代里,加拿大太平洋铁路既是整个加拿大西部开拓和发展的主导性因素,又是对抗美国扩张势力,支持麦克唐纳国家政策的重要力量。它成功地用一条大锁链把东西加拿大紧紧地连在一起,实现了麦克唐纳国家政策的梦想,并置身于西加拿大发展的各项事业之中,为西部的开拓做出了重要贡献。加拿大太平洋铁路是加拿大的一级铁路之一,由加拿大太平洋铁路公司营运。其网络横跨西部温哥华至东部蒙特利尔,并设有跨境路线,通往美国的明尼阿波利斯、芝加哥、纽约市等大型城市。公司总部设于阿尔伯塔省卡尔加里。

the Trans-Canada Highway 加拿大横贯公路　又称横加公路,是一条横贯加拿大的公路。

主线西起不列颠哥伦比亚省的维多利亚，东至纽芬兰省的圣约翰斯，全长 8 030 千米。另一条连接温尼伯和夏洛特皇后群岛马塞的耶洛黑德公路也是本公路的一部分。工程于 1950 年开始，1962 年启用，1971 年完工。由于加拿大国内公路系统的建造和管理都是由各省政府负责（而非联邦政府），横加公路的干线编号亦因省而异。横加公路沿线设有绿底白枫叶的标记以作识别。横加公路有别于美国的州际公路系统，横加公路并未全线达到高速公路的标准，部分路段只是双线行车。

John A. Macdonald 约翰·亚历山大·麦克唐纳爵士 全名 Sir John Alexander Macdonald，生于 1815 年 1 月 11 日，卒于 1891 年 6 月 6 日。麦克唐纳爵士是加拿大第一任总理，加拿大联邦重要领导人。他的政治生涯跨越近半个世纪，任加拿大总理 19 年。

Yellow-head Pass 黄头山口　加拿大洛基山脉的山口，位于加拿大阿尔伯特省和不列颠哥伦比亚省的交界处，在贾斯珀国家公园和罗布森山省立公园内。

National Parks Act《国家公园法》　加拿大联邦政府颁布的旨在管理自然资源的环境保护法律。1930 年加拿大首次颁布《国家公园法》。这部法律要求公园管理局设计并保护自然公园的自然风貌。此外，其他有一定生态意义的荒野也可被圈定为自然保护地。保护自然资源是为加拿大子孙后代提供福祉，具有重大的教育意义。各国家公园的发展要先征得公园管理局的准许，但首要依据则是这部《国家公园法》。

curling 冰壶　又称掷冰壶、冰上溜石，是以队为单位在冰上进行的一种投掷性竞赛项目，冬奥会比赛项目，设男女两个小项。冰壶为圆壶状，由不含云母的苏格兰天然花岗岩制成，且世界上所有的制造优质冰壶用的天然花岗岩均产自苏格兰近海的一个小岛，也只有苏格兰人掌握着制作世界顶尖水平冰壶的技术。冰壶周长约为 91.44 厘米，高（壶的底部与顶部）11.43 厘米，重量（包括壶柄和壶栓）最大为 19.96 公斤。刷冰的原因是减小摩擦。

UNESCO World Heritage Site 联合国教科文组织世界遗产　指被列入联合国教科文组织具有特殊文化或自然意义名单的森林、山川、湖泊、沙漠、历史遗迹、建筑物或城市等，由文化遗产和自然遗产的政府间委员会，即世界遗产委员会负责《保护世界文化和自然遗产公约》的实施。世界遗产委员会成立于 1976 年 11 月，由 21 名成员组成。委员会每年召开一次会议，决定哪些遗产可以录入《世界遗产名录》，对已列入名录的世界遗产的保护工作进行监督指导。

Canadian Parks and Wilderness Society（CPAWS）加拿大天然公园与荒野协会　成立于 1963 年，旨在保护加拿大的野生动植物资源。2007 曾位列加拿大十大组织之首，13 个分会的 1.3 万名会员遍布加拿大各个地区。协会的目标是保护加拿大至少一半的公共土地和水中的野生动植物资源不被人类活动所伤害。

Translation of Selected Sentences

1. Provincial forests and Yoho National Park are neighbors to the west, while Kootenay National Park is located to the south and Kananaskis County to the southeast.

西边是省有森林和优鹤国家公园,库特奈国家公园坐落在它的南面,而班夫国家公园的东南面是卡那那斯基斯郡。

2. As Banff is one of the world's most visited national parks, the health of its ecosystem has been threatened. In the mid-1990s, Parks Canada responded by initiating a two-year study, which resulted in management recommendations, and new policies that aim to preserve ecological integrity.

由于班夫国家公园是世界上游客最多的国家公园,它的生态系统已经遭到了破坏。20世纪90年代中期,加拿大公园管理局进行了为期两年的研究,最终决定加强管理、制定新的政策,以维护班夫国家公园的生态完整性。

3. The park was established in 1885, in response to conflicting claims over who discovered hot springs there, and who had the right to develop the hot springs for commercial interests.

1885年班夫国家公园成立,平息了当时关于谁最早发现了温泉和谁有权利开发温泉牟利的争端。

4. Bowing to pressure from grazing and logging interests, the size of the park was reduced in 1911 to 4,663 km^2, eliminating many foothills areas from the park.

1911年迫于人们要求放牧及采伐以增加收入的压力,公园的面积减少到了4 663平方千米,舍弃了一些山麓的面积。

5. During World War I, immigrants from Austria, Hungary, Germany and Ukraine were sent to Banff to work in internment camps.

第一次世界大战期间,来自奥地利、匈牙利、德国和乌克兰的战俘来到班夫的俘虏收容所做劳工。

6. Bowing to pressure, Jean Chrétien, then the Minister of Environment, the government department responsible for Parks Canada, withdrew support for the bid, which was eventually lost to Sapporo, Japan.

当时的环境部长、加拿大公园管理局的负责人让·克雷蒂安迫于压力,不再支持申请,最终日本札幌市获得了该届奥运会的主办权。

7. With public sentiment tending towards environmentalism, Parks Canada issued major new policy in 1979, which emphasized conservation.

公众舆论支持保护公园的自然环境,所以1979年加拿大公园管理局出台政策,重点保护班夫国家公园的自然环境。

8. During the 1980s, Parks Canada moved to privatize many park services such as golf courses, and added user fees for use of other facilities and services to help deal with budget cuts.

20世纪80年代,加拿大公园管理局允许公园的一些服务设施(如高尔夫球场等)私有化,以维护公园的设施维护和资金预算困境。

9. Elk, mule deer, and white-tailed deer are common in the valleys of the park, including around (and sometimes in) the Banff town-site, while moose tend to be more elusive, sticking primarily to wetland areas and near streams.

在公园的山谷地带,随处可见麋鹿、长耳鹿、白尾鹿在悠闲散步,有时甚至在镇子上也能看到这些动物的身影。驼鹿则仅在沼泽地和溪边徜徉,很少涉足其他地方。

10. In 2005, a total of five caribou were counted, making this species one of the rarest mammals found in the park.

2005 年在班夫国家公园仅发现了 5 头驯鹿,因此驯鹿便成为公园内迄今发现的最稀少哺乳动物种群了。

11. Due to the harsh winters, the park has few reptiles and amphibians with only one species of toad, three species of frog, one salamander species and two species of snakes that have been identified.

由于公园地处气候严寒的北方,园内发现的爬虫类和两栖类动物仅有一种蟾蜍、三种青蛙、一种火蜥蜴和两种蛇。

12. Since the early 1980s, Parks Canada has adopted a strategy that employed prescribed burns, which helps to mimic effects of natural fires.

20 世纪 80 年代开始,加拿大公园管理局采取有规划火烧策略,模仿山火的作用,以达到维护公园生态和谐的目的。

Questions

1. Why did the Canadian government believe that the ecosystem of Banff National Park has been threatened?

2. What did the government do to deal with the problem of the issue of ecosystem of Banff National Park?

3. Who helped to resolve the issue of hot springs in Banff National Park?

4. Who were the early tourists visiting Banff? How did they get to Banff?

5. How did Banff National Park get its name?

6. Why was the 1972 bid to host Olympics the most controversial in Banff National Park? What's the result of the bid?

7. How did Banff National Park deal with its budget cuts in 1990?

8. How does Banff manage fire in the park?

Chapter 3

Bruce Peninsula National Park

布鲁斯半岛国家公园

【简介】

在休伦湖和乔治亚湾之间，安大略省的布鲁斯半岛因其壮观的海岸线和尼亚加拉的悬崖岩层而闻名，它也是加拿大最古老和最长的徒步旅行路径，即著名的布鲁斯步道的北部终点站。布鲁斯半岛国家公园是加拿大 39 座国家公园之一，成立于 1987 年。公园以尼亚加拉断崖的奇特地貌和乔治亚湾清澈的湖水著称，是公园的标志景观之一。此外公园内还有安大略西南地区最大的一片原始森林，森林内物种丰富，动物出没频繁，较为珍贵的包括黑熊、响尾蛇以及 43 种兰花品种。在布满碎石的岸边行进，著名的尼亚加拉断崖地貌在眼前不断变幻，公园里的一切都以自然为本，以自然的发展需要为主。

Bruce Peninsula National Park is located on the northern tip of the Bruce Peninsula, between Georgian Bay and Lake Huron, in eastern Ontario. The park is a patchwork（拼凑物）of limestone cliffs, beach and ancient forest scenery, a place of global significance in the heart of the World Biosphere Reserve.

The shoreline cliffs in the park are part of Ontario's "Great Wall", the Niagara Escarpment, a World Biosphere Reserve that runs from Niagara Falls to Tobermory. The park features a 20 – km segment of the Bruce Trail, which follows the Niagara Escarpment for 782 kms. Plant life in the park includes 43 species of orchids（兰花）and more than 20 varieties of ferns, including the rare northern holly fern（蕨类）. At Tobermory, the cliffs run underwater, surfacing as the Fathom Five Islands, such as Flowerpot and Bears Rump, a separate national marine park. The dramatically varied landscape and rich natural and human history are part of the West St Lawrence

Lowlands Natural Region preserved by the park. The Lake Huron shore is alternately rimmed (为……之边) by rock, sand and fen wetlands which, coupled with the mild climate, provide a greatly diversified habitat for animals and plants. The boundaries of its 155 square kilometres encompass the former Cyprus Lake Provincial Park and lands purchased from private owners as they become available.

During theSilurian Period and the Paleozoic Era, 400 million years ago, this area was covered by an inland sea where the skeletons of underwater corals and other sea creatures formed layers of calcium-based sediments on the sea floor. As the water began to recede (退去), the dolomite (白云石) bedrock of the peninsula was formed by the compression of these calcium layers to form limestone, which then combined with magnesium (镁) from the sea. The harder dolomite limestone forms much of the rock of the escarpment cliffs along Bruce Peninsula National Park's Georgian Bay shoreline. At Niagara Falls, the dolomite "caprock" erodes more slowly than the rock below it, creating the sculptured cliffs for which the area is famous. After thousands of years of water erosion, the softer lower rock beneath the caprock was undermined and huge segments broke across straight fault lines to create the topography (地形) of the Niagara Escarpment, the backbone of the peninsula. Similarly, the much smaller flowerpot and pillar formations along the shoreline, and the coastline and karst caves are molded by wave action. The largest shoreline cave on the peninsula, the Grotto, is accessible underwater. Over 7 kilometres of karst caves have been mapped in the park. Ontario's "Great Wall" is now a World Biosphere Reserve, an international designation that promotes a harmonious relationship between people and their environment.

The shipping industry around Tobermory was well underway during the 1850's, when demand for lumber (木材) in the southern cities of Ontario and some of the border states was at its peak. Wasteful logging inhibited the growth of many trees, especially pine, leaving tree stumps and forests charred (烧焦的) by fire. Settlements and roads were determined by the logging industry and the first lumbermill was built in Tobermory in 1881. Three lighthouses were constructed to guide ships: at Cove Island in 1858, Big Tub in 1885 and Flowerpot in 1897, in response to the alarming frequency of shipwrecks. By the 1890's propeller (螺旋桨) driven steamers were replacing the two- and three-masted schooners on the Great Lakes. The steamers also carried passengers to the little towns along the shore when travel overland was slow and difficult. Eventually the service evolved from these early vessels to today's Chi-Cheemaun, the largest vessel of its kind on the Great Lakes. Commercial fishing developed in the mid-1800's and flourished with the advent (来临) of steam-powered fishing tugs. Ice taken from the bay and kept year-round in icehouses was used to preserve the fish. In 1932, the lamprey eel (鳗鱼) entered the Great Lakes through the newly completed Welland Canal. Overfishing of some species, in combination with the devastating effect of the eel preying on the Lake Trout, led to the decline of the fishery.

Commonly seen wildlife on the Bruce includes chipmunk, squirrel, raccoon, porcupine, snowshoe hare, skunk, white-tailed deer, snakes and frogs. The Upper Bruce Peninsula ecosystem has one of the very few forests large enough and sufficiently free of roads and farms to support typical interior forest species such as the black bear, fisher, long-eared bats, northern flying squirrel, fox, martin, Massasauga rattlesnake, red-shouldered hawk, barred owl, hermit thrush, black-throated blue warbler, scarlet tanager and yellow-spotted salamander. Countless caves and crevices in the barren cliff face provide homes for ravens, turkey vultures, swallows and bats. Black bears, whose population is currently being monitored by the park to determine their number, size, habitat use and overall health, create problems only when they have been foolishly encouraged to interact with humans who feed them or leave garbage behind. Black bears are smaller than their grizzly and polar cousins weighing anywhere from 100 to over 250 kilograms. They eat literally anything that is available. Solitary by nature, they mate briefly and the female raises the young alone. Bears reach maturity at age three and a half. The Eastern Massasauga rattlesnake, now an endangered species, was once found throughout southern Ontario. They are poisonous, but avoid encounters with humans and rarely present a danger. Other species include garter, ring-necked, De Kay's, ribbon, smooth green, red-bellied, milk and water snakes.

The Bruce has a uniquely diverse range of habitats from rugged cliffs to the dry alvars to the swampy wetlands. The massive, rugged cliffs of the park are inhabited by thousand year old cedar trees, overhanging the crystal clear waters of Georgian Bay. It is the largest remaining chunk of natural habitat in southern Ontario. The broad-leafed forests on the peninsula are made up of mainly of sugar maple, beech, basswood, white oak, red oak, shag-bark hickory, black walnut and butternut. White pine, red pine and red oak dominate the forest along dry ridge tops. The Dorcas Bay Nature Reserve is a fragile ecosystem that begins with wet bogs（沼泽）that are home to some of the peninsula's 43 species of orchid and insectivorous plants including pitcher plants, sundew and butterworts. Of the orchids the most amazing are the Calypso Lady's Slipper, the Dwarf Lake Iris, and the Alaskan Rein. About half the world's Dwarf Lake Iris, most of Canada's Indian plantain and over 20 varieties of fern, including the rare Northern Holly Fern, are found on the peninsula. Beyond the fen, is sandy forest with stands of red and white pine and dense cedar bush interspersed（穿插）with lichen-covered rocky knolls called alvars which may harbour mixed forest of cedar, spruce, birch and aspen. Alvars are flat open areas of limestone bedrock with little or no soil where primarily vascular（血管）plants, mosses and lichens create a unique habitat. Of the 54 vascular plants known to grow mostly on alvars, 17 are designated as provincially rare. In winter, alvars have little snow coverage but in spring, growth begins early as the rock warms quickly. In summer a prairie-type flora often gives the alvar a grassland appearance. Ferns and moss also grow in the cracks of rocks where small amounts of soil have collected. Most surprising are the dwarfed and twisted cedars that grow on ledges and out of cracks in sheer rock faces far from any soil. In 1988 a 511 year old cedar was discovered growing from one of these cliffs.

In the Bruce Park, people can take a series of activities, including hiking, camping, swimming, fishing, canoeing, kayaking, cross-country skiing, snowshoeing, wildlife and wildflower viewing, interpretive and educational programs. For hiking lovers, there are several self-guided trails: Cyprus Lake Trail, Georgian Bay—Marr Lake Trail, Horse Lake Trail, and Singing Sands. For the less rugged experience, follow this trail around Cyprus Lake. A watchful eye will note the charred stumps from forest fires of the early 1900s. There are many access trails from the Cyprus campground and campers are encouraged to use this trail as their path to the Head-of-Trails. The Georgian Bay Trail is the easiest and quickest path to the park's scenic cliffs and shore. At the shoreline this trail meets the Bruce Trail, giving the option of two different routes. Camping in the park is popular during summer, while in spring and autumn, there are fewer visitors. Good but unsupervised swimming is available in several locations. Cyprus Lake has a shallow beach. A 30 – minute walk from the Head-of-Trails parking lot gets you to Indian Head Cove and the Grotto. This intricately carved rocky shoreline is very popular with swimmers who like cold, deep water. There are acres of shallow sandy shoreline at Singing Sands Beach at Dorcas Bay on Lake Huron. Motorboats are prohibited on Cyprus Lake. Non-powered craft such as canoes and sailboards may be used on the lake anytime. Canoeing gives visitors a unique perspective on the park. Most of the Bruce Peninsula National Park's inland lakes can easily be explored in day trips. During the winter months visitors may use the park for cross-country skiing, snowshoeing and camping. Cyprus Lake campground is open during the winter but the services are limited. There are also many excellent opportunities for photography within Bruce Peninsula National Park whose magnificent landscape and unique diversity of plant and animal species attract wildlife photographers from around the world. However, sometimes visitors are obliged to observe the rules of "no trace photography" within the national park.

There are also learning programs provided by the park for those who have desires to get a betterknowledge of the park. In July and August the park service offers interpretive programs to help visitors enjoy the special character of Bruce Peninsula National Park. The schedule of events is posted at various locations around Tobermory and is available at any park office. The regular events include naturalist demos, guided hikes and evening programs. School programs mainly are organized by an interpreter, beginning with the Georgian Bay / Marr Lake Trail. The hike will focus on past and present factors influencing the formation of the Niagara Escarpment. Emphasis will be placed on the special ecology of the cliff edge. During the hike, students will see cedar trees which are part of the oldest forests in Ontario, experience the magnificent view over Fathom Five National Marine Park's islands, scramble over rock beaches formed thousands of years ago, and visit sink holes, ancient coral reefs, and sea caves. But student programs are only available during spring and fall.

In 2006, a new visitors' centrewas opened to serve Fathom Five National Marine Park and the Bruce Peninsula National Park. It features an information centre, reception area, exhibit hall and theatre. A 20 – meter viewing tower was also constructed to provide visitors with aerial views of the surrounding park and Georgian Bay. The centre was designed with environmental sustainability in mind, receiving $224,000 from the Federal House in Order initiative for implementation of innovative greenhouse gas reduction technology.

With long time of efforts made to keep this region clear and new devises implemented to help this, the park is the ultimate paradise throughout year. The shores of Georgian Bay, the cliffs created by the Niagara Escarpment, the Indian Head Cove, and the Grotto from Cyprus Lake Campground, all of them are wonderful choices.

Notes

Bruce Peninsula 布鲁斯半岛　位于加拿大安大略省的休伦湖和乔治亚湾之间，地处尼亚加拉陡坡北末端，是一片大约80千米长、裸露地表的石灰岩地区。半岛的一边是寒冷、清冽的乔治亚湾，而另一边是温暖得多的休伦湖。布鲁斯半岛形状狭长，三面环水，湖岸是著名的尼亚加拉断层。风化的岩层姿态多变，依水蜿蜒，映着蓝天碧水，风景绝佳。与此同时，岛上有丰富的植被和动物。著名的布鲁斯徒步路线，从这里一直连接到尼亚加拉瀑布，绵延八百千米，是徒步旅行爱好者经常出没的地方。

Lake Huron 休伦湖　北美洲五大湖之一，北美五大湖中第二大湖，其位置居中。美国和加拿大共有。它由西北向东南延伸，长330千米，最宽295千米。面积5.96万平方千米，在五大湖中居第二位。湖面海拔177米。平均水深60米，最大深度229米。蓄水量3 540立方千米。湖岸线长2 700千米，较曲折，东北部有乔治亚湾。湖区铀、金、银、铜、石灰石和盐等矿产资源丰富，是美、加两国重要的工业区。圣克莱尔河东岸多炼油厂和石油化工厂，被称为加拿大的“化工谷”。湖泊水质好，盛产鱼类。湖区伐木业和捕鱼业亦很发达。主要经济项目有伐木业和渔业。沿湖多游览区。

Ontario 安大略省　简称安省。安大略北至哈得逊湾，东邻魁北克，西接马尼托巴，南部与美国的明尼苏达、密歇根、俄亥俄、宾夕法尼亚和纽约诸州为界。安大略同美国的边界基本上都是河流或者湖泊，自西面的森林湖直到东部康沃尔的圣劳伦斯河。这其中包括五大湖中的苏必利尔湖、休伦湖、伊利湖和安大略湖。安大略全省大约有2.5万个湖泊以及全长超过10万千米的河流。省会为多伦多，也是该省最大城市，是安大略湖西岸所谓"金色马蹄"城市带的主要成员

Niagara Escarpment 尼加拉瓜悬崖　位于美国和加拿大之间，西起美国的纽约州，穿过安大略省、密歇根省、威斯康辛州和伊利诺斯州。这条绵延数千里的悬崖包括志留纪的地质风貌。该悬崖因尼加拉瓜大瀑布而闻名于世。

Niagara Falls 尼亚加拉瀑布　位于加拿大和美国交界的尼亚加拉河中段，号称世界七大奇景之一，与南美的伊瓜苏瀑布及非洲的维多利亚瀑布合称为世界三大瀑布。它以宏伟的气势、丰沛而浩瀚的水汽震撼了所有的游人。从伊利湖滚滚而来的尼亚加拉河水流经此地，突然垂直跌落50多米，巨大的水流以银河倾倒之势冲下断崖，声及数里之外，场面震人心魄，气势磅礴。瀑布的河水从高处的伊利湖流入安大略湖，湖水经过河床绝壁上的羊岛，分隔成两部分，分别流入美国和加拿大。大瀑布因其外表形成一个马蹄状而被称为马蹄瀑布。

Silurian Period 志留纪　早古生代的最后一个纪，也是古生代第三个纪。本纪始于距今4.38亿年，延续了2500万年。由于志留系在波罗的海哥德兰岛上发育较好，因此曾一度被称为哥德兰系。志留纪可分早、中、晚三个世。植物登陆成功和有颌类的出现是发生在志留纪的最重要的生物演化事件。由于剧烈的造山运动，地球表面出现了较大的变化，海洋面积减小，大陆面积扩大。这个时期最大的特点是植物开始登上陆地，在海中也出现了有颌骨的鱼类——棘鱼类，海中有成群的珊瑚聚集生活，最后形成珊瑚礁。

Welland Canal 韦兰运河　加拿大南安大略的水道，它使南方的伊利湖和北方的安大略湖之间通航大船成为可能，并成为圣罗伦斯航路的重要纽带。1829年开凿。20世纪初因人口激增和工业发展需要而重加疏浚，于1932年完工。全长44.4千米，深8米，宽61米，从克伯尼港至威乐港共有8个水闸，可供较大轮船通行。货运以铁矿、小麦、煤、玉米、钢铁、大麦为主。由于冬季河道结冰，每年停航3个月，此停航期内进行设备、设施的维修。圣劳伦斯海道管理公司在韦兰运河设有一个现场交通控制中心，布设工业电视监视系统和船舶自动识别系统，可以随时了解船舶的方位和过闸情况，减少了船舶滞留并提高了航运安全度，航行安全保障率持续保持在99%以上。船舶通过圣劳伦斯海道需要交纳通行费，安装导航设施及GPS船用终端，必须遵守加拿大航海机构和圣劳伦斯海道管理公司制定的制度和规则。

Federal House in Order initiative 联邦政府自制行动　此行动方案下，加拿大联邦政府通过控制政府行动所产生的有害物质排放，减少有害物质排放，并解决各种污染物排放问题。

这份行动方案受到加拿大环境保护局、加拿大自然资源保护局和加拿大公共事务局等的支持。这一行动牵制 11 个政府部门和机构。据估计，这些部门和机构占政府总排放量的 95%左右。

Translations of Selected Sentences

1. During the Silurian Period and the Paleozoic Era, 400 million years ago, this area was covered by an inland sea where the skeletons of underwater corals and other sea creatures formed layers of calcium-based sediments on the sea floor.

志留纪和古生代时期，也就是400 万年前，这个地区被一片内海所覆盖，水下珊瑚和其他水生生物的骨骼组成了海底的钙基沉积物。

2. Commonly seen wildlife on the Bruce includes chipmunk, squirrel, raccoon, porcupine, snowshoe hare, skunk, white-tailed deer, snakes and frogs. The Upper Bruce Peninsula ecosystem has one of the very few forests large enough and sufficiently free of roads and farms to support typical interior forest species such as the black bear, fisher, long-eared bats, northern flying squirrel, fox, martin, Massasauga rattlesnake, red-shouldered hawk, barred owl, hermit thrush, black-throated blue warbler, scarlet tanager and yellow-spotted salamander.

布鲁斯公园内常见的野生动物有金花鼠、松鼠、浣熊、豪猪、雪靴野兔、臭鼬、白尾鹿、各种蛇和青蛙。上游的布鲁斯半岛生态系统拥有广阔的树林，没有马路和农田，足以维持典型的森林生物的生存，如黑熊、食鱼动物、长耳蝠、北美鼯鼠、狐狸、燕科小鸟、侏响尾蛇、红肩鹰、大林鸮、隐士夜鸫、黑喉蓝林莺、红唐纳雀和黄点火蜥蜴。

3. The Bruce has a uniquely diverse range of habitats from rugged cliffs to the dry alvars to the swampy wetlands. The massive, rugged cliffs of the park are inhabited by thousand year old cedar trees, overhanging the crystal clear waters of Georgian Bay. It is the largest remaining chunk of natural habitat in southern Ontario.

布鲁斯的栖息地较为独特，其类型众多，从崎岖不平的山崖峭壁，到干燥的矮灌木丛草地，再到潮湿的沼泽地。公园内大块崎岖的山崖峭壁上栖居着千年古松，悬挂于乔治亚湾透明的清水之上。这是南安大略湖现存的最大的自然栖息地。

4. There arealso many excellent opportunities for photography within Bruce Peninsula National Park whose magnificent landscape and unique diversity of plant and animal species attract wildlife photographers from around the world. However, sometimes visitors are obliged to observe the rules of "no trace photography" within the national park.

布鲁斯半岛公园给大家提供了很多精彩的摄影机会。这个公园内壮丽的景色和独特的动植物吸引了世界上众多的野生生物拍摄者。但是，有时候游客有义务遵守公园内"禁止入内拍照"的规定。

5. During the hike, students will see cedar trees which are part of the oldest forests in Ontario, experience the magnificent view over Fathom Five National Marine Park's islands, scramble over rock beaches formed thousands of years ago, and visit sink holes, ancient coral reefs, and sea caves. But student programs are only available during spring and fall.

远足时,学生们将会看到安大略省最古老的森林之一——雪松林,可以在五浔国家海洋公园内的岛上欣赏壮丽的自然景色,可以攀登千年之前形成的岩石,可以参观灰岩坑、古老的珊瑚礁和海蚀洞。但是这些学生活动方案仅在春秋两季实施。

Questions

1. What are the features of the Bruce Peninsula National Park in terms of its topography?
2. What is Ontario's "Great Wall" and how about its features?
3. What led to the decline of the fishery in the Bruce Peninsula region?
4. What are the suggestions given by the article for hiking lovers in this park?
5. What are the wonderful choices for visitors to this park according to the last paragraph?

Chapter 4

Cape Breton Highlands National Park

布雷顿角高地国家公园

【简介】

布雷顿角高地国家公园位于加拿大新斯科舍省布雷顿角岛北部。该国家公园成立于1936年,是加拿大第一个位于大西洋沿岸的国家公园,主要野生生物有驼鹿、黑熊、土狼等。公园东临大西洋、西濒圣劳伦斯湾,面积958平方千米。公园内海岸岩壁陡峭,森林树木茂密,溪流峡谷纵横,禽兽飞行出没,游客在公园内游览,仿佛置身于原始森林之中。通向加拿大布雷顿角高地国家公园的是世界著名的卡伯特小径。在这里,人们可以看到鲸鱼、秃鹰和驼鹿。漫步在风景如画的山谷和海洋洞穴,享受阿加底亚、苏格兰和爱尔兰的丰富的遗产宝藏。园内有6个露营地和25条登山步道,也有提供游泳、骑自行车或打高尔夫球等活动的场所。每年9月下旬或10月初的时候,公园到处是一片让人眼花缭乱的秋天的颜色。冬天,这个公园则是人们滑冰、滑雪橇或者是进行越野速度滑雪的理想之地。

Cape Breton Highlands National Park is the first national park designated in Atlantic Canada. The Cabot Trail, a world-famous scenic highway, runs along parts of the coastal borders on both sides of the park and crosses the highlands. Renowned（闻名）for its hiking trails, the park is home to a diverse mix of boreal（寒带）and temperate species not found elsewhere in Canada.

Cape Breton Highlands National Park is one of the tourist attractions in Canada, known for its spectacular highlands and ocean scenery, such as high mountainous areas with lovely waterfalls and abundant forests and the low ocean fantastic view. Steep（陡峭的）cliffs and deep river canyons carve into a forested plateau bordering the Atlantic Ocean. The park is often referred to as the place "where the mountains meet the sea". The dominant feature of this region is the elevated plateau, divided by steep-walled river canyons; northern species and habitats on the plateaus coexist with the more temperate habitats and species of the lowlands. Consequently, there is much diversity.

Established in 1936, the National Park covers 950 square kilometers, protecting about 20% of northern Cape Breton. It is one of the largest protected wilderness areas inNova Scotia and is one of a system of national parks protecting outstanding Canadian landscapes. Approximately 88 percent of the park is forested. The plateau, or upper reaches, is dominated by both boreal and taiga vegetation and is part of the worn-down Appalachian Mountain chain that stretches from Georgia to Newfoundland. The boreal land region of this plateau features large swaths (大片) of coniferous trees, sprinkled with barrens and wetlands. Most of the entire population of Nova Scotia's endangered Canada lynx (猞猁) lives here, as well as moose, hare, grouse, and marten. The taiga land region of the plateau features a tundra-like landscape characterized by scrub forest, barrens, and bogs. Within the park, several dozen species of rare or threatened plants and animals can be found, as well as old growth forests of international importance. Small populations of arctic-alpine plants left over from the last ice age can also be found here.

The dominant feature of northern Cape Breton is the Cape Breton Plateau, averaging 350 meters at its edges but rising to more than 500 meters atits centre! Extending over 70% of the park, the plateau appears flat-topped but actually consists of broad, gently rolling hills, deeply cut by steep-walled river canyons and broad valleys. On the west coast the plateau meets the Gulf of St. Lawrence in steep cliffs. On the east, the highlands border the Atlantic Ocean in a more gently sloping coastal plain with low headlands and a few long sandy beaches.

Cape Breton Island is made up of three main groups of rocks: the Bras d'Or terrane (地块), the Avalon terrane and the Blair River inlier (英利尔山区). Some of the oldest rocks in the Maritime Provinces are found on Cape Breton Island, in the Blair River inlier. The plateau as we see it today is the product of millions of years of geological change, including mountain building, erosion, and glaciation. Faults (断层), glaciers and rivers have all worked the landscape over millions of years to produce spectacular river canyons. There are a number of faults in Cape Breton Highlands National Park but noticeable earthquakes are rare. The faults were probably created long ago when two continental plates met and pushed the sea floor between them upwards to form the Appalachian Mountains. Nowadays, the North American continent is moving westwards, so most of the tectonic (构造) activity occurs on the western side of the continent where the Rocky Mountains are still being pushed up today.

Cape Breton Highlands National Park protects part of the Maritime Acadian Highlands

Natural Region, a Parks Canada designation, which is foundin the Acadian Forest Region of Canada. This natural region in conjunction with the Great Lakes—St. Lawrence Forest Region is part of a mixed hardwood-softwood forest that stretches from the Great Lakes to New England and Maritime Canada.

The climate and landscape of northern Cape Breton have created an unusual mix of southern and northern plant species. Southern temperate(温带) plants such as sugar maple, yellow birch and Dutchman's breeches usually occur farther south in New England, but thrive in the rich woodlands of Cape Breton's sheltered valleys. Most plants on the Cape Breton plateau are cold-loving northern boreal species such as balsam fir, white birch, black spruce and bunchberry. Northern Cape Breton has 631 native plant species and 118 introduced, or "exotic", species. These species were generally brought to the area by humans, sometimes on purpose as garden plants, and sometimes accidentally. They include plants like ox-eye daisy, lupine, and black-eyed susan, as well as dandelions and purple loosestrife. Although part of the Maritime Acadian Highlands Natural Region, the vegetation in northern Cape Breton can actually be divided into three distinct forest regions: Acadian, composed of mixed woods; Boreal, of softwood; and Taiga, made up of scrub forest, barrens and bogs.

The rich tapestry (如织锦般的栖息地) of mountains, forests, rivers, lakes and coastlines provides habitats for both northern and southern species of animals. Because the park is a protected wilderness area, these animals can live out their lives in a natural state, as close to being unaffected by humans as possible in the modern world. Mammals, birds, amphibians, reptiles, fish and invertebrates (无脊椎动物) are all found within Cape Breton Highlands National Park.

Some species of animal in thepark are particularly sensitive to human activity and require extra protection. Several species that are separated from the rest of their populations in North America are also found here. There are about 40 species of land mammals in northern Cape Breton, including moose, coyote, and snowshoe hare. Land mammals of particular interest include Canada lynx, rock vole and Gaspé shrew (加斯佩鼩) because they are at risk. Marine mammals such as harbor seals and pilot whales are found in the water adjacent to the park. Along the coasts, seabirds such as gulls, cormorants and guillemots nest. In the woods there are many forest songbirds as well as grouse, hawks, bald eagles and owls. Cape Breton Island isn't home to very many species of amphibians and reptiles because it is isolated and cold. The most common frog in the park is the green frog; the garter snake is the most common reptile. The lakes and rivers are home to brook trout, gaspereau and Atlantic salmon as well as a few other species of freshwater fish. The Gulf and the Ocean next to the park are inhabited by ground fish such as

Atlantic cod and sharks like the blue shark and Basking shark. Invertebrates are extremely important to the ecosystem. Thousands of species of invertebrates can be found in the park, including spiders, beetles, dragonflies, periwinkles, crabs and slugs.

As Cape Breton Highlands National Park is located on an island of protected land isolated by mountains and ocean, island populations are thought to be more vulnerable to extinction becauseof their small size and isolation. The plants and animals of northern Cape Breton include a number of species at risk. Nine of these are now extinct or extirpated (灭绝) such as the woodland caribou. Population studies for some species at risk in the park have been or are being conducted. Many plant species, although not yet officially listed at risk are rare and probably at risk. As assessment of plant populations increases, it is likely that more species will be protected.

There are several ways Parks Canada helps species at risk in Cape Breton, including implementing research and monitoring, management plans, developing partnerships and through legislation. Several research projects are ongoing at the moment, including work on the wood turtle and the Bicknell's thrush. New projects may also be started in the future. Observation of the population trends to detect whether a problem exists. At present the park is monitoring herpetiles (amphibians and reptiles) and birds of prey (owls, eagles, hawks and falcons). Park staff as well as volunteers walk various trails in the park, actively looking for these particular species. Zoning in parks protects habitat where species at risk live. The Canada National Parks Act and Regulations, the Species at Risk Act and the Nova Scotia Endangered Species Act give special protection to species at risk. Then the Accord is an agreement between the federal, provincial and territorial governments to help prevent species in Canada from becoming extinct as a consequence of human activity. It is a key component of the Canadian Biodiversity Strategy because it ensures that the federal and provincial governments work together to protect species at risk. Parks Canada, a federal government agency, actively helps protect species as part of the Accord. The last one is the Endangered Spaces Campaign in Canada which began in 1987. It was a 10 year campaign that concluded in 2000 and added 38 million hectares to Canada's parks and protected areas. Parks Canada is dedicated to preserving representative natural habitats of Canada's wilderness. Thirty nine natural regions in Canada have been identified and 25 of these are included in Canada's National Parks system, leaving 14 unrepresented.

A series of exhilarating adventures are available in the hiking capital of Canada. The park has six campgrounds, 26 hiking trails, several beaches, and a world-class golf course. Bordering communities have seasonal amenities (季节性设施), recreational programs, and cultural activities. Cape Breton's magnificent northern shore and the incredible plateaus on the world-renowned Cabot Trail are worthy a visit.

Geocaching is a modern treasure hunt—a new sport that is growing in popularity with people of all ages. Given the latitudinal and longitudinal co-ordinates, participants use their GPS (Global Positioning System) unit to locate the hidden "cache", a small box with a logbook to record their visit. Cape Breton Highlands is pleased to offer four geocaches: Ecocache find all five official park caches to earn a limited-edition geocoin; Middle Head Multi-cache explore this historic peninsula, site of the Keltic Lodge; Boomer's Campground Kid Caches discover geocaching in the Chéticamp and Broad Cove Campgrounds; Earthcache, for the Earth Science buffs. Besides, visitors can also take other activities, including camping, hiking, sightseeing, swimming, fishing, picnicking.

The park also provides learning programs for visitors to be more acquainted with the environment here. From late June to August, park interpreters offer fun, informative programs about the park's wildlife, plants and history. All interpretation programs are included free with your park entry permit. The park offers extensive programming for schoolchildren from their first year in primary through to grade twelve. Park interpreters offer field trips in the park as well as class sessions. Programs such as puppet shows and beginner field trips introduce the younger children to their national park and build continually on their experience until high school at which time the students are encouraged to use their knowledge and critical thinking skills. Working closely with the curriculum, interpreters bring to the classroom the significance of Canada's protected heritage areas and the importance of working together to ensure their continued protection.

The headlands and cliffs of Cape Breton Highlands tower over the rich, natural heritage that is all around. Home to the famous Cabot Trail, the land is blessed with spectacular scenery, abundant wildlife and a human history that stretches back to the last Ice Age. The park offers many accessible treasures and experiences remarkable in their diversity, beauty, and wonder. The possibilities at Cape Breton Highlands are endless. Enjoy the park and enjoy yourself is not impossible.

Notes

Atlantic Canada 加拿大大西洋省区　指加拿大在大西洋海岸的四个省:新斯科舍、纽奔驰域、爱德华王子岛和纽芬兰与拉布拉多。而其中新斯科舍、纽奔驰域和爱德华王子岛又合称为加拿大海洋省份(又译作加拿大滨海诸省或加拿大沿海诸省)。这四个省比加拿大其他省份落后和贫困,但自然环境优美。经济以渔业和旅游业为主。四省内最大城市是哈利法克斯,也是加拿大唯一在大西洋的港口城市。近年在新斯科舍和纽芬兰近岸发现丰富油田,因此区内石油业正在加速发展.

Atlantic Ocean 大西洋　世界第二大洋,原面积 8 221.7 万平方千米,在南冰洋成立后,面积调整为 7 676.2 万平方千米,平均深度 3 627 米,最深处波多黎各海沟深达 8 605 米。从赤道南北分为北大西洋和南大西洋。北面连接北冰洋,南面则以南纬 66 度与南冰洋连接。大西洋中的海洋资源相当丰富,已勘探和利用的资源主要是矿产资源和水产资源。大西洋中的矿产资源主要有石油、天然气、煤、铁、重砂矿和锰结核等。大西洋生物资源丰富,最主要的是鱼类,其捕获量约占大西洋中海洋生物捕获量的 90% 左右。大西洋的渔获量曾居世界各大洋之首,后低于太平洋,退居第二位。大西洋在世界航运中处于极为重要的地位,它西通巴拿马运河连太平洋,东穿直布罗陀海峡、经地中海、苏伊士运河通向印度洋,北连北冰洋,南接南极海域,航路四通八达、十分便利。同时大西洋沿岸几乎都是各大洲最发达的地区、经济水平较高的资本主义国家,贸易、经济交往频繁,是世界环球航运体系中的重要环节和枢纽。

Nova Scotia 新斯科舍省　加拿大大西洋四省之一,面积 55 491 平方千米,人口约 93 万人,早期欧洲人移民到加拿大的登陆点,也是历史上英法殖民者利益争夺的焦点地区之一。1605 年法国殖民者建立的第一个移民区——新法兰西殖民地即位于罗耶尔港(Port Royal)。直到今天,在新斯科舍仍能发现苏格兰的传统文化。由于靠近海洋,新斯科舍的气候暖湿而多雨,每年有 140 日无霜期,在加拿大算是暖和的省份。省会为哈利法克斯(Halifax),是大西洋沿岸最大的城市及世界第二大深水港,是大西洋省区的政治、经济和文化中心。因处于欧美两洲之间,该省一直以来都是加拿大的海军基地,也是加拿大大西洋舰队司令部的所在地和重要的军事基地,在第二次世界大战期间发挥了重要的功用。

Newfoundland 纽芬兰岛　北美大陆东海岸的大西洋岛屿,面积 11.1 万平方千米,人口 466 172 人(2001)。岛上人口主要分布在沿海地区,文化艺术风格独特。经济素以渔业著称,20 世纪以来,经济趋于多样化。全岛约 1/3 的人口直接或间接同渔业生产有关。纽芬兰浅滩石油和天然气储量丰富。与 12 亿年形成的大自然景观相辅相成的是纽芬兰近 5 000 千年的人文景观。早在哥伦布发现美洲大陆之前,纽芬兰就已经是欧洲人探索新大陆的门户。纽芬兰曾经是一个独立的国家,1949 年通过公民投票决定加入加拿大联邦。这一决定使它成为加拿大最年轻的省份,同时也给加拿大带来了北美新大陆最悠久的历史。

Gulf of St. Lawrence 圣劳伦斯湾　位于拉布拉多半岛、纽芬兰、新伯伦瑞克与布雷顿角岛之间,是加拿大东南部的大西洋海湾,是进入整个北美大陆腹地的一个门户。该海湾是全球五个最佳赏鲸地点之一。

Canada National Parks Act and Regulations《加拿大国家公园法令》　该法案于 1930 年通过,旨在为加拿大国家公园管理局提供法律条文来管理国内各国家公园以及各种自然资源,最新版本为 2000 年修订的《国家公园法》。1988 年《加拿大国家公园法》做出了具有前瞻性的修改:除了再次确认“保护生态完整性”是第一任务之外,还规定,每个国家公园都要公布管理计划,要有公众广泛参与,非政府组织可以在法庭上挑战其主管单位——加拿大公园管理局的管理流程。

Species at Risk Act《濒危物种法》 2002年12月12日经过对加拿大境内大多数公民和各种机构、部门的意见征求,议会通过了这一旨在保护加拿大地区内各种动植物的法案。这一法案的目的是帮助联邦政府阻止各种濒危动物的消失,为各种濒危动物恢复正常生活提供保证。这一面向未来的法案列举了各种濒危物种,同时也提供了各种拯救濒危物种的措施。

Nova Scotia Endangered Species Act《新斯科舍省濒危物种法》 新斯科舍省立法机构于1999年通过的濒危动物保护法,旨在保护该省内各种动植物资源,包括保护各类动植物保护地、保护并拯救因人类破坏活动而濒危的动植物等,同时响应《国家濒危物种法》的各项法案,保护动植物栖息地,减少人为破坏。

Canadian Biodiversity Strategy《加拿大生物多样性战略》 加拿大是发达国家中最早签署生物多样性条约的国家。继条约批准后,1995年又公布了《加拿大生物多样性国家战略》,其中不仅明确了各种生态系统,特别是森林生态系统的保护和管理,还明确了生物资源可持续利用的目标和方向。该国家战略为加拿大各省和特区政府制定各自独立的生物多样性战略和行动计划提供了依据和参照。这一战略也明确了保护生物多样性需要举全国之力:保护生物多样性不仅是政府单一方面的责任,同时也是每一位公民的责任和义务。

GPS (Global Positioning System) 全球定位系统 随着现代科学技术的迅速发展而建立起来的新一代精密卫星定位系统。20世纪70年代由美国陆海空三军联合研制的新一代空间卫星导航定位系统。其主要目的是为陆、海、空三大领域提供实时、全天候和全球性的导航服务,并用于情报收集、核爆监测和应急通信等一些军事目的。经过20余年的研究实验,耗资300亿美元,到1994年3月,全球覆盖率高达98%的24颗GPS卫星星座已布设完成。由于GPS技术所具有的全天候、高精度和自动测量的特点,作为先进的测量手段和新的生产力,已经融入了国民经济建设、国防建设和社会发展的各个应用领域。GPS技术市场的应用前景非常可观。

Translations of Selected Sentences

1. Established in 1936, the National Park covers 950 square kilometres, protecting about 20% of northern Cape Breton. It is one of the largest protected wilderness areas in Nova Scotia and is one of a system of national parks protecting outstanding Canadian landscapes.

该公园成立于1936年,覆盖面积达950平方千米,占北部布雷顿角保护地的20%左右。这是新斯科舍省最大的野生生物保护地之一,也是众多保护美丽的加拿大风景的国家公园之一。

2. Extending over 70% of the park, the plateau appears flat-topped but actually consists of broad, gently rolling hills, deeply cut by steep-walled river canyons and broad valleys. On the west coast the plateau meets the Gulf of St. Lawrence in steep cliffs. On the east, the highlands border the Atlantic Ocean in a more gently sloping coastal plain with low headlands and a few long sandy beaches.

高原占据公园的面积超过70%,它看上去平坦,但实际上包含了广袤的、绵延的丘陵地,中间形成了陡壁峡谷和地形宽广的山谷。在其西边,高原与圣劳伦斯湾相接,连接它们的是陡峭的悬崖。在东边,高地与大西洋接壤,形成了地势更为平坦的海岸平原、岬和绵延的沙滩。

3. Although part of the Maritime Acadian Highlands Natural Region, the vegetation in northern Cape Breton canactually be divided into three distinct forest regions: Acadian, composed of mixed woods; Boreal, of softwood; and Taiga, made up of scrub forest, barrens and bogs.

尽管北部布雷顿角地区是阿卡迪亚高地海岸自然区的一部分,但这里的植被实际上分布在三种地势截然不同的区域:阿卡迪亚,由混交林组成;北部针叶林,由针叶树组成;以及泰加林,由灌木丛林、荒林和沼泽组成。

4. There are several ways Parks Canada helps species at risk in Cape Breton, including implementing research and monitoring, management plans, developing partnerships and through legislation. Several research projects are ongoing at the moment, including work on the wood turtle and the Bicknell's thrush.

加拿大公园管理局在布雷顿角公园内采取了一系列保护濒危动物的措施,包括调查和控制、管理、共同发展和完善环保法律。如今,公园内实施了一些研究计划,包括对木纹龟和比氏夜鸫的研究。

5. Geocaching is a modern treasure hunt—a new sport that is growing in popularity with people of all ages. Given the latitudinal and longitudinal co-ordinates, participants use their GPS (Global Positioning System) unit to locate the hidden "cache", a small box with a logbook to record their visit.

地理寻宝是一项现代寻宝游戏——这种新游戏正受到各个年龄层的喜爱。告知经纬度坐标之后,参与者使用手中的全球定位系统来定位隐藏的藏匿品——一个小箱子,其中有记录簿子来记录他们的发现结果。

Questions

1. What is the dominant feature of Cape Breton Highlands?
2. What are the three main groups of rocks in Cape Breton Island?

3. Is Cape Breton Island home to very many species of amphibians and reptiles? If not, what are the reasons?

4. What are the ways that Parks Canada takes to help species at risk in Cape Breton?

5. What are the extensive programs the park offers for schoolchildren from their first year in primary through to grade twelve?

Chapter 5

Elk Island National Park of Canada
麋鹿岛国家公园

【简介】

从埃德蒙顿向东驱车30分钟,便可到阿尔伯塔省首屈一指的麋鹿岛国家公园。麋鹿岛成立于1913年,是为保护加拿大濒危、脆弱的白杨山林生态而设,也是观察野生动物的最佳地区。成群的野牛、麋鹿、红鹿自由自在地生活在公园之内,许多野牛是从北美洲各地被送来此地保育。麋鹿岛目前有250种鸟类,是绝佳的赏鸟天堂。目前游览区内共有1 700只麋鹿、600只草原野牛、375只森林野牛,以及梅花鹿、驼鹿和海狸等动物。麋鹿岛国家公园是一座天然的"岛屿",它的周围被壮美的自然景观环绕。一望无际的草原、郁郁葱葱的森林、连绵起伏的群山,令人叹为观止的景致吸引了无数游人来此观光游览。每年冬季,许多城市居民来此越野滑雪,夏季则选择健行或打高尔夫球。

It may come as a surprise to travellers on the Yellowhead Highway to find one of Canada's 42 national parks bisected（一分为二）by a major highway and located just 45 kilometres from the capital city of a province. Aren't national parks supposed to be located in remote scenic areas with mountains, pristine lakes, panoramic（全景）views?

Located less than an hour away fromEdmonton, Elk Island National Park of Canada is a beautiful, unique and historically rich national park, protecting one of the most endangered habitats in Canada, the Southern Boreal Plains and Plateaux Natural Region. It is one of the few places in North America where one can see all of the original grazing species coexisting.

Elk Island National Park is situated in the Beaver Hills—an area, as the name suggests, abundant in beaver at the turn of the 19th century. It is thought that the Sarcee were the first people to inhabit the area. Prior to the 1800s large bands of Cree forced the Sarcee into the surrounding plains. German, English, and Ukranian settlers arrived in the area during 1881 – 82. They quickly began clearing land and farming the area around the Beaver Hills. The Beaver Hills were considered too hilly, wet and generally less suitable for agriculture. For this reason, very little serious land clearing and agriculture was attempted. Most, but not all homesteads on the land which would later become Elk Island, were primarily filed for speculative purposes. For centuries, aboriginal peoples travelled through the western landscape. The Beaver Hills provided relief along the journey; a place to hunt, gather and rest. Aspen thickets surrounded by prairie provided forage, protection and plenty of water for plains bison, elk, moose and deer.

On March 28, 1906 Canada's first wildlife sanctuary was established. Five local men pledged by bond to the Dominion Government of Canada to protect one of the last remaining herds of elk in Canada. William Andrew Dickson Lees, Johnson Carscadden, Ellsworth Simmons, William H. Cooper and Francis Austin Walker signed a $5000 bond to establish Elk Park, a 41 square kilometre fenced enclosure around Astotin Lake. In 1908, Elk Park was renamed Elk Island Park.

The plains bison in Elk Island National Park originated from a private herd owned by Charles Allard and Michel Pablo of Ronan Montana. For $245 a head, the Dominion Government of Canada purchased Pablo's share of the herd. In 1907, approximately 400 plains bison were shipped to the Park by railway, via Lamont, en route to their final destination, Buffalo National Park at Wainwright. Forty-eight animals eluded (躲避) capture; their offspring remain in Elk Island to this day. This herd flourished and over the next century thousands of plains bison from the Park were sent to other natural areas in North America to re-establish herds where they once roamed. Their return represents one of the most significant events in Canadian conservation history.

By 1922, the Park was increased in size to allow for growing wildlife populations. Development for visitor use began in 1923 and by 1924 the Elk Island Parkway was constructed from the north to the south.

In 1930, the Park was renamed Elk Island National Park as declared under the new National Parks Act and was obligated to preserve the natural environment for the enjoyment and use by the people of Canada and for future generations. New recreation facilities were constructed after World War II including two bungalow camps, a dance hall, a band shell, tennis courts and a service station. Sandy Beach Campground opened in 1947. In keeping with the preservation mandate, a program was initiated, in 1941, to bring back the beaver to the Beaver Hills after centuries of overtrapping. Today more than 1000 beaver reside in the national park.

In 1951, Prime MinisterLouis St. Laurent officially opened the Ukrainian Pioneer Home, a replica (仿制品) of a typical home built by early Ukrainian settlers. The Ukrainian Pioneer Home was federally designated as the oldest purpose-built Ukrainian museum in Canada. The home remains unoccupied, as it does not meet today's building standards.

At the turn of the millennium, theBeaver Hills Initiative was established in collaboration with Elk Island National Park, municipal partners, environmental groups, the provincial government and industry. Its mission is to support cooperative efforts to sustain the quality of water, land, air, natural resources and community development.

Thought to be extinct, a small herd of pure wood bison were located in 1957 in a remote corner of Wood Buffalo National Park. For conservation purposes, 23 animals were transported to Elk Island National Park in 1965. Most wild populations of wood bison that currently exist in North America can be traced back to the Elk Island herd.

In 1979, fire was reintroduced into the Park through carefully planned prescribed burns. After decades of fire suppression, the aspen forest in Elk Island National Park began to dominate the grassland areas. Fire became recognized as an essential tool tore-establish the balance of forest, grassland and other types of vegetation, thus maintaining healthy ecosystems.

In 1987, Elk Island National Park, together withthe Canadian Wildlife Service, Parks Canada Agency, and the Friends of Elk Island, initiated the Elk Island National Park Trumpeter Swan Reintroduction Program. The trumpeter swan population declined in the 1800's when the settlement of the prairies resulted in over-hunting and loss of critical habitat. In 1998, eleven years after the implementation of the reintroduction program, a milestone was reached—a pair of reintroduced trumpeters successfully raised four cygnets (小天鹅), becoming the first breeding pair in the Park in over a century. There were four other sub-adult (non-breeding) trumpeters in the Park at that time, bringing the total number of swans in the Park to ten. After an absence of over a hundred years, the largest waterfowl in North America has once again firmly established itself in the Park.

Throughout the history of the park, that of the elks in this area seems to be needed to pay more attention to. Elk management is not an easy task which herd management, monitoring of other animals, and co-operative agreements should be taken into consideration.

Populations of elk were abundant in areas around present day Elk Island National Park at the end of the 1700's. The depletion of elk numbers came with European settlement and the need for subsistence and commercial meat. By 1894, elk were practically eliminated from much of the area. As mentioned above, a fence was made to restrict the movement of animals into and out of the Park.

Then, elk and bison herds must be managed in the Park to prevent over-population of these species. This intervention is necessary as continuous, annual, intensive grazing in the very long-term impacts both plant and animal diversity and predisposes large herbivores to disease and large "die-offs".

The elk management program in Elk Island involves intensive monitoring to determine the number of elk that are to be surplused each year. These elk are then trapped and handled during the winter months for relocation. Close co-operation between Elk Island National Park, the Wild Elk Federation (Canada), the Rocky Mountain Elk Foundation (USA), the Canadian Food Inspection Agency, and several other agencies is necessary to relocate these animals to the wild.

An ungulate (有蹄类动物) aerial survey is flown every winter. Total elk, moose, deer, and bison numbers are recorded, which includes the age and sex of animals. Through this survey, the rate of increase and other related population dynamics are recorded. Habitat utilization is assessed annually by monitoring browse, grass, and sedge use on established transects. From this information, an assessment of the number of elk to be surplused from the Park each year is calculated.

All of the Park's surplus elk are released to the wild through co-operative elk transplants with federal, provincial, or state governments. Elk Island elk are used as a source herd for re-establishing elk populations in native range throughout North America. The Park has become renowned for its conservation role in repopulating areas where elk populations used to exist or in supplementing existing herds that have been drastically reduced and are too small to increase on their own. Over the last two decades, an average of 160 elk each handling year has been live-trapped and transplanted for this purpose. Surplus elk have been allocated to the Canadian provinces of British Columbia, Alberta, Saskatchewan, Ontario, the Yukon, and Northwest Territories, and to Kentucky, Tennessee, and North Carolina in the United States.

Elk Island National Park has an extensive cultural history dating back to the receding of the glaciers. There are 227 Native sites recorded, mostof which were campsites and stone tool-making sites, and 13 non-native sites which include two refuse pits, remains of homesteader cabins, and the site of Alberta's first Forest Ranger station.

Memories and buildings are the most obvious reminders of Elk Island's rich history. In Canada, any federal government building more than 40 years of age is evaluated by the Federal Heritage Buildings Review Office (FHBRO) to assess the historic nature of the structure and the appropriate level of protection. In Elk Island National Park, the Superintendent's residence, the Pavilion overlooking the beach area, the horse barn, and the Ukrainian Pioneer Home have FHBRO protection. These buildings are maintained and managed in accordance with FHBRO direction.

The Ukrainian Pioneer Home was designated as a "Classified" federal heritage building because of its commemorative history, its stature as the oldest purpose-built Ukrainian museum in Canada, and its association with important individuals and events related to the commemoration of Ukrainian immigration to Canada. Two Prime Ministers, Louis St. Laurent and Lester Pearson, have participated in events at this site.

All historic buildings undergo regular maintenance to protect their historic qualities. The Ukrainian Pioneer Home requires special efforts. Recent maintenance includes the re-thatching of the roof during the summer of 1995. The Home will be closed until restoration of the building——specifically repairs to the foundation by replacing rotting logs—is complete.

During this restoration period, a portion of the collection of tools, clothing and textiles, and other artifacts related to the Ukrainian settlement in the Elk Island Park area—formerly displayed in the Home—has been lent to the Basilian Fathers Museum in Mundare, Alberta, for public viewing.

Of course, Elk Island National Park offers exciting opportunities for visitors and students to learn about ecological processes, the Family of Parks Canada, and the significance of Elk Island National Park. Onsite summer interpretation programs, school group environmental education programs, exhibits, self-guided trails, special events and the Information Centre all offer learning opportunities for visitors of all ages. Elk Island National Park is also an ideal place to give children the opportunity to explore nature. There are three trails under 4 km in length: Amisk Wuche, Lakeview and Beaver Pond. The Living Waters Boardwalk is a 300 m floating sidewalk off the shore of Astotin Lake. In the Astotin Lake Area, children can enjoy the playground and the beach. Children can build sand castles and watch the amusing Richardson's ground squirrels.

The year 2006 marked the 100th anniversary for Elk Island as the first wildlife preserve in Canada. The Park has evolved in many ways but the vision remains the same. Today it continues to be a place for nature and for people. A hundred years have come and gone, and with that vision, an incredible legacy began. Imagine what the next hundred years will bring.

Notes

Edmonton 埃德蒙顿　阿尔伯塔省的省府，该省的文化、政治和教育中心，加拿大第五大城市。城市拥有轻轨运输，拥有北美最大的室内购物中心——西部埃德蒙顿购物中心，和加拿大最大的历史公园——埃德蒙顿堡垒公园。埃德蒙顿是阿尔贝塔省北方的经济中心和石油、天然气工业的中心。20世纪40年代起就获得了"加拿大石油之都"的美称。供应与服务工业的驱动和新科技发展的支撑使得阿尔伯塔省大量的石油、天然气和油沙储量增值很多。有报道称,其储量在世界上排名第二,仅次于沙特阿拉伯。

Beaver Hills 比弗山　位于加拿大萨斯喀彻温省,距离该省的约克顿市约 35 英里,是白沙河的发源地,山上布满了灌木丛和各种野草。该山所在地区主要有两个县城,分别为艾蒂尤纳和凯利赫。

Dominion Government of Canada 加拿大自治领政府　加拿大立国初期的官方全名。1867 年英国政府通过了《英属北美法案》,同意在 1867 年 7 月 1 日成立加拿大自治领联邦政府。同年,安大略、魁北克、新斯科舍(新苏格兰)和新不伦瑞克四个省份同意加入联邦政府,组成英帝国内的加拿大自治领。到 1910 年,加拿大已经有 9 个省份(纽芬兰省是在 1949 年加入联邦的)。1931 年的《威斯敏斯特条例》扩大了加拿大政府自主权,包括外交。英国国会在 1982 年 通过的《加拿大法案》,把修改宪法的权力从英国国会移交给加拿大。现在的加拿大由 10 个省及两个地区组成。

Lamont 拉蒙特　加拿大阿尔伯塔省的一个县,位于该省中部。总面积 2 402.1 平方千米,总人口 4 167 (2001),包含 27 个镇区,主要景点为赤鹿岛国家公园。

Louis St. Laurent 路易斯 · 圣劳伦特　加拿大魁北克省康普顿人,加拿大第 12 任总理。圣劳伦特于 1882 年 2 月 1 日出生在魁省东部城镇的一个小商人之家,父亲是法裔加拿大人,开店为生;母亲是爱尔兰人,这使得圣劳伦特在流利的英法双语环境中成长。1905 年圣劳伦特从魁北克城的拉瓦尔大学(Laval University)法律系毕业。大学毕业后一直从事律师工作。1894 年被任命为对外事务部部长。在担任对外事务部长期间,路易斯 · 圣劳伦特代表加拿大参加了联合国代表大会的成立,他指出联合国在战时没有军队,导致对各国战争监管不力,因此建议成立一个联合国部队,用以维持战争国家的和平。1948 年起担任总理一职。期间,圣劳伦特的内阁在国际舞台上扩大了加拿大在战后全球秩序重组上的所扮演的角色。1949 年加拿大加入北大西洋公约组织(NATO),参与了美国的朝鲜战争及联合国的军事事务。在加拿大国内,则促成纽芬兰省加入加拿大联邦,并建立了新的社会和工业政策。圣劳伦特政府对加拿大进行了谨慎的改革,开始实行征税、建立了支持艺术发展的加拿大委员会、扩大社会福利计划,如家庭福利和老年退休金等的发放范围、引入了医疗保险的最初雏形。政府同时参与了多项大型公共建设,如横贯加拿大高速公路、横贯加拿大输油管线、圣劳伦斯河海运等。1956 年加拿大掀起了对横贯加拿大输油管线建设的争议,导致公众对政府耗用巨资不满。

Beaver Hills Initiative 比弗山倡议　旨在保护加拿大比弗山地区生态完整性的倡议。在这一倡议之下,结合其他的经济、社会、文化等组织,对这一地区进行科学调查、管理和保护,并为该团体提供有效的环境保护措施,以保护这一地区的生态完整和多样性。

Wood Buffalo National Park 伍德布法罗国家公园　也译为"木水牛城国家公园",位于加拿大中北部的平原上,占地 44 807 平方千米。公园里栖息着北美数量最多的野牛,同时这里还是美洲鹤的天然巢穴。公园另一个引人入胜的景点是位于皮斯河和阿萨巴斯卡河之间世界上最大的内陆三角洲。这里随处可以看到广阔的大草原、森林、湿地、盐沼,以及不多见的内陆大三角洲。1983 年联合国教科文组织将伍德布法罗国家公园作为自然遗产,列入《世界遗产名录》。公园包括北部辽阔的森林及大草原,还有一些北美洲幸存的未受破坏的巨大草场和莎草牧场。这些水草肥美的地方维持着世界上最大的流动野牛群的生活,如众所周知的水野牛。

Canadian Wildlife Service 加拿大野生动物局　加拿大环境部下属的一个分支机构,主要负责保护各种野生动植物,如候鸟、濒危物种等。此外,也负责调查、研究国内重要的野生生物,杜绝重要物种的国际性贸易等,并掌管着国内 140 个国家野生动物保护区。

Parks Canada Agency 加拿大公园管理局　加拿大内务部的一个分支机构,成立于 1911 年,主要根据《加拿大国家公园法》负责保护加拿大境内的重要资源,如自然和文化遗产,维护生态环境,是世界上第一个国家公园保护机构。加拿大公园管理局管理着加拿大 36 个国家公园、7 个国家公园保护区、2 个国家海上保护区、1 个国家风景区和 157 个国家历史遗迹。

Lester Pearson 莱斯特·皮尔逊　加拿大第 18 任总理,曾经成功地化解了苏伊士运河危机,1957 年诺贝尔和平奖得主。在任职期间,皮尔逊政府取得了不少的业绩,例如,正式建立了全民医疗体系、引入学生贷款、全国养老金、创设加拿大勋章,以及成立皇家双语及双文化委员会等。皮尔逊政府还有一项值得称道的成绩,就是把枫叶旗正式确立为加拿大国旗。新国旗于 1965 年 2 月 15 日正式启用。

Translations of Selected Sentences

1. The Beaver Hills were considered too hilly, wet and generally less suitable for agriculture. For this reason, very little serious land clearing and agriculture was attempted. Most, but not all homesteads on the land which would later become Elk Island, were primarily filed for speculative purposes.

人们认为比佛山地区小山太多不平整,太潮湿,不是种植农产品的理想之地。由于这个原因,很少有土地被清理干净用于农业耕作。大多数的土地(不是所有的农地)曾经主要用于投机活动,后来则成为麋鹿岛。

2. In 1930, the Park was renamed Elk Island National Park as declared under the new National Parks Act and was obligated to preserve the natural environment for the enjoyment and use by thepeople of Canada and for future generations.

1930 年《国家公园法》将这个公园重新命名为麋鹿岛国家公园。为加拿大现在和未来的公民提供休闲娱乐和使用,此公园有义务保护自然环境。

3. The elk management program in Elk Island involves intensive monitoring to determine the number of elk that are to be surplused each year. Theseelk are then trapped and handled during the winter months for relocation. Close co-operation between Elk Island National Park, the Wild Elk Federation (Canada), the Rocky Mountain Elk Foundation (USA), the Canadian Food Inspection Agency, and several other agencies is necessary to relocate these animals to the wild.

麋鹿岛公园内的麋鹿管理方案包括对麋鹿数量的集中监察，以此来决定每年麋鹿被判"多余"的数量。这些多余的麋鹿然后被安置在某个地方，在随后的冬季被遣送至其他区域。为将这些麋鹿放生，麋鹿岛国家公园、加拿大联邦野生麋鹿管理部、洛矶山麋鹿基金会、加拿大食品检验局和其他一些机构的紧密合作非常必要。

4. The Park has become renowned for its conservation role in repopulating areas where elk populations used to exist or in supplementing existing herds that have been drastically reduced and are too small to increase on their own. Over the last two decades, an average of 160 elk each handling year has been live-trapped and transplanted for this purpose.

这个公园因以下两个举动而闻名，一是保护麋鹿再生之地（麋鹿曾经在此生活），二是增加麋鹿数量（因该地区麋鹿数量急剧下降，在自然情况下无法繁殖）。在过去的 20 多年中，平均每一个麋鹿数量调整年就有 160 只麋鹿被重新安置。

5. The year 2006 marked the 100th anniversary for Elk Island as the first wildlife preserve in Canada. The Park has evolved in many ways but the vision remains the same. Today it continues to be a place for nature and for people. A hundred years have come and gone, and with that vision, an incredible legacy began.

2006 年标志着麋鹿岛国家公园成为野生生物保护地已有 100 年之久。这个公园在很多方面发生了转变，但是其远景不变。如今，它继续是一个服务于自然和人类的地方。100 年已过去，带着这个远景，它将书写新的传奇。

Questions

1. Who did contributions to the establishment of the Elk Island National Park?

2. What program was initiated in Elk Island National Park in 1941? And what is its purpose?

3. How does the park deal with all of the Park's surplus elk?

4. In Elk Island National Park, which buildings are protected by Federal Heritage Buildings Review Office, and are maintained and managed in accordance with FHBRO direction?

5. What are the three trails available in the park to explore nature for visitors?

Chapter 6

Fundy National Park of Canada
芬迪国家公园

【简介】

芬迪国家公园是加拿大国家公园系统中的一座，坐落在新不伦瑞克的芬迪湾，靠近阿尔玛村。芬迪国家公园呈现了芬迪湾的海岸栖息地及高地阿卡迪亚森林景观。面积206平方千米，长满森林的小山和河谷遍布公园，海湾险峻的峭壁高达61米。从卡其布黄色、红色的砂条岩，到灰绿夹杂黑色熔岩流的花岗石，间或有河谷穿过这峭壁围墙留下大而深的凹痕，汹涌抢攻的河水形成瀑布、湖泊，绿荫地上野花丛生，有野生动物出没，海岸外芬迪湾以浪高流急的海潮而闻名。由于沿岸的独特地形，潮水可上涨21米，是世界上潮位最高、潮差最大的海湾，是观赏世界大潮汐的最理想处。退潮后，海滩上留下大量海螺、海贝，颇吸引游人。园内除观赏大潮汐外，还可游泳、海浴、垂钓、泛舟，游览森林和野生动物保护区。

Parks Canada has identified 39 natural regions in Canada. Fundy National Park represents the Maritime Acadian Highlands region, encompasses some of the last remaining wilderness in southern New Brunswick. Here, the conifer dominated Caledonia Highlands roll down to meet the fog-generating Bay of Fundy. The tidal fluctuation of the Bay of Fundy is the highest in the world. Watch fishing boats come and go to the rhythm of the bay. Inland, explore lush forests and deep stream valleys.

The park has two major environmental systems: the marine coastal environment of the Bay of Fundy and the Caledonia Highlands plateau (part of the Appalachian Mountain range) with its deeply cut river valleys. These are divided into several ecosystems: marine coastal system, Caledonia Highlands plateau systems.

The intertidal zone—the part of the beach that lays between high and low tide—is brimming (充满) with life. The plants and animals here have to be hardy. They live in salt water at high tide. At low tide they live in the fresh water of rain or runoff, or they get completely dried out and baked in the sun. Temperatures can rise to over 30 degrees Celsius (摄氏) in the summer, or drop to minus 25 degrees Celsius in winter. Waves crash upon them, battering them and covering them with silt. Winter ice-cakes crush them. Yet life thrives here. In fact, an organism's tolerance to exposure dictates where species live in the intertidal zone.

On rocky shores brown seaweeds and barnacles find anchorage. Some of the animals living between boulders and cobbles are periwinkles, side swimmers, gumdrop sea-slugs, dog whelks, limpets, butterfish and crabs.

Fundy is renowned for its vast tidal flats. The difference between the view of Alma Beach at high and low tide is dramatic. The Alma Beach tidal flat may extend more than one kilometre on a very low tide. Peculiar animals live in this thick mud substrate, especially mud shrimps, clams and a multitude of worms. The spectacular colours of the clam worm change from pink to brown to green iridescence. The bamboo worm lives in a vertical mud tube. These hidden organisms constitute a succulent (多汁的) feast for sandpipers and plovers in late summer migration through the Bay of Fundy.

Above the normal high tide, but inundated (淹没) regularly on very high tides, are salt marshes. The salt marsh at the mouth of the Upper Salmon River and those on the Point Wolfe and Goose Rivers are kept fertile and lush by the giant tides.

Ducks, geese, herons and other birds feed or nest near these marshes, or use them as migration stops. Fish (mummichog and threespine stickleback) forage and grow in the briny (海水) creeks, and smaller organisms prosper in the rich mud and marsh waters, where nutrients from the land and sea come together to produce excellent growing conditions. Water flowing out of the marshes carries plant and animal matter into the bay. This is a rich nutrition source for animals in the upper bay. The salt water cordgrass dominates this brackish ecosystem. This vigorous tough grass forms dense stands close to the water and does not mind being inundated occasionally. It has air passages in the stem that permit the transportation of oxygen to the often submerged roots, while glands excrete (分泌) excess salt. The roots and rhizomes (根茎) help prevent erosion.

In most ecosystems, the food chain begins with plants. For example, at the mouth of the Bay of Fundy, phytoplankton (浮游植物) grows in nutrient-rich upwellings of water. This is the beginning of the food web for zooplankton, shrimps, fishes and whales. In contrast, at this end of the Bay of Fundy, the water is too turbid and murky for the growth of phytoplankton. Here the food chain begins with bacteria constantly stirred up by tides and fed by detritus from the salt marsh.

Seaside cliffs are always impressive. This specialized habitat is often noticed for its striking beauty, but overlooked as a place for plants and animals. Fossils of 300 million-year-old plants lie embedded in giant blocks of sandstone, indicating the past lifetimes of Fundy. Even the ripple marks of ancient beaches are preserved in this rock. Today the endangered peregrine falcon（游隼）nests on these cliffs. Extreme slope, salt spray and wind support only sparse and specialized vegetation, such as roseroot（蔷薇根景天）. The rarest plants in the park, which grew in the cold of retreating glaciers, like to grow on these steep cliffs along the Bay of Fundy.

The Caledonia Highlands quickly rise 300 metres from the nearby coast. Fundy National Park is in the transition zone between the strictly coniferous boreal forest to the north and deciduous-dominated forest to the south.

Generally the park's forest is a mix of red spruce, balsam fir, yellow birch, white birch and maples. It is known as the Acadian Forest. While maples contribute the most brilliant autumn colours from mid-September to mid-October, balsam fir（冷杉）gives a wonderful fragrance to the forests of Fundy. The forest floor is covered with carpets of moss, woodfern and bunchberry. By mid-late summer, the bright red clumps of the bunchberry fruit are very common.

Representing the southern element of the Acadian Forest and covering only 5.4% of the park are pure hardwood stands. Although common as part of the mixed forest that blankets the park, hardwood species grow in pure stands only on well-drained rocky hillsides with southern exposure. Yellow birch, the largest of Fundy's hardwoods, and white birch are the most abundant. Other species include sugar maple, red maple and beech. A unique group of plants, called spring ephemerals（短命）, grows in the hardwood forest. They bloom before the trees grow their leaves in late May. In a few short weeks, species such as spring beauty and trout lily have bloomed and disappeared underground, waiting for next year's spring rush.

The coniferous forest represents the boreal element of the Acadian Forest. With the rarity of fire along the Bay of Fundy coast, spruce budworm has been the most important agent of change in the spruce-fir forest and is responsible for maintaining the forest in a state of dynamic equilibrium. Pure stands of conifer are rare in Fundy National Park, but it is there that one can appreciate the majesty of the tall red spruce, which can live 200 years or more. The park harbours some of the last pure stands of red spruce in eastern North America.

Fundy National Park protects a wide variety of plants. There are 658 species of vascular（血管）plants here. These include ferns and clubmosses, which reproduce by spores. Flowering plants such as the rare bird's eye Primrose—found only in Fundy National Park in New

Brunswick—is a cliff-dwelling plant remnant of a colder climate.

Less known are the bryophytes（苔藓植物）represented by mosses and liverworts. These small plants are often found growing on rocks and tree trunks. Many also grow on the forest floor. There are 276 species of bryophytes in Fundy National Park.

More than 400 species of lichens have been found in Fundy National Park. These small plants, which few people know by name, cover the forest floor, rocks and tree trunks. One may find as many as 30 species on a single tree! Lichen do not harm the trees they grow on, they simply use them to avoid competition from other taller plants on the ground.

In wet, poorly drained areas such as along the Caribou Plain trail, black spruce and larch live in a thick sphagnum moss carpet. Some of these stunted（发育不良的）trees have lived here for more than a century, yet are no taller than a person. A lack of nutrients in naturally acidic soil is a major constraint to their growth. Some bog plants have found ways to supplement（补充）their meagre diets by catching and consuming insects. Three species of insectivorous plants may be seen at Caribou Plain bog; they are the pitcher plant, sundew and bladderwort.

The shallow waters of ponds and streams surrounding the bog support much animal life. At dusk or dawn, moose and beaver become active, feeding on succulent yellow pond lilies. Although the bog pond is too acidic for most frog and fish eggs to hatch, many adult frogs can be seen in and near the pond.

Fundy is well positioned on the Atlantic migration route, and over 260 bird species have been identified in the park or on the adjacent bay. Of those, approximately 95 species have nested in the park. A common resident of Fundy's forests is the ruffed grouse. Early spring resonates with the deep thumping sounds of the male grouse's drumming.

Spring and early summer nights vibrate（颤动，摇晃）with the mating calls of frogs and toads. The earliest performers are the wood frogs and the tiny spring peepers whose forceful whistles echo in shrill massed chorus across the dark ponds. These sounds are soon augmented（增强）by the long trills of American toads. Marshy ponds and lakes are the home of leopard and pickerel frogs whose roars and growls mix with the "plunks" of green frogs, and later in the summer the bass droning roar of bullfrogs.

Salamanders（蝾螈）, the other amphibians in the park are mute. Seven species live in moist places: the yellow-spotted, red-backed and four-toed in mossy glades and inside rotten logs; the dusky, blue-spotted and two-lined under rocks near springs and brooks, and adult newts in ponds and lakes.

Of these seven species, three are considered rare. Four-toed salamanders have not been found anywhere else in New Brunswick, and Fundy is the only Canadian national park known to contain northern dusky salamanders. The third rare salamander is the blue-spotted salamander.

Opportunities exist in Fundy National Park to observe mammals in their natural habitat, especially along trails and streams where openings are common. Of the 38 species in the park, those most commonly seen include snowshoe hare, chipmunk, red squirrel, little brown bat, eastern coyote, white-tailed deer and moose. Moose are the largest animals in the park and may weigh 1000 kilograms. They feed primarily in lakes and wetlands, but may browse along roadsides at dusk. Caution signs along the highway are meant to be heeded!

Fundy National Park lies between the Saint John River system and the Petitcodiac River system. Easy inland portage routes used by native peoples joined these waterways. It appears that the exposed coastal paddle was not a favourite route of travel. Humans left little evidence of their passing in what is present-day Fundy National Park until European settlement began in this area around 1825.

This is a late date for settlement considering that, nearby, Acadians had arrived in the 1600s, and Mi'kmaq and Maliseet long before that. The rugged coastline and quickly rising highlands were not particularly inviting. Rocky, acidic soils combined with cool summers and harsh winters were not conducive to prosperous homesteading. It was a good place for trees, however.

The unexploited wilderness soon became attractive as a source of lumber for the rapidly growing city of Saint John, eighty kilometres to the west, and for export to England, the Caribbean Islands and New England. Today you can still find pieces of brick on Point Wolfe beach from the 20 metre tall smokestack that once stood at the Point Wolfe sawmill.

Several small communities grew within the existing park boundaries, populated by immigrants from Scotland, Ireland and England. The more desirable coastal land was the first to be settled. Many immigrants received land grants on the highlands where conditions were not good for farming, and many abandoned their land within a generation. The coastal settlements fared better. Small sawmills, shipbuilding and fishing operations thrived during the 1800s. Point Wolfe village, with its sawmill at the mouth of the Point Wolfe River, became the main settlement in the area.

Jobs were seasonal in nature. Most men worked in winter lumber camps and on spring log drives as primary sources of income. Summer and autumn work focused on their farms. Women remained on the homestead year-round, left to manage the complexities of the household, farm and child rearing.

As with any resource-based industry, times change. The easily accessible trees had been cut by the late 1800s. Sawmill refuse had clogged rivers and nearby coastal areas, harming fish populations and preventing Atlantic salmon from entering rivers to spawn. The fishery declined. Wooden boats and sail gave way to steel and steam. By 1922 the large mills, river drives and shipbuilding were no more. People emigrated, seeking better opportunities in cities elsewhere in Canada and the United States. A period of decreasing population and declining economic activity followed. Years later, in 1948, this area was chosen as New Brunswick's first national park, both to stimulate the economy and to preserve the area's natural beauty.

Fundy National Park was proclaimed in the Canada Gazette on April 10, 1948. The minister responsible, the Honourable James Allison Glen, headed the Department of Mines & Resources from April 18, 1945 to June 10, 1948 under William Lyon Mackenzie King.

Be rich in natural and cultural heritage, Fundy National Park is a place for learning and entertaining. The tides here are especially spectacular. Twice daily, the most dramatic tides in the world cover the ocean bed with up to 38 feet of water—the equivalent of a four-story building. Over the next six hours, the water sluices back to reveal the sea bottom: mucky, covered with seaweed, and bursting with intertidal wildlife.

In addition to the tides, the park offers a nine-hole golf course designed byStanley Thompson, a heated saltwater swimming pool just off Point Wolfe Road, summer hiking, camping, and shore exploration, as well as winter cross-country skiing, hiking, and snowshoeing. It's also liberally peppered with iconic Canadian imagery for photographers lucky enough to catch sunlight instead of frequent morning fogs in summer.

Notes

New Brunswick 新不伦瑞克省　加拿大三大沿海省份之一。北临沙勒尔湾(Chaleurs Bay),东以圣劳伦斯湾(Gulf of St. Lawrence)和诺森伯兰海峡(Northumberland Strait)为界,南边一直到芬迪湾(the Bay of Fundy)。该省的西部与美国的缅因州接壤,西北部则和魁北克省毗邻。总面积 73 440 平方千米,大致呈长方形,东西相距 242 千米,南北相距 322 千米。首府为弗雷德里克顿,最大的城市为圣约翰(Saint John)。人口约 75 万(2004),是加拿大以英法双语同为官方语言的省份之一,也是除魁北克外的最大法语区。新不伦瑞克的天然资源丰富,为强大的纸浆和造纸业提供了有利的保障。森林产品和食品加工是该省的主要工业。新不伦瑞克是加拿大绿色的沿海省,省内拥有 85% 的常绿森林,乡村田园风光与美丽的海岸线相互交映,青草、碧海与蓝天所构成的独特自然风景使该省成为旅游爱好者以及露营者的天堂。

Bay of Fundy 芬迪湾　大西洋西岸主要海湾之一,位于加拿大新不伦瑞克省和新斯科舍省之间,三面为陆地包围,仅西南与缅因湾连通,两者以大马南岛为界。芬迪湾以世上最为大型潮汐而闻名。湾内海底沉积物以砾石最为广泛，近岸狭长地带沉积了多种类型相互混合的沉积物，在开阔海岸的波蚀岩石台地上，覆盖着种类单调来源于当地的沉积物；小湾尽头和河口两侧，则有许多沼泽和由泥和粘土组成的潮滩。芬迪湾湾顶处还有两个狭窄的小湾：北面为齐尼克托湾，南面是米纳斯湾。通常在齐尼克托湾大潮差达14米，而在米纳斯湾可达16米以上，甚至曾观测到最大潮差为21米的记录。

Acadians 阿尔卡迪亚人　17世纪定居在加拿大阿尔卡迪亚地区殖民者的后裔。阿卡迪亚曾是法国的殖民地，范围覆盖北美洲的东北部，包括现魁北克东部、整个加拿大海洋省份和新英格兰，往南一直延伸到费城；而实际上法国政府指明的是与大西洋接壤，在大约纬度40度与46度之间的沿岸地带。其后，阿卡迪亚被划入英国殖民地的一部分，并发展成现在数个美国州和加拿大省份。

Mi'kmaq 米克马克人　加拿大东部沿海各省（新斯科舍省、新伯伦瑞克省、爱德华王子岛省）最大的印第安部落。米克马克人是季节性的游牧民族。冬季猎捕驯鹿、驼鹿及一些小动物；夏季捕鱼、拣拾贝壳动物并在沿海猎捕海豹。冬季住房为覆以桦皮或兽皮的圆锥形棚屋，夏季住所则式样不一，一般为长方形的露天棚屋。衣着为东部林地印第安人的典型式样：男人缠腰布，女人著紧身衣，均穿罩袍，缘饰甚多。男女善驾独木舟。米克马克人礼仪较为繁缛。

Maliseet 马勒塞特人　印第安部族，过去多居住在今日加拿大之新布仑兹维克及美国之缅因州东北部。

Saint John 圣约翰城　加拿大一个古老的工业城市，建市已有200多年，人口约14万，是加拿大东部重要通商港口城市，是纽宾士域省的贸易、工业和商业中心，以及加拿大生产糖浆的重要基地，拥有大规模石油、化工和造纸等企业。圣约翰临近芬迪湾，为大西洋省区一个重要的港口，冬季时圣约翰常有浓雾，故有"雾城"之称。天然资源丰富，土地为树林覆盖，纸浆与造纸业十分蓬勃，而树林底下蕴藏着大量铅、铜、铁等矿物，所以采矿是圣约翰主要的经济活动，另外，捕鱼业发达，盛产龙虾，每年夏天，龙虾又便宜又新鲜，除供应加拿大各省外，还有外销，运往欧洲各大城市。

Caribbean Islands 加勒比群岛　泛指西印度群岛，西印度群岛中安的列斯群岛东部和南部的岛群，包括维尔京群岛、背风群岛、向风群岛和委内瑞拉以北海面上的许多岛屿。面积1.29万平方千米。人口约300万（1981），以黑种人和黑白混血种人为主。大部分岛屿通行英语。东部岛群南北走向，分内、外两列岛弧，内弧为火山岛，外弧是石灰岩岛；气候湿热，降水丰沛，多飓风。南部岛群东西走向，系安第斯山北段支脉的延伸，多为低丘或山地；除特立尼达岛外降水较少，干湿季分明。经济以农业和旅游业为主。主要农作物为甘蔗、香蕉、海岛棉、葛薯、蔬菜等。有小规模家畜饲养业和渔业。工业较薄弱，主要有制糖、酿酒、轧棉、水果加工、石油提炼等产业。岛上多冬季疗养地。

James Allison Glen 詹姆斯·埃里森·格伦　加拿大议员，1940—1945年间加拿大众议院议长。格伦1877年出生于苏格兰兰顿,毕业于英国格拉斯哥大学法律系,1911年移居加拿大,定居于温尼伯。1926年初次进入加拿大众议院。1930年落选。1935年重新进入议会。1940年被选为众议院议长,主要解决当时的战争事宜。1945年格伦被任命为当时政府的内阁成员,担任矿产资源部部长一职。1948年因心脏病突发而辞职。随后被麦肯齐·金任命为加拿大国际联合委员会主席。1950年逝世。

William Lyon Mackenzie King 威廉·莱昂·麦肯齐·金　加拿大第11、13及15任总理,出生在安大略省柏林市,后改名为基切纳市(Kitchener)。父亲是一位律师,母亲则是加拿大起义者威廉·莱昂·麦肯齐的女儿。金在1895年和1897年先后获得多伦多大学学士和硕士学位。从芝加哥大学毕业后,他进入哈佛大学,取得经济学硕士(1898年)和哲学博士(1909年)。1908年应洛里埃之邀担任了洛里埃内阁的劳工部长。1911年自由党竞选失败后,金暂时离开政界,研究劳资关系。1919年洛里埃去世以后,金出任自由党领袖职务。他在1921—1926年、1926—1930年、1935—1948年三度担任加拿大总理,在位时间长达21年,是英联邦历史上在位时间最长的一位总理。1999年麦肯齐·金被历史学家评选为最伟大的加拿大总理;他的肖像现被印在加币50元纸钞上。

Stanley Thompson 斯坦利·汤普森　著名加拿大高尔夫球场设计师。汤普森出生于多伦多,曾就读于安大略农业学院,第一次世界大战期间参军。战后回国之后,汤普森成为全职的高尔夫球场设计师。20世纪20年代正是高尔夫流行年代,因此汤普森的知名度不断提高。其作品多集中于1912—1952年,作品中的设计理念为保护自然草原和土地。汤普森的经典作品包括费尔蒙特班夫泉城堡酒店高尔夫球场(Fairmont Banff Springs Golf Course),有"恶魔之炉"(Devil's Cauldron)之称的4号洞(标准杆3杆)则被认为是世界最佳球洞之一;另一项经典之作位于费尔蒙特贾斯珀公园酒店(The Fairmont Jasper Park Lodge),曾连续六年被"Score Golf"评选为加拿大最佳高尔夫度假村。

Translations of Selected Sentences

1. The park has two major environmental systems: the marine coastal environment of the Bay of Fundy and the Caledonia Highlands plateau (part of the Appalachian Mountain range) with its deeply cut river valleys. These are divided into several ecosystems: marine coastal system, Caledonia Highlands plateau systems.

这个公园有两个主要的环境系统:芬迪湾海洋海岸环境和加利多高地平原(阿巴拉契山脉的一部分),后者有着深邃的河谷。这两个大环境分为几个生态系统:海洋海岸生态系统和各加利多高地平原生态系统。

2. Fundy is renowned for its vast tidal flats. The difference between the view of Alma Beach at high and low tide is dramatic. The Alma Beach tidal flat may extend more than one kilometre

on a very low tide.

芬迪公园以巨大的潮滩出名。在阿尔玛海滩上,高潮和低潮的落差相当大。低潮时,阿尔玛海滩的潮坪可以长达近千米。

3. The Caledonia Highlands quickly rise 300 metres from the nearby coast. Fundy National Park is in the transition zone between the strictly coniferous boreal forest to the north and deciduous-dominated forest to the south.

加利多高地平原比旁边的海岸突然高出300米。芬迪国家公园处于北部完全意义上的松叶林和南部落叶林的过渡地段。

4. Fundy is well positioned on the Atlantic migration route, and over 260 bird species have been identified in the park or on the adjacent bay. Of those, approximately 95 species have nested in the park. A common resident of Fundy's forests is the ruffed grouse.

芬迪公园正处于大西洋迁徙路线之上,在公园内或附近的海湾地区发现的鸟类超过了260种。在这些鸟类中,大约95种在公园内筑巢。公园的森林内,常见栖息鸟为环毛松鸡。

5. Twice daily, the most dramatic tides in the world cover the ocean bed with up to 38 feet of water—the equivalent of a four-story building. Over the next six hours, the water sluices back to reveal the sea bottom: mucky, covered with seaweed, and bursting with intertidal wildlife.

一天两次,世界上潮差最大的潮水覆盖了这片海域,最高可达38英尺——相当于四层楼的高度。在随后的六个小时内,潮水退去,露出海岸底部的模样:满是粪肥,充满了海藻,里面的潮汐带生物活蹦乱跳。

Questions

1. What are the park's major environmental systems and its sub-ecosystems?

2. What is Fundy National Park renowned for? Why?

3. What is the difference of food chain at the end of the Bay of Funday and in other ecosystems?

4. What are the three rare Salamanders in Fundy National Park?

5. What does "jobs were seasonal in nature" (in the last but 5 paragraph) mean?

Chapter 7

Glacier National Park of Canada

冰川国家公园

【简介】

加拿大冰川国家公园原是布莱福特部族的印第安保留地的一部分，1910 年被辟为国家公园。因这里有 50 余条冰川，故得名冰川国家公园。公园占地 41 万公顷，被称作“北美大陆分水岭”的洛矶山脉从北到南贯穿公园中心。冰川国家公园是北美特有物种的大观园，有 1 000 多种树木花草。在比较干燥的东部山坡上，恩氏云杉、亚高山冷杉、小干松、花旗松和大枝松，苍劲挺拔，亭亭玉立。在气候较暖、比较湿润的西部山坡上，落叶松、冷杉、云杉茂密葱茏、郁郁苍苍。冰川公园里动物种类也很多，有大角山羊、美洲豹、山狗、山猎、驼鹿、美洲大角鹿、黑熊、灰熊、白尾鹿等。冰川退缩后留下许许多多美丽的山川、湖泊、瀑布等，风景从大到小变化万千，非常精致。冰川国家公园是一个奇特而美丽的自然风景区。

Glacier National Park of Canada protects, for all time, a portion of theColumbia Mountains Natural Region, in the interior wet belt of British Columbia. The steep, rugged mountains, warm, moist climate and wide variety of plant and animal life are typical of this natural region. The park protects unique stands of old-growth cedar and hemlock and critical habitat for threatened and endangered wildlife species such as the mountain caribou, mountain goat and grizzly bear. The Rogers Pass National Historic Site is located in Glacier National Park. Rogers Pass was so designated for its importance in the construction and development of the country's first major national transportation route.

Glacier National Park was established in 1886, simultaneously (同时) with Yoho National Park to the east. The Canadian Pacific Railway had just completed its transcontinental line, linking together the scattered（疏散的，零落的）settlements of the young nation of Canada. The spectacular scenery along the rail line suggested a potential for tourism to the railway company. Lodges and hotels were built to entice（诱惑）travellers to what had been uncharted wilderness only a few years before. The famous Rogers Pass is located in the center of Glacier National Park. The Pass was named for its discoverer, Major A. B. Rogers, Engineer-in-Chief for the railway. It has been designated a National Historic Site in commemoration of its role as an essential, yet perilous, link in the building of the transcontinental railway.

Very little is known about the use of the area by First Nations peoples before the arrival of the railway route finders in the 1860s and 1880s. Archaeological research has not yet discovered any occupation sites, although few such sites would have survived subsequent development for highway and railway operations. The long, snowy winters with their frequent deadly avalanches, the tangled and thick underbrush in summer, and poor hunting and fishing opportunities may have made most of the interior Selkirks, including what is now Glacier National Park, less attractive than the neighbouring valleys of the resource-rich Columbia River. However, at least four aboriginal groups are known to have used the region.

The nearby "Great Glacier" was a major attraction. Annual visitors from Philadelphia, the Vaux family, created a valuable photographic record of the progressive retreat of this glacier up-valley. Following the closure of Glacier House in 1925, facility development in the park was limited primarily to trail construction, until the early 1960s.

With the opening of the Rogers Pass section of the Trans-Canada Highway in 1962, additional facilities intended for the new automobile-oriented visitor were developed, including campgrounds, picnic areas, viewpoints and commercial services. Today's explorers will find heritage resources of national significance, both historical and natural. The abandoned right-of-way, stone bridges and trestle（支架）supports, equipment, snow sheds and other structures associated with the construction of the railway may all be visited on self-guiding trails.

Mount Revelstoke and Glacier National Parks of Canada together represent the Columbia Mountains Natural Region within Canada's system of national parks. The Columbia Mountain ranges form the first tall mountain barrier east of the Coast Mountains.

They are geologically and climatically distinct from the Canadian Rockies, found east of Glacier National Park. Glacier National Park lies within both the Selkirk and Purcell Ranges of the Columbia Mountains.

This region's weather originates over the Pacific Ocean. Wet, mild westerly air masses are intercepted by the wall of the Columbias, dumping heavy rains in summer, and heavy snowfall in the winter. High annual precipitation, deep snow accumulations and relatively moderate winter temperatures are characteristic of the Columbias and their two national parks, Mount Revelstoke and Glacier.

Glacier National Park contains three distinct life zones created by elevation. The three life zones can be thought of as "Rainforest, Snowforest and No Forest". The higher the elevation, the more often precipitation (降水) occurs as snow rather than rain. Deep, lingering snow and colder temperatures at high elevations shorten the growing season, until an elevation is reached beyond which no trees can grow.

At the lowest elevations in the park, the region's high annual precipitation falls mostly as rain. This is the interior rainforest: dense vegetation and large trees remind many visitors of the coastal Pacific Northwest. Western red cedar, western hemlock, and western white pine tower over shrubs such as devil's club, Pacific yew, and mountain box. A rich variety of ferns and mosses carpets the forest floor.

Foresters call this zone the Interior Cedar-Hemlock forest. The once-vast forests of British Columbia's interior wetbelt constitute (组成) the world's only inland temperate rainforest. One can experience this magnificent forest type along the Hemlock Grove Nature Trail. The Beaver Valley hiking trail leads through some of the most extensive contiguous old-growth forest left in southern British Columbia.

At the elevation of the Illecillewaet Campground, cedars thin out and more subalpine fir is seen in the forest. You are entering the interior subalpine life zone, known to foresters as the Engelmann Spruce/Subalpine fir zone. The lower subalpine forests are dominated by Englemann spruce, subalpine fir and mountain hemlock. Trees begin to thin in the upper subalpine near treeline, creating park-like, open meadows. The meadows explode into colour in August, with wildflowers including paintbrush, glacier lily, fireweed, lupine and monkeyflower.

At about 2000m elevation on average, the growing season becomes so short that trees can no longer survive. Fully 50% of the area of Glacier National Park is found above treeline. Much of this area is rock. About 12% of the park is permanent ice and snow. Tundra vegetation includes heather and sedge meadows, but in many places only lichen survives, encrusting (包裹) the rock. Areas above the treeline provide year-round or seasonal habitat for such wildlife as grizzly bear, mountain goat, white-tailed ptarmigan, hoary marmot, golden-mantled ground squirrel, mountain caribou and pika.

Specialized habitats found in Glacier National Park include avalanche paths and wetlands. Wetlands are very rare in the steep, narrow valleys of the Selkirk Mountains. The Beaver Valley contains many fens and wetlands along the Beaver River. Avalanche paths sweep from ridgetop to valley bottom. These areas green up early in spring and are important early season feeding areas for grizzly bears. Many park trails cross avalanche paths.

In order to ensurethe park remain healthy, and unimpaired（不受损害）for future generations, Parks Canada is working to maintain or restore the ecological health of this national park. Next are two of success stories:

Grizzly bears（熊）require large home ranges, have a relatively low reproductive rate and are easily displaced by human activity. Two of Glacier National Park's popular hiking trails, Balu Pass and Sir Donald, are located in prime grizzly bear habitat. It is recognized that there is an need to improve human use on these trails in order to minimize the impact on the grizzly and reduce the potential for bear/human encounters. Bear/human conflict data and information are gathered through grizzly bear monitoring. Different portions of the trails are marked so hikers and bears would be less likely to cross paths. In addition, interpretive signs are installed along the trails to inform visitors about the reasons for trail realignment. Early results indicate that grizzly bears continue to use habitat near the decommissioned（退役）trail segments, while hikers are using the new trail sections. This project will inform future trail mitigations and ensure that grizzly bears and people can safely share the alpine landscape in Glacier National Park.

Many bird species are in danger of disappearing throughout the world as more and more of them are found on rare and threatened species' lists. With theirsmall size, hours of activity and mobility, monitoring birds is no easy feat. Park staff and volunteers set up sound stations throughout Glacier national park. Microphones record the sound of the environment allowing researchers to record and identify birds at any time, in many locations. The recordings are collected and analysed and the results are music to the ears. This sound technology provides more accurate and complete data sets about where different species prefer to live, eat, mate, raise their young and how they respond to disturbances like noise from human activity. This technology to monitor these species are still to be used to determine(确定,决定) whether improvements made to their habitats could increase populations of threatened birds in Glacier national park.

Under considerable carefulness and great efforts, this park becomes more clear and attractive. At the same time, a variety of activities are provided for visitors, including ski touring, hiking, caving, mountaineering, wildlife viewing, backcountry, huts, fishing, and trail conditions report. Besides, one can also enjoy education programmes in this beautiful site.

There are a number of different ways to learn more about Glacier National Park. Educational and interpretive programs offered by national park staff are seasonally available. The Friends of Mount Revelstoke and Glacier offer courses every year to help members learn more about the history and ecology of the area and to teach recreational skills. The Columbia Mountains Institute offers professional development workshops and organizes conferences on a variety of subjects related to the North Columbia Mountains. Past conferences have discussed grizzly bear ecology, issues around caribou（驯鹿）survival in the Columbias, and the use of DNA in wildlife research, among many other topics.

Situated in the Columbia Mountains, Glacier National Park's 1,349 square kilometres of peaks, glaciers and forest straddle（跨越）the crest of the Selkirk Range, an area of legendary（传奇的）snowfall. Visitors to the park will find ski touring terrain including glades, alpine bowls and icefields where descents of more than 1,500 metres are possible. Of course, all ski destinations in the park require knowledge of travel in avalanche terrain. Skiers are urged to wear avalanche transceivers（收发器）and be prepared for self rescue.

In general, many wild lives can be seen here. Mid-May through early June is the best time to view bears in the parks. They are just out of hibernation and often can be seen from the highway. Scan the avalanche slopes in Rogers Pass for both grizzlies and blacks. When the dandelions bloom, black bears often consume a floral salad at roadside. During July and August bears are harder to spot because of the dense vegetation. Try an early morning drive along the Trans-Canada keeping a sharp lookout for black bears. By this time, grizzlies have usually moved up slope into the upper subalpine/alpine tundra. Goats may be seen all year above the highway snowsheds on the eastern side of Rogers Pass. May and June are the best months for bird watching.

As far as fishing is concerned, anational park fishing licence is required. However, the park's glacier-fed rivers and small lakes do not support many game fish. Non-native species such as Brook Trout and Rainbow Trout live in Mount Revelstoke's Eva and Jade Lakes. To protect Bull Trout, recently placed on the endangered species list, a fishing ban is in effect on all streams (not lakes) in both Mount Revelstoke and Glacier National Parks.

In this rugged section of the northern Rockies, exhilaration（兴奋）is the universal language.

Notes

Columbia Mountains Natural Region 哥伦比亚山脉自然景区　位于加拿大不列颠哥伦比亚省东南部，东界落矶山沟，南临哥伦比亚河，西接内陆高原，北滨弗雷泽河，与加拿大境内的落矶山脉平行。哥伦比亚山脉由四条显然分开的山脉卡里布（Cariboo）、莫纳希（Monashee）、塞尔扣克（Selkirk）和珀塞尔（Purcell）组成，海拔均超过3 000米。该山脉是包括国立和省立公园在内的主要游览区。山中有金、银、铜、铅和锌等重要矿物资源。

Yoho National Park 优鹤国家公园　坐落在加拿大洛矶山脉，沿大陆分水岭位于不列颠哥伦比亚省东南部的西坡，落矶山五大国家公园之一，以壮观、瑰丽、险峻的景观而闻名。

优鹤国家公园的三大景观是碧绿的翡翠湖、塔卡考瀑布和踢马河。翡翠湖因湖底的岩石粉经阳光照射而产生翠绿色彩,塔卡考瀑布以254米落差的壮阔水瀑景观引人入胜,奔腾咆哮的踢马河以水滴石穿之功洞穿巨石而形成令人叹为观止的天然桥。

Canadian Pacific Railway 加拿大太平洋铁路公司　苏格兰移民、蒙特利尔银行董事长乔治·史狄芬于1881年创建该铁路公司。100多年来,加太铁路担负着政府和公众的重托,是整个加拿大西部开拓和发展的主导性因素,又是对抗美国扩张势力,支持麦克唐纳国家政策的重要力量。它成功地用一条大锁链把东西加拿大紧紧地连在一起,实现了麦克唐纳国家政策的梦想,并置身于西加拿大发展的各项事业之中,为西部的开拓做出了重要的贡献。加太铁路公司的第一个工程是一条横贯加国、连接西部口岸的长达1 900英里的铁路。除经营铁路外,加太铁路公司的章程中还包括可以经营电报、电话、码头,以及任何“必要和有用”的业务。1981年加太铁路公司成立百年之时,该公司已发展为一个强大的联合体,业务包括所有主要的运输业、自然资源业和房地产业,年收入达126亿加元。

Rogers Pass 罗格斯通道　英属哥伦比亚塞尔扣克山脉的高山通道,是通往哥伦比亚河大本德地区的捷径。1881年5月加拿大太平洋铁路公司的调查员罗格斯少校发现了这条通道。它是冰川国家公园的中心所在,周围的群山适合滑雪、野营、远足等。现在,这一地区已成为旅游中心,设有罗格斯通道博物馆、国家公园服务中心和一个酒店。这一通道是加拿大历史遗址之一。

Major A. B. Rogers 罗格斯上校　因发现加拿大英属哥伦比亚的罗格斯通道而闻名。1834年罗格斯出生于美国马赛诸塞州,曾就读布朗大学,一年后转入耶鲁大学,获得工程学学位。印第安战争期间在美国骑兵部队服役,获得少校军衔。1881年加拿大太平洋铁路公司任命罗格斯在塞尔柯克和洛矶山脉之间发掘一条铁路路径。罗格斯对早期调查员的调查结果进行了仔细研读,终于发现了罗格斯通道。

Engelmann Spruce 恩氏云杉　北美西部原生云杉,生长于洛矶山脉高海拔山区。常与白云杉混合交织在一起,被称为内陆云杉。常规用途是锯材,制作胶合板和纸浆等。

Friends of Mount Revelstoke and Glacier 雷夫尔斯托克山和冰川保护协会　1986年由一批环保人士组建的非营利性环境保护协会,1987年正式获得政府承认。经过近20年的发展,现在成员多达200多人,遍布世界各地,其主要任务在于保护或保留自然环境,在国家公园内开展教育和环保娱乐项目等。该组织涵盖了一系列的活动,如环保教育、室外活动等。其经济来源主要是位于冰川国家公园罗格斯通道内冰川书店的营业收入。

ColumbiaMountains Institute 哥伦比亚山区研究所　雷夫尔斯托克地区的非营利性组织,致力于各项应用生物学的研究和实际应用,如资源保护、管理、调查研究、学术交流,以及其他关注英属哥伦比亚南部地区生态环境的事项。此研究所的亮点在于举行各种基于科学研究的学术会议、开办特殊环保教育课程、向成员或其他非成员等提供最新的研究进展情况。

Selkirk Range 赛尔扣克山脉　美国爱达荷州北部山脉、华盛顿州东部山脉和英属哥伦比亚东南部山脉组成的山脉，属于哥伦比亚山脉的一部分，长度大约为 320 千米。山脉西面、东北面和背面被哥伦比亚河围绕。

Translations of Selected Sentences

1. The Pass was named for its discoverer, Major A. B. Rogers, Engineer-in-Chief for the railway. It has been designated a National Historic Site in commemoration of its role as an essential, yet perilous, link in the building of the transcontinental railway.

这条路径以其发现者——铁路总工程师罗杰斯上校而命名。当时设立了一处国家岩石历史遗迹，以此纪念它作为建设横跨大陆的铁路时必要却又险要的通道。

2. This region's weather originates over the Pacific Ocean. Wet, mild westerly air masses are intercepted by the wall of the Columbias, dumping heavy rains in summer, and heavy snowfall in the winter. High annual precipitation, deep snow accumulations and relatively moderate winter temperatures are characteristic of the Columbias and their two national parks, Mount Revelstoke and Glacier.

这个地区的气候源于太平洋。潮湿、温和的西风因受哥伦比亚省的阻隔而不能汇聚，夏季降雨量大，冬季降雪量大。丰富的年降水量、常年的积雪和相对温和的冬季气候是哥伦比亚地区的气候特点，同时也是该地区瑞佛史托克山脉公园和冰川国家公园的气候特点。

3. The Beaver Valley contains many fens and wetlands along the Beaver River. Avalanche paths sweep from ridgetop to valley bottom. These areas green up early in spring and are important early season feeding areas for grizzly bears. Many park trails cross avalanche paths.

比佛谷内，比佛河沿岸有许多的沼泽地和湿地。雪崩发生的地方可从脊顶至河谷底部。春天，这些地方很早就披上绿装，对黑熊来说，这是它们早期的觅食之地。许多公园的路径穿过这些雪崩之地。

4. Bear/human conflict data and information are gathered through grizzly bear monitoring. Different portions of the trails are marked so hikers and bears would be less likely to cross paths. In addition, interpretive signs are installed along the trails to inform visitors about the reasons for trail realignment.

大灰熊控制系统收集了熊/人冲突的数据和信息。路径的不同区域都有标志，因此远足者和熊不易跨过禁区。此外，路边也安装了解释性的标志牌，告知游客路径重新安排的原因。

5. In general, many wild lives can be seen here. Mid-May through early June is the best time

to view bears in the parks. They are just out of hibernation and often can be seen from the highway. Scan the avalanche slopes in Rogers Pass for both grizzlies and blacks. When the dandelions bloom, black bears often consume a floral salad at roadside.

总的来说,这里可以见到许多野生生物。从五月中旬至六月上旬,是公园内看熊的最好时间。这些熊刚脱离冬眠期,人们可以在公路上看到它们。在罗杰斯隘口看雪崩山坡,同时也可以看灰熊和黑熊。蒲公英开放的时候,黑熊经常在路边享用这些植物沙拉。

Questions

1. What is the Rogers Pass designed for?

2. What are the characteristics of the weather in the Columbias where the park is in?

3. What are the distinct life zones in the park? And what are the causes of this feature?

4. What does Parks Canada do to maintain or restore the ecological health of the Glacial National park?

5. According to the information given by the last but 2 paragraph, if one wants to fish in the park, what does he/she have to pay attention to?

Chapter 8

Grasslands National Park of Canada
大草原国家公园

【简介】

成立于1981年的萨斯喀彻温省的草原国家公园是加拿大最新的公园之一。覆盖面积约为375平方英里，公园的高地是荒野和没有任何记号的；到这里的徒步旅行者可以享受到风景的自然美，也可以自由自在地到处漫游。这个自然公园内生活着大量的动物，如北美敏狐、穴鸮鸟、岩鸻、大艾草榛鸡、高山矢嘲鸫、伯劳鸟、游隼和斯普氏鹨等濒危物种，还有黑尾草原犬鼠、红头啄木鸟、铁质鹰、短耳鸮、大杓鹬、东部贲腹蛇和大短角蜥等稀有物种。当然，这里也生长着大量美丽的植物，如蓝色格兰马草、冠毛针茅草、刺梨仙人掌、秋葵月见草等。任何一种不常见的植物和低矮的草丛都展示了这个生态系统的存在，以及它对于维持地球上的生命的重要性。

Did you know that Grasslands National Park is the only national prairie park in Canada? The East Block Badlands are the richest resource for dinosaur fossils in Canada? Grasslands National Park is the only place in Canada where one can see the Black-tailed Prairie Dog and the Black-footed Ferrets in their native habitat? And, there are over 12,000 ancient teepee rings within the park? Yes, Grasslands National Park is the only national park in all of Canada to protect the mixed-grass prairie ecosystem. It is a land of rolling hills, rugged coulees, and steep ravines. The West Block centres on the Frenchman River Valley and the East Block features the Killdeer "Badlands" and the Wood Mountain "Uplands". Grasslands National Park of Canada is where sky meets prairie at the northern extension (延伸) of the great plain

With over 70 different species of grass and over 50 different species of wildflowers,

Grasslands has an important role in protecting the prairie ecosystem. We often hear about problems in the rainforest, but we don't realize that one of the most endangered ecosystems is right in our backyard—the native prairies! At least 80% of our native prairie has been lost. In our southwest corner of Saskatchewan, Grasslands National Park contains the most intact and greatest example of the remaining native prairie in Canada.

The grasses are the heart and soul of the prairies and without them, there wouldn't be a home for such unique animals. This is one place in Canada where you can see the Buffalo roam and the Deer and the Antelope play. Grasslands are home to the Black-footed Ferret—often considered North America's most endangered mammal. Also, the park is the only place in Canada where you can see Black-tailed Prairie Dogs in the native habitat!

Grasslands National Park also tells the amazing history of the prairie, a story of survival and persistence. Ancient teepee rings and bison drive lanes a just a few remnants (残余) of the First Nations people who once called the prairie wilderness home. Old corrals using river willow fenceposts and remnants of early ranching homesteads dot the landscape to remind us of the end of the 'wild west' and the transition (过渡) to settlement.

Thenatural history at Grasslands is a story going back millions of years—to the Dinosaurs. The thousands of fossils found in the park tell us that both land and aquatic (水生的) dinosaurs once roamed here. The prairie landscape has been a sea, a tropical jungle and has survived the glaciers. And all of these influence what the prairies look like today.

Native habitation dates back to 10,000 years ago. By the 1600's, theGros Ventre followed the bison herds in this area. More recently, the Assiniboine, Cree, Sioux, and Blackfoot also inhabited this grassland area on a seasonal basis. Campsites, tipi rings, vision quest sites, medicine wheels, and bison drive lanes are some of the cultural heritage.

By the 1880's, Euro-Canadian settlement had pushed farther west, bison herds were declining and so were other native species. Cattle herds replaced bison on the open range. Large ranches, like the 76 Ranch, held lease (租) to thousands of acres of grazing lands. The cowboys who worked these ranches were the cowboys of the old west, their lifestyles romanticized on the silver screen.

MacDonald's national policy and the Homestead Act had a noticeable impact on this area by 1908. The Act allowed for grants of crown lands to potential farmers. Farmers took up the homestead challenge and put up fences, which changed a part of ranching forever. Today, farming and ranching coexist to form a unique blend of economies and south-western hospitality.

Efforts to protect the grasslands of this area and its native species have been going on for over 50 years by various individuals and groups. During the 1950's and 60's prairie conservationists

（自然资源保护者）promoted the protection of a significant area of natural grasslands. In October 1963, members of the Saskatchewan Natural History Society passed a resolution urging the federal government to establish a national park in south-western Saskatchewan. A study of potential park areas was conducted in 1965 in southern Saskatchewan and southern Alberta, concluding that the Killdeer Badlands-Frenchman river area was the most suitable.

In 1975, a Memorandum-of-Intent setting out some of the terms and conditions for establishment of Grasslands National Park of Canada (GNPC) was signed by both governments. In accordance with the Memorandum, an independent Public Hearings Board was appointed in 1976 to determine the degree of support for the proposed Grasslands National Park. The Board reported that the Park proposal had the public's support and should proceed.

On June 19, 1981, Canada and Saskatchewan signed an agreement to establish the Park. Subsequently, Parks Canada purchased two ranches totaling 140 km^2 in the Frenchman river area. Acquisition of additional park lands ceased when the conditions in the agreement pertaining to oil and gas exploration and water resource management proved unworkable. Negotiations to resolve the impasse went on for five years.

In 1988, with help of a coalition of non-governmental conservation organizations, an agreement was reached between Canada and Saskatchewan to revise the 1981 accord（协议）and proceed with establishment of Grasslands National Park of Canada. The proposed park boundary encompasses（包围）approximately 900sq. km. in two blocks. Currently, Parks Canada has acquired over 55% of the land. The agreement allows Parks Canada to acquire land on a willing-seller, willing-buyer basis: allows Saskatchewan to uphold existing water use agreements and uphold international commitments: and, allows Parks Canada 30 years before Saskatchewan can terminate its obligations to the 1988 agreement. Grasslands National Park of Canada was officially proclaimed a national park under the new Canada National Parks Act on February 19, 2001.

The Saskatchewan Natural History Society (now Nature Saskatchewan) has played a significant role in advocacy of the park and in pro-active conservation measures. The Nature Conservancy of Canada has also assisted with establishment of the park. In 1991, Amoco Canada sold its title to mineral rights to Parks Canada.

Grasslands National Park of Canada has been established to preserve and present a representative portion of the Canadian mixed grass prairie ecosystem. It is part of a larger Canadian system of protected heritage areas. Grasslands National Park represents the Prairie Grasslands Natural Region of Canada. It is the only park of these 39 national parks that does so. Water is one of the primary shapers of prairie ecosystems. Low precipitation（降雨）in Grasslands National Park area influence many of the conditions that create prairies. Waterways such as the Frenchman River in the West Block and Rock Creek in the East Block add significantly to the biodiversity (生物多样性) and are important habitats. The unique combinations of landscape and climate create niches (生态位) for specially adapted plants, animals and endangered species.

Identify flora such as blue gramma grass, historically a favourite of the bison; needle-and-thread grass, whose sharp awns lend its name; prickly pear cactus, that breathes only at night; and gumbo evening primrose, whose flowers change colour in 24 hours. The non-domination of any single species of flora and combinations of short and medium sized grasses illustrate the character of this ecosystem, and its importance in sustaining life of this planet.

The dominant landscape found in the park is rolling uplands and the river valley. During the retreat of the last glaciation, over 10,000 years ago, the ice melted and moved in two divisions toward the area known as the Frenchman River Valley. The northern division was higher in elevation than the southern division; consequently, the meltwater flowed south. The flow of water from the north was stopped by the southern lobe. Where the water hit the ice and ran along its margin (边缘) an "ice stream" was cut into the land and became the present Frenchman River Valley. The huge amount of runoff changed the landscape, creating coulees (熔岩流), buttes and creeks. The Frenchman River Valley flows south and is part of the Missouri watershed.

The plants found in the coulees, on buttes and along creeks are indicative of the amount of moisture and type of soil present. The natural grasslands of south-western Saskatchewan are called mixed-grass prairie. In the park area the dominant plant forms are grasses. Common grass species are spear grass, wheat grass, and blue gramma grass. Grasses are only one element that makes up the natural cover of grasslands. Trees and shrubs such as aspen, green ash, wolf willow and buffalo berry take hold on the valley floors and in the coulees where there is more moisture. In drier locations, sage, rabbit brush, greasewood, mosses, lichens and cacti make up a significant part of the plant community. There is a succession of colours and aromas in the grasslands as the wildflowers bloom. Crocus, prairie onion, cinquefoil, rose, vetch, locoweed, violets, asters, fleabanes, goosefoot, and buttercups are just a few of the many wildflowers that are found in the area.

A fairly complete range of prairie fauna is still present, although several species have been extirpated (灭绝) from the area. Whether on purpose or through incidence, black-footed ferret, greater prairie chicken, bison, plains grizzly bear, prairie wolf and wolverines are species that have been extirpated. Endangered species in the area include swift fox, burrowing owl, mountain plover, greater sage grouse, and sage thrasher. The loggerhead shrike, peregrine falcon and Sprague's pipit are classified as a threatened species. Species of special concern are black-tailed prairie dog, red-headed woodpecker, ferruginous hawk, short-eared owl, long billed curlew, eastern yellow-bellied racer, and greater short-horned lizard. Common mammals in and near the proposed park are pronghorn antelope, mule deer, white-tailed deer, beaver, coyotes, red fox, skunks, porcupines, badgers, and bobcats. Birds include golden eagles, hawks, owls, grouse, songbirds, ducks and geese. For more specifics, see our species lists.

Sources of water in the Park area are limited. Surface waters are derived mainly from spring runoff and vary seasonally. Many creeks flow in the spring and early summer but dry up during the hot, windy summer. The usefulness of surface water sources is greatly limited because much of the substrata consists of marine shales or soft clays with high concentrations of salts or mud in suspension. Potable water is not available within the park.

Catfish and carp are two of the most common fish species that inhabit the Frenchman River, Rock Creek and a few tributaries that maintain water for a part of the year. Painted turtles occur in some numbers along the fresh and standing waters of the Park. Leopard frogs and chorus frogs can also be found.

Many people find the prairies a difficult landscape to appreciate. There are many who find it a challenge to see the beauty in what is apparent emptiness. To appreciate the prairies demands patience, a different eye to the world, an open heart and a keen sense of space.

Other environments seem easier to appreciate. The overwhelming drama of Rocky Mountain vistas, the size and epic grandeur rivets you to yourseat; it grabs you and says "pay attention!" Who would disagree with a mountain? The seashore mesmerizes (入迷) with rhythm of surf and wind; it induces a nature trance. The northern forests present mystery. The silence, the warmth of darkness and the subdued green invites you, which draws you like a fly drawn to a pitcher plant. The prairies present a challenge. A challenge to the eye, to the imagination and to the soul.

Space, the final frontier... certainly there are two major themes underlying the prairielandscape: subtlety and space. The space is overwhelming; huge!

Wallace Stegner, when writing of the landscape, said here was: "a distance without limits, a horizon that did not bound the world but only suggested endless space beyond". To the explorers who came, the land offered nothing beyond furs. It was a physical and an aesthetic desert. Its landscape was overwhelming in its emptiness, "... its awesome, inhuman scale" said Ronald Rees, and it was lonely. The land and sky offered nothing to comfort the traveler, especially the European, so accustomed to streams, trees, forests and hills. The prairie was, to William Butler, "The Great Lone Land". Alexander Sutherland, speaking of the prairie isolation wrote "... as we go floating by, we are completely isolated as if we were out on the sea".

In recent years, we have come to approach the prairies not as pioneers, industrialists, settlers, trappers, explorers or Europeans but solely as people; prairie people. We have come to see the land, not in terms of its immediate use, but for what it is, "the least common denominators (共同特性) of nature, the skeleton requirements simply, of land and sky—Saskatchewan prairie", wrote W. O. Mitchell. This bare bones approach to the land was extended to the individual and his or her unique connection to the land—it shaped identity. In the broad circle of prairie sky the individual was, "the question mark in the circle" to Wallace Stegner.

One man, Captain Butler, was able to view the new land for what it was: something clean, crisp (清新的), new and immense, "a view so vast that endless space seems for once to find embodiment, and at a single glance the eye is satiated with immensity... [the land is] reduced to its own nakedness, space stands forth with almost terrible grandeur". A land such as this must instill (逐步灌输) certain traits and feelings within prairie people and visitors. Wallace Stegner was well aware of this: "It is a country to breed mystical people, egocentric people, perhaps poetic people, but not humble ones. At noon, the total sun pours on your single head; at sunrise or sunset, you throw a shadow a hundred yards long. It was not prairie dwellers who invented the indifferent universe or impotent man. Puny you may feel there, and vulnerable, but not unnoticed".

Subtletyis the element of the prairies that makes the greatest demands on one's patience; but patience is a key to unlocking some of the most glorious secrets of the prairie. Settlers, travelers and explorers, when first confronted with the prairie, felt as though they were standing on the shore of the ocean. From their feet to the horizon stretched an endless sweep of grass that shifted and danced like waves in the wind. Here is subtlety. The wind paints a canvas of changing colors, not mountainous black and white contrasts of glacier and peak, but infinite shades of green, brown, and blue that dance against a backdrop of a sky alive with color. "A land of wide open spaces, of suave flowing lines, of harmonies of color," noted C. W. Jeffries. Hidden in the grasses, a gift to the patient and curious, are multitudes of flowers, delicate and shy. Here also is the sound of birds, infinite and invisible; making the world increasingly abstract (抽象的) and infinitely beautiful.

"Desolate? Forbidding? There never was a country that in its good moments was more beautiful ... You don't get out of the wind, but learn to lean and squint (斜视) against it. You don't escape sky and sun, but wear them in your eyeballs and on your back. You become acutely (强烈地) aware of yourself. The world is very large, the sky even larger and you are very small..." wrote Wallace Stegner.

Take the time to explore; the wind and sky will ensure that no one place is ever the same, the birds will be your company, the wail of the coyote will be your mystery, the stars will be your guardian and the space will be your freedom.

Notes

Black-tailed Prairie Dog 黑尾草原犬鼠　又称黑尾土拨鼠,一种小型穴栖性啮齿目动物,原产于北美洲大草原,当地人称之为“草原犬”。如果算上短尾巴,土拨鼠身长平均约为30－40厘米。身体长大而肥壮,头部短而阔。颈部粗短,耳朵短小。体毛主要为灰褐色。腹部为灰黄色。尾巴末端为黑色。土拨鼠栖息在美国、加拿大和墨西哥。家族群生活。它们的社会组织性很明显,既有抵抗捕食兽的功用,还能够调整本种的种群密度。它们的社会基本单位是由雄性1～2只、雌性2只以上,再加大批幼仔组成。每一群体各据家园,不让邻居侵犯,以维持食物供应和种群数量之间的平衡。

Black-footed Ferrets 黑足雪貂　唯一原产于北美的雪貂。成年黑足雪貂平均体长0.5米,尾长0.15米,体重约1公斤。毛色通常呈浅黄色,头顶、眼圈、腿和尾尖呈深褐色或黑色,具有夜行性和穴居的生活习性。分布在加拿大南部,沿洛矶山脉向东到美国的奥克拉荷马、堪萨斯州和内布拉斯加广大地区。黑足鼬是北美著名的濒危物种。1937年加拿大野生黑足鼬灭绝。1967年在美国被列为濒危物种。2008年美国网站《生活科学》评出黑足鼬为全球十大最濒危的稀有动物物种之一。

Saskatchewan 萨斯喀彻温省　简称萨省,位于加拿大中心地带,东西与曼尼托巴省和阿尔伯塔省为邻,南部与美国的蒙大拿州和北达科他州接壤。“萨斯喀彻温”的名称被认为是由早期印第安土著民提出的,意思指遍布境内的湍急河流。被誉为加拿大的“产粮之篮”,以牧场和麦田而闻名。萨省面积65.19万平方千米,占加拿大面积的6.5%,其中森林面积约35.2万平方千米,淡水面积约8.16万平方千米。人口2002年约102万,60%以上的人口居住在城市。萨省居民具有多种族背景。无边的土地和开阔的天空使得这个省份以美丽的日落而闻名。

Assiniboine 阿西尼博因人　美洲平原印第安民族,操苏族(Sioux)语系语言。17世纪前,从扬克托奈·达科他人中间分裂出来。鼎盛时期住在加拿大温尼伯湖以西,沿阿西尼本河与萨斯喀彻温河一带地区。阿西尼博因人是杰出的水牛猎手,常在哈得逊湾及密西西比河上游同欧洲商人进行贸易,用水牛肉干换取火器及其他商品。后由于英国人和法国人的入侵,不断向西移动,进入加拿大平原、蒙大拿平原及北达科他平原,与一直控制着大草原的黑脚人相互对抗。阿西尼博因人分成若干群组,各有首领和议会。各群组经常移动,以追踪流动的水牛群。18世纪引进马匹之前,专靠步行,使用狗拖撬。妇女担任与住房(圆锥形帐篷)有关的全部劳动,包括搭拆帐篷。战争中的骁勇表现为取敌头皮,夺取马匹,与敌人短兵相接。最有重要意义的宗教礼仪是太阳舞。

Blackfoot 黑脚人　居住在加拿大亚伯达省和美国的蒙大拿州，是阿尔冈昆人中从林地西迁大草原的第一批人。18 世纪初期，徒步猎取水牛，住在落矶山东约 644 千米的萨斯喀彻温(Saskatchewan)谷地。1750 年前获得马匹和火器，将一些弱小部落驱走，向西推进到落矶山脉，向南则深入到现在的蒙大拿州。19 世纪初势力达到鼎盛，疆土辽阔，北起萨斯喀彻温，南达密苏里河源头。黑脚人是西北平原上力量最强大、最富于军事侵略精神的一支。每一部落都分为若干狩猎群组，由一名或多名首领领导。这些群组在可藏身的河谷分散居住，度过冬天；夏天则聚集在一个大型宿营地点，举行太阳舞。太阳舞是主要的部落宗教仪式。20 世纪末，有 6 000 多人居住在蒙大拿州黑脚印第安人保留地，另有 2600 人居住在亚伯达的黑脚人、布卢德人及皮埃甘人等保留地。

MacDonald 麦克唐纳　加拿大政治家，政府总理(1867—1873，1878—1891)。1815 年 1 月 11 日生于英国的格拉斯哥。1820 年由苏格兰移居金斯顿。1836 年取得律师资格。1844 年作为保守党成员被选入联合的加拿大省议会。1854 年在美国圣路易斯的麦克唐纳-道格拉斯公司的飞机工厂领导并建立自由保守党。1857 年担任加拿大省总理。1864 年领导自由保守党与自由党达成协议，共同组成加拿大省联合政府。在促进实现《英属北美法》、创建联邦的过程中，麦克唐纳在夏洛特敦会议上起了重要作用。他是魁北克会议的主要组织者和 1866 年伦敦会议的主席。1867 年 7 月加拿大自治领建立后，他担任第一任总理，被封为爵士。麦克唐纳执政的前 6 年实现了自治领的扩张。1870 年自治领获得原属英国哈得孙公司在西部、西北部的大片领土，建立了马尼托巴省。1871 年和 1873 年不列颠哥伦比亚省、爱德华岛相继加入联邦。为巩固自治领，发展国内经济，麦克唐纳提出了三项国家政策：大力组织向西部移民，开发西部；修建横贯东西的加拿大太平洋铁路；实行保护关税。这三项政策为加拿大早期经济的繁荣奠定了基础。在对外政策方面，他对美国保持独立，主张加强与英国的联系，支持英帝国的统一外交并竭力争取加拿大在统一外交中的平等地位。

Saskatchewan Natural History Society 萨斯喀彻温省自然历史协会　加拿大自然资源保护局之下的一个非盈利性地方环境保护组织，成立于 1949 年，意在宣传自然环境的审美作用和保护自然环境和资源等自然生态系统的重要性。该协会提供了一系列的服务，如雨林保护、针叶林保护等。协会的目标是实现人与自然的和谐相处。

Alexander Sutherland 亚历山大·苏瑟兰　知名教育家、作者和哲学家。1852 年出生于苏格兰。1864 年随父母移居澳大利亚。1874 年毕业于墨尔本大学。随后两年执教于墨尔本的苏格兰学院，1877 年创建卡尔顿学院。1892 年退休，开始全身心致力于文学。但 1893 年的经济危机使苏瑟兰不得不为《阿古斯》报撰写文稿，以获得一定的经济收入。1890 年出版《三十首短诗》(Thirty Short Poems)。1898 年出版了其最重要的著作《道德本能的起源和生长》(The Origin and Growth of the Moral Instinct)，共两卷，获得了英国哲学界的好评。随后，与乔治·萨瑟兰合写了《澳大利亚简史》。此外，萨瑟兰也曾参加政治活动。1902 年去世。

Translations of Selected Words

1. We often hear about problems in the rainforest, but we don't realize that one of the most endangered ecosystems is right in our backyard—the native prairies! At least 80% of our native prairie has been lost.

我们经常听到雨林的问题,但我们没有意识到最濒临危险的生态系统之一就在我们身后——本地的草原系统!至少我们地区内80%的草原已经消失。

2. In accordance with the Memorandum, an independent Public Hearings Board was appointed in 1976 to determine the degree of support for the proposed Grasslands National Park. The Board reported that the Park proposal had the public's support and should proceed.

与这份报告一起,一个独立的大众听证委员会于1976年委派成立,以决定对提议中的大草原国家公园的支持程度。这个委员会的报告称:公园的提议受到公众的支持,应该肯定。

3. Identify flora such as blue gramma grass, historically a favourite of the bison; needle-and-thread grass, whose sharp awns lend its name; prickly pear cactus, that breathes only at night; and gumbo evening primrose, whose flowers change colour in 24 hours. The non-domination of any single species of flora and combinations of short and medium sized grasses illustrate the character of this ecosystem, and its importance in sustaining life of this planet.

寻找以下植物:蓝色格兰马草,它曾是水牛最喜欢的食物;冠毛针茅草,它那尖尖的芒刺给了它这一名字;刺梨仙人掌,它只在晚上呼吸;秋葵月见草,它的花色在24小时之内会发生变化。任何一种不常见的植物和低矮的草丛都展示了这个生态系统的存在,以及它对于维持地球上的生命的重要性。

4. Endangered species in the area include swift fox, burrowing owl, mountain plover, greater sage grouse, and sage thrasher. The loggerhead shrike, peregrine falcon and Sprague's pipit are classified as a threatened species. Species of special concern are black-tailed prairie dog, red-headed woodpecker, ferruginous hawk, short-eared owl, long billed curlew, eastern yellow-bellied racer, and greater short-horned lizard.

这个地区濒临物种包括:北美敏狐、穴鸮鸟、岩鸻、大艾草榛鸡、高山矢嘲鸫。伯劳鸟、游隼和斯普氏鹨被划为受胁物种。需要特别关心的物种有黑尾草原犬鼠、红头啄木鸟、铁质鹰、短耳鸮、大杓鹬、东部黄腹蛇和大短角蜥。

5. Subtlety is the element of the prairies that makes the greatest demands on one's patience; but patience is a key to unlocking some of the most glorious secrets of the prairie. Settlers, travellers and explorers, when first confronted with the prairie, felt as though they were standing

on the shore of the ocean. From their feet to the horizon stretched an endless sweep of grass that shifted and danced like waves in the wind. Here is subtlety.

精细是这个草原的特点，它使人们要用耐心去观赏草原；发现这个草原最大的秘密的关键是耐心。当定居者、旅游者和探索者第一次与草原邂逅时，觉得自己似乎站在海岸边上。从他们的脚下到天边之间，绵延着无穷尽的草地，如波浪一般在风里舞蹈。这就是精细所在。

Questions

1. What does the topography like in Grasslands National Park?
2. Who have inhabited this grassland area in history?
3. When did Canada and Saskatchewan sign an agreement to establish the Park? And when was the accord revised and proceeded? What did the new agreement say?
4. What are the endangered species in this grassland area?
5. How to appreciate the park according to the article?

Chapter 9

Gros Morne National Park
格罗莫讷国家公园

【简介】

格罗斯莫讷国家公园始建于1970年。它位于纽芬兰州的纽芬兰岛的西北部，面积1 805平方千米。它西起圣劳伦斯湾，东至长岭山脉的支脉。整个国家公园自然景观优美奇异。格罗斯莫讷地区的景色被认为是加拿大东部最引人入胜的，这里有高耸的山脊与陡峭的大悬崖、海岸沼泽与高地苔原、多变的海滩与湖泊。1987年格罗斯莫讷国家公园被联合国教科文组织作为自然遗产，列入《世界遗产名录》。然而这个地区之所以在世界遗产名录中占有一席之地，并不是因为这里独特而且美丽的自然风光，而是由于这个地区特殊的地质构造。公园有两种不同的地貌，一种是濒临圣劳伦斯海湾的低地，一种是山地高原——长岭山。多样的地形给动植物提供了各种生存地，各种各样的动植物包括了中温带、寒带和极地的各种类型。

Gros Morne National Park is a world heritage site located on the west coast of Newfoundland. At 1,805 km^2 (697 sq mi), it is the second largest national park in Atlantic Canada (surpassed by Torngat Mountains National Park at 9,600 km^2/3,700 sq mi). The Gros Morne National Park Reserve was established in 1973. It wasn't until October 1, 2005 that the National Parks Act was applied to the reserve, thereby making it a Canadian national park. Gros Morne National Park of Canada was designated a UNESCO World Heritage Site in 1987. The park is an area of great natural beauty with a rich variety of scenery, wildlife, and recreational activities. Visitors can hike through wild, uninhabited mountains and camp by the sea. Boat tours bring visitors under the towering cliffs of a freshwater fjord (峡湾) carved out by glaciers(冰川). Waterfalls, marine inlets(海湾), sea stacks (岩柱), sandy beaches, and colorful nearby fishing villages complete the phenomenal natural and cultural surroundings of Gros Morne National Park of Canada.

The park takes its name from Newfoundland's second-highest mountain peak (at 2,644 ft/806 m) located within the park. Its French meaning is "large mountain standing alone", or more literally "great sombre". Gros Morne is a member of the Long Range Mountains, an outlying range of the Appalachian Mountains, stretching the length of the island's west coast. It is the eroded remnants(残余) of a mountain range formed 1.2 billion years ago. The park provides a rare example of the process of continental drift, where deep ocean crust (地壳) and the rocks of the earth's mantle(地幔) lie exposed. More recent glacial action has resulted in some spectacular scenery, with coastal lowland, alpine(高山的) plateau, fjords, glacial valleys, sheer cliffs, waterfalls and many pristine(原始的) lakes.

The Gros Morne National Park illustrates some of the world's best examples of the process of plate tectonics(构造地质学). Within a relatively small area are classic, textbook examples of monumental earth-building and modifying forces that are unique in terms of their clarity, expression, and ease of access. The property presents the complete portrayal of the geological events that took place when the ancient continental margin of North America was modified by plate movement by emplacement of a large, relocated portion of oceanic crust and ocean floor sediments (沉积物). The park also presents an outstanding demonstration of glaciations in an island setting. The fjords, waterfalls and geological structures of the park combine to produce a landscape of high scenic value.

The park comprises part of the Long Range Mountains on the Gulf of St Lawrence. The park includes coastal lowlands and an alpine plateau. The marine areas included in the park covers the inner portion of St Paul's Inlet, intertidal(潮间的) zones and estuaries(河口湾). The shoreline features beaches, steep cliffs, and dune formations up to 30 m in height.

An upland alpine plateau, with perched lakes, bare rock and valleys, covers a large proportion of the eastern central park. The serpentine(蜿蜒的) hills in the south-west comprise ultra-basic igneous rocks(火成岩), which, due to high heavy metal content, inhibit most plant life. A number of steep sided, glacial valleys cut through the Long Range scarp(陡坡) face, forming deep, oligotrophic(寡营养的) fjords, with vertical cliffs up to 685 m high. A number of waterfalls are fed in the summer by snow-melt at higher altitudes.

The park isgeologically diverse with areas of Ordovician sedimentary rocks, Precambrian granite(花岗岩) and gneiss(片麻岩), Palaeozoic serpentinized ultra-basic rocks, gabbros(辉长岩), volcanic and Lower Palaeozoic sedimentary rocks. Exposed oceanic crust, mantle, a section of ancient Mohorovicic discontinuity(莫霍间断面), and other distinctive geological features are also found. There is also an unusually complete paleontological sequence which has been proposed as the world stereotype for the Cambrian-Ordovician boundary.

There are up to 36 distinct vegetation types and communities, with some vascular(维管的) species and bryophytes(苔藓植物), representing about 60% of Newfoundland's insular flora(植物群).

The coast includes typical shoreline communities, active dunes(沙丘) with white spruce(云杉), and cliffs with prostrate spruce and balsam fir(胶冷杉). This is a mountain valley forest of well-spaced, old balsam fir trees. Their tops are wind pruned, and their lower branches are bowed downward by the weight of snow. The coastal plain further inland has a number of plant communities including a mosaic of sedges(莎草科植物) in meadows with American larch scrub. Larch Scrub is an open forest of twisted larch trees rooted in mucky soil. Beneath the larch grow shrubby dwarf birch and sweet gale. Black spruce dominates wet, oligotrophic sites and balsam fir is found in more protected and mesic(长于湿地的) areas. Tundra(苔原) vegetation has developed on the plateau above this and varies from small areas of coniferous forest and stunted forest to bare rock.

Faunal diversity resembles an oceanic rather than continental-shelf island and is markedly reduced compared with the mainland. However a number of species scarce in Canada are found, including lynx(山猫), caribou(北美驯鹿) and arctic hare.

Woodland caribou are much smaller than moose(驼鹿); a male adult can weigh up to 270kg, and females are about a quarter smaller. They are usually seen on the Long Range traverse(横断物) and sometimes on Gros Morne Mountain. In recent winters, they have been seen in large numbers on the coastal lowlands north of Berry Hill and St. Paul's.

Occasionally seen along certain park trails: Stuckless Pond , Lomond , Gros Morne Mountain , and Snug Harbour. The Newfoundland black bear is generally larger than its mainland relatives, ranging in size from 90 to 270 kg and averaging 135 kg. It also has one of the longest hibernation (冬眠) periods of any bear in North America! Black bears are wary of humans; however they sometimes wander into campsites(营地) in search of food.

Common in Gros Morne National Park and are often seen in areas frequented by humans. Red foxes resemble a small slender dog and range in size from 3. 6 to 6. 8 kg. Newfoundland's Red Fox comes in a range of colors—red, black, silver, yellowish, and a mixture of colors called a patch phase.

These large hare (up to 6kilograms) can be seen on the Long Range traverse and on Gros Morne Mountain. Arctic hares in Newfoundland are the southernmost in the world, and have the lowest reproductive potential of any hare or rabbit in the world—one litter per year with an average of three young! Because of their low numbers, restricted habitat, and low reproduction rate, the park monitors these animals closely. The more common marine mammals that can be observed from the park, albeit with a diminishing frequency in recent years include pilot, minke (缟鰛鲸) and finback whales and harbor seals.

The avifauna(鸟类) comprises arctic, boreal(北方的) and pelagic(浮游的) species, with strays from mainland, the north-west Atlantic and Europe. A total of 239 species of birds have been placed on the Gros Morne National Park of Canada checklist. Out of the 239 species, 105 have nested and 11 others have possibly nested in the park. The park is a significant breeding site for harlequin duck, blackpoll warbler, common tern and arctic tern(燕鸥), a nesting site for bald eagle, rock ptarmigan(雷鸟) and American tree sparrow, and a stopover for migrating shore birds. Anadromous(海鱼为产卵而游入江河的) Atlantic salmon and arctic char(嘉鱼) are found in park waters and also in permanent freshwater form in certain landlocked lakes on the Long Range Mountains.

The most abundant mammals in Gros Morne National Park are introduced species, while the threatened and endangered animals are native to the island. This may be due to different lifestyles. Most of the introduced animals are generalists—they can live in a wide range of conditions. The native species at risk are more specialized, and we humans are profoundly changing the habitats and special conditions that they require.

The Newfoundland subspecies of marten(貂) was once common, but now only a few hundred survive in three locations on the Island. Marten live in mature forest. Much of the Island's mature forest has been cut for lumber and pulp(纸浆) production. Another threat is that marten are easily caught in snares set for snowshoe hares. Because of these threats, this subspecies is now listed as endangered.

There are a number of archaeological sites in the park and human habitation can be traced back to the MaritimeArchaic Indians (4,500 - 3,000 years ago) and the Dorset Eskimos (1,800 - 1,200 years ago). Europeans settled the area from the late 18th century.

Maritime Archaic Indians who crossed over from Labrador first settled this land some 5,000 years ago. The earliest evidence for their fully maritime lifestyle comes from L'Anse Amour in southern Labrador, which is also the site of the oldest known burial mound in the Americas. The major Maritime Archaic site discovered so far in Newfoundland is at Port au Choix, 160 kilometres north of Gros Morne National Park.

Cooler times brought an arctic folk, the Palaeo-Eskimos to these shores. These people specialized in hunting marine mammals and intensely used whatever resources were abundant. Seals were their most important food, and when seals were scarce starvation came. For 16 centuries they hunted these shores, and then left no further trace.

The term Recent Indian Cultures encompasses all Indian occupation of Newfoundland since the end of the Palaeo-Eskimo period. Unfortunately this occupation is not well represented in the archaeological record of the park, and the people who left the remains are unidentified. There are traces of Indian occupation within Gros Morne National Park at Cow Head and Broom Point about a thousand years ago, but so far nothing more recent has been found.

The ocean's bounty is not endless—the over-exploited fishery failed. People turned to the woods. In the 1920s, the St. Lawrence Timber, Pulp, and Steamship Company set up in a place called Lomond, named by the mill manager, George Simpson from Scotland. Logging brought cash to a society based on barter. Fishermen took to the woods for the winter working out of lumber camps.

During World War II, Canada recognized Newfoundland's strategic importance, and was worried by American ambitions in the colony. After two referendums, Newfoundland and Labrador agreed to Confederation in 1949. Canada's social programs and the development of new industries completed the switch to a cash economy. Roads linked communities, and new schools were built. Electricity and television brought a very different way of life.

The national importance of the Bonne Bay area was recognized in 1973. By agreement with the Province of Newfoundland and Labrador, the Government of Canada established Gros Morne National Park to protect and present an outstanding example of Newfoundland's western highlands. In 1987, the United Nations declared Gros Morne National Park a World Heritage Site for its exceptional geological features and natural beauty.

Notes

Newfoundland 纽芬兰岛　北美洲东海岸外大西洋岛屿，属加拿大纽芬兰省。在圣劳伦斯湾外。北以贝尔岛海峡与拉布拉多半岛相望，西南以卡博特海峡和新斯科舍半岛相隔。主要河流均朝向东北流入大西洋。海岸曲折，多半岛、港湾。它属寒冷大陆性气候。大部地区生长针叶林，以云杉、松为主。有铁、铜、银、石棉、石油等矿藏，资源丰富。东南面的纽芬兰浅滩，正当寒暖流交汇处，是世界最大渔场之一。

Long Range Mountains 长岭山脉　加拿大纽芬兰岛最高山脉。从雷伊角(Ray)沿西海岸向北绵亘约400千米，平均高度近670米，路易斯山为最高点814米。森林茂密，为纸浆和造纸厂提供丰富原料。为亨伯河(Humber)发源地。南部多强烈海风，风速每小时达193千米。

Appalachian Mountains 阿巴拉契亚山脉　又译阿帕拉契山脉，是北美洲东部的一座山系。南起美国的阿拉巴马州,北至加拿大的纽芬兰和拉布拉多省。最北部余脉则延伸到魁北克的加斯佩地区。最高峰在北卡罗莱纳州的米切尔峰(2037 米)。构成阿巴拉契亚山脉的有纽芬兰省的长岭山、魁北克的圣母山、缅因州的朗费罗山、新罕布什尔州的怀特山、佛蒙特州的格林山脉、塔库尼克山、马萨诸塞州的勃克夏山、跨宾西法尼亚州、马里兰州和西佛吉尼亚州三州的阿莱干尼山、跨宾西法尼亚州、马里兰州、西佛吉尼亚州以及佛吉尼亚州四州的阿巴拉契亚岭谷。还有从宾西法尼亚州南部到佐治亚州北部的蓝岭。

Gulf of St Lawrence 圣劳伦斯湾　它位于加拿大东南部的大西洋海湾。在圣劳伦斯河口东南,是一个宽广而几乎被陆地包围的水域,仅东端由贝尔岛海峡和卡博特海峡与大西洋相通。海湾西起圣劳伦斯河河口,北到贝尔岛海峡,南至卡伯特海峡。面积 23.8 万平方千米。平均水深 127 米,最大水深 572 米。海湾是在地质构造运动中发生沉降作用后形成,因而湾岸曲折,多暗礁、浅滩,不利于航行。圣劳伦斯湾是进入整个北美大陆腹地的一个门户。

Ordovician 奥陶纪　古生代第二个纪,约开始于 5 亿年前,结束于 4.4 亿年前。在此期间形成的地层称奥陶纪,位于寒武纪之上,志留纪之下。奥陶纪是地质学上古生代中 5.05—4.38 亿年前的这段时间,可分为三个时期——奥陶纪早期 (5.05—4.78 亿年前),奥陶纪中期(4.78—4.53 亿年前)和奥陶纪晚期(4.53—4.38 亿年前)。奥陶纪亦分早、中、晚三个世。奥陶纪是地史上海侵最广泛的时期之一。

Precambrian 前寒武纪　寒武纪以前全部地质时期的总称。又称先寒武纪,简称前寒武或先寒武。这一时期形成的地层称前寒武系或先寒武系。一般认为地球的年龄是46 亿年,地球演化中的地质阶段约从 40 亿年前开始,寒武纪的起点时间是 5.7 - 6 亿年前。故前寒武纪通常就是指大约 40 亿年前最早地质阶段起到寒武纪开始,时距 34 亿年,约占地质历史 85% 的时间。

Palaeozoic 古生代　地质时代中的一个时代,开始于同位素年龄 542 ± 0.3 百万年(Ma),结束于 251 ±0.4 白万年。古生代属于显生宙,上一个代是新元古代,下一个代是中生代。古生代包括了寒武纪、奥陶纪、志留纪、泥盆纪、石炭纪、二叠纪。

caribou 北美驯鹿　产于北极地区,有些生物学家把北美驯鹿分成两种,一种叫北方驯鹿,栖息在北方的冻土地带和针叶树林地带。另一种叫森林驯鹿,栖息在加拿大的森林之中。

black bear 美洲黑熊　大量分布在北美地区。居住范围北起阿拉斯加,向东横穿加拿大,直至东海岸的纽芬兰—拉布拉多省;向南则经美国部分地区,一直延伸到墨西哥的那亚里特和塔毛利帕斯州。美洲黑熊主要在这些分布地的山区密林中活动,海拔约 900 ~ 3000 米。美洲黑熊是独居动物。它们的活动时间根据居住地和季节的不同而有所变化。在春季,它们常在拂晓或薄暮时分外出寻找食物,到了夏季它们会花大量时间在白天活动。进入秋季,它们不论白天黑夜都会出来觅食游荡。

harlequin duck 丑鸭　一种体型小巧而结实的深色海鸭。栖息于山溪，以水生动物为主。喜欢成对或成群活动。它们分布于北美地区（包括美国、加拿大、格陵兰、百慕大群岛、圣皮埃尔和密克隆群岛及墨西哥境内北美与中美洲之间的过渡地带），欧亚大陆及非洲北部（包括整个欧洲、北回归线以北的非洲地区、阿拉伯半岛以及喜马拉雅山－横断山脉－岷山－秦岭－淮河以北的亚洲地区。

blackpoll warbler 黑顶白颊林莺　一种北美鸣禽，黑顶白颊林莺林莺属，雄性有黑冠。

Translation of Selected Sentences

1. The park provides a rare example of the process of continental drift, where deep ocean crust and the rocks of the earth's mantle lie exposed.

公园的地貌展现了罕见的大陆漂移过程，在这里，深海地壳和地球地幔的岩石裸露出来。

2. The Gros Morne National Park illustrates some of the world's best examples of the process of plate tectonics.

格罗斯莫讷国家公园是证明地质构造学的最好例子。

3. The park includes coastal lowlands and an alpine plateau. The marine areas included in the park covers the inner portion of St Paul's Inlet, intertidal zones and estuaries.

公园包括海洋低地和高山高原。公园的海洋区域覆盖了圣保罗内部的海湾、潮间区域和河口湾。

4. Because of their low numbers, restricted habitat, and low reproduction rate, the park monitors these animals closely.

由于数量少、栖息地有限、再生育率较低，公园密切监控着这些动物。

5. These people specialized in hunting marine mammals and intensely used whatever resources were abundant. Seals were their most important food, and when seals were scarce starvation came. For 16 centuries they hunted these shores, and then left no further trace.

这些人善于捕猎海洋动物，利用各种丰富的资源。海豹是他们最主要的食物。当海豹匮乏时，饥荒也就随之而来。在将近16个世纪里，他们在这些海岸捕猎，没有去过更远的地方。

Questions

1. What is the general condition of the Gros Morne National Park according to this passage?
2. What kind of animals does the Gros Morne National Park have?

3. By what kind of examples does the Gros Morne National Park illustrate the process of plate tectonics?

4. Why was themarten listed on the endangered subspecies?

5. When did Canada recognize the importance of the Gros Morne National Park?

Chapter 10

Gulf Islands National Park Reserve
海湾群岛国家公园

【简介】

海湾群岛国家公园是一座新兴的公园，它始建于2003年。这里是太平洋遗迹中一处别具意义的保留地。海湾群岛国家公园为游人们提供了一处近距离接近海岸，并理解其深藏内涵的机会。这里共有15座大小不一、形态各异的岛屿，以及7座小型的群岛。同时，该公园还汇集了拥有最甜美田园风光的赛特纳岛、充满历史古迹的梅恩岛和以宁静海滩著称的片打岛。游客们可以搭乘著名的渡轮依依到达这些岛屿，尽情地观赏和探险。气候温暖干燥的海湾群岛国家公园融合了惊世的美景、悠久的历史以及当地独一无二的风俗习惯。壮观秀丽的风景、细软无垠的沙滩、陡峭嶙峋的海角，还有别具特色的奥里根白栎和杨梅林，这些都是此公园吸引络绎不绝的游人来此观光的亮点。

Gulf Islands National Park Reserve is the 40th National Park in a system of 43 parks and park reserves across Canada. The Gulf Island National Park was established in 2003. Off the southern coast of British Columbia, adjacent(毗邻) to the Strait of Georgia, is a collection of islands, islets and reefs(礁) with rocky shorelines and steep headlands. Known as the Southern Gulf Islands, these islands are blessed by Canada's only Mediterranean-type climate—warm and dry in the summer and mild and wet in the winter. About 15 of these islands and over 30 smaller islets comprise the Gulf Islands National Park Reserve. It is an environment popular for many activities like ocean kayaking, wildlife watching, wilderness camping, fishing, sailing, boating and scuba diving(用水肺潜水). The new national park reserve offers a variety of opportunities for Canadians to learn about and experience an exceptional coastal island landscape and the cultures of the people who live there.

Today this area is a beautiful place to explore. The park protects "one of the most ecologically important and most at risk natural areas in southern Canada". All of the larger islands share a similar past. They were inhabited by semi-nomadic(半游牧的) First Nations people and settled by Europeans in the mid 1800s. In the 1900s, the islands were known for bountiful fruit production.

Many animals and plants in the area are not found elsewhere in Canada. A total of 28 species are listed as Endangered, Threatened or of Special Concern, ranging from Orcas(虎鲸) to the Sharp-tailed snake. Some of the protected species include Garry oak trees(奥里根白栎), arbutus trees(杨梅树), wild lilies, shellfish and whales. Wildlife watching and sightseeing are popular activities when exploring the ocean channels in the park or when visiting one of the main islands. There are opportunities to view sea birds and marine wildlife like seals, Sea Lions, porpoises(鼠海豚), Blue Herons(鹭), Bald Eagles and Red Tail Hawks. Human impact on the islands since the turn of the last century—logging, commercial fishing, residential development, and increased visitor traffic has threatened and continues to threaten, these rich island ecosystems. The creation of the national park helps protect the islands in perpetuity(永恒) and, hopefully, encourages all of us to better appreciate the natural and cultural uniqueness of this area.

The park includes 4 geographical zones—the Outer Islands, Prevost Island, the Penders and the Inner Islands. The Outer islands region includes Saturna Island, Tumbo Island, Cabbage Island, Mayne Island, Anniversary Island, Samual Island and the Belle Chain Islets. The Prevost Island region includes such highlight destinations as James Bay, Selby Cove, Portlock Point, Channel Islands and the Nearby Islets. The Penders region includes North and South Pender Islands. The highlight destinations include Roesland Lake, Loretta's Wood, Mount Norman, Greenburn Lake and Blunden Islet. The Inner Islands region includes Russel, Island, Portland Island, Brackman Island, Isle-de-Lis, Sidney Spit and D'Arcy Island.

Fronting on the open waters of the Strait of Georgia, the Outer Islands are the last bastion(棱堡) against the urban metropolis of mainland British Columbia across the waters to the north. Facing the silhouettes of skyscrapers by day, and the sparkling lights of a big city by night, the Outer Islands are a serene contrast—so near to and yet so far from the frenzy (疯狂) of modern life. These islands have always been important to those who travel the waters of the inland Salish Sea. They were used by First Nations as a launching area before they travelled into open water heading to the Fraser River. Similarly, miners headed for the goldfields of British Columbia interior made Mayne Island their last stop, giving rise to the name Miners Bay.

The numerous small islets paralleling Mayne and Saturna Islands support an abundance of marine life and waterfowl(水鸟). In days past, First Nations people made stinging nettle nets to hang across narrow gaps between islets to intercept migrating waterfowl codfish and their eggs were harvested in shallow areas. The herring(鲱鱼) fishery has also played an important role in the culture of First Nations. The method of fishing was a ritual undertaken with grace and skill. Paddling a canoe(独木舟), the fisher would lower a rake which had pins of bones and in a gentle swooping motion would prick the herring and catch it on the sharp ends.

Saturna Island was and still remains an important and special place for First Nations. The island had affluent permanent village sites particularly on the east and west shores of the inlets. People lived in large cedar(雪松) houses built using corner cedar posts. In some places, these posts still stand. Narvaez Bay on Saturna Island is a beautiful, fairly undisturbed bay that is surrounded by regenerating Douglas fir (黄松) forest. The area is a great place to go for a hike and enjoy a picnic.

Tumbo Island has recovered well from the fur farming, timber harvesting and coal mining activities that highlighted its rich and varied past. Today, the island, located off the southern tip of Saturna Island, is largely forested with old growth Douglas fir and Garry oak meadows(草地). Tumbo Island derives its name from the landform that gives it its distinctive shape—a tombolo. A tombolo is a sandbar(沙洲) extending outward from shore connecting with an island, or from island to island as it does here. First Nations found a safe harbor on Tumbo Island when hand trolling for fish offshore in their canoes or on their journeys across the Strait. The island remains an important spiritual place to this day.

The Channel Islands in Captain Passage, located between Prevost Island andSalt spring Island, are important as seal and sea lion haul outs and are also used for nesting by various bird species. The Channel Islands were used by First Nations as landmarks when navigating between islands, and for harvesting of marine mammals, shellfish, barnacles(藤壶), chitons(石鳖) and sea urchins(海胆). The Red Islets, Bright Islet and Hawkins Islet located adjacent to Prevost Island feature relatively undisturbed coastal bluff and Garry oak/arbutus woodland. The sensitive ecosystems on these islands and islets are being afforded the highest level of protection within the national park: authorized access only.

Safeguarded by its inaccessibility, Blunden Islet off Teece Point on South Pender Island is a relatively undisturbed islet, with typical coastal bluff vegetation including arbutus, Garry oak and old growth Douglas fir, and no evidence of invasion by exotic species.

Like most of North Pender Island, Roesland was once a farm operation. In 1905, Scottish immigrant Robert Roe purchased land that stretched from Otter Bay to Shingle Bay and included

Roe Lake. Robert, Margaret, and their four children cleared more than 30 acres for pasture and a garden and built a home, barn and several outbuildings. In 1910, they added a cabin for visiting family and friends. Its popularity spurred the idea for a holiday farm resort.

Each winter they built one or two more cabins and by 1950 had seventeen. The cabins were basic: wood floors, shake roofs, two double beds and wood-burning stoves. Guests brought their own bedding, dishes, cutlery and cooking utensils, and had to carry their own water in buckets from the well. By 1927, the resort had a store for guests and commercial fishermen, and a marine fuel station on an island in Otter Bay.

Robert Roe died in 1969, and the resort was purchased by David and Florence Davidson who would operate it for almost 20 years. In the fall of 1991, Roesland Resortceased operation. Six years later it was acquired to become part of the new Gulf Islands National Park Reserve (GINPR). As well as protecting the natural environment, the national park reserve is committed to protecting the park reserve's cultural heritage. A variety of initiatives, including an archaeology (考古学) program, an oral history program, a conservation program and ongoing research, help the Gulf Islands National Park Reserve to identify cultural sites needing protection. Through research undertaken so far, the GINPR recognizes the important story to be told at Roesland.

GINPR began its job of interpreting this site with an assessment of the existing buildings. Only select buildings were identified for conservation, including the 1908 Roe house, which has been painstakingly and beautifully restored by the Pender Island Museum Society in partnership with the GINPR. Six cabins and Bert and Irene Roe's house have been removed. Great care was taken in their removal so that the natural landscape was not scarred. The remaining cabins, which span the time from the 1930s to the 1960s, will stay in place to tell the story of the resort. The long-term vision for the site is to restore the buildings, including the repair or replacement of windows, doors, siding and logs, and to furnish them to reflect the way they looked in their respective time periods.

The Gulf Islands area is one of the most popular marine recreation areas in Canada. Sailing, power boating and kayaking are all great ways to visit the nationalpark reserve. Mooring buoys (系泊浮筒) are located at Sidney Spit (Sidney Island), Beaumont Marine Park (South Pender Island) and Cabbage Island. Limited dock space is also available at Sidney Spit, and dinghy (小舟) docks are located at Tortoise Bay on Portland Island during peak season and at Royal Cove year round. Only one boat is permitted per buoy and vessel size restrictions are printed on the buoys. Fees for mooring buoys are collected from May 15 to September 30 and the maximum total stay is 14 days per calendar year.

Kayaking is a fun and challenging way to explore the Gulf Islands. Whether kayaking with friends or a tour operator, be sure to check the weather and tide tables in advance. Kayakers will need to bring adequate water and pack out all their garbage.

Owing to the strength of the tidal currents coursing through Active Pass between Mayne Island and Galiano Island, there are plenty of nutrients in the water to attract marine life. Salmon feed in these waters, although in recent years the large Chinook has all but disappeared. It's not as easy to purchase fresh seafood on the islands as you might imagine. Most islanders do their own crabbing and shrimping, but at Horton Bay on Mayne Island, you may find such delicacies for sale. Although the availability is seasonal, stop by the federal dock between 4 and 6pm and look for a truck with the personalized license plates CRAB 4U, which says it all.

Bird watching can be a good motivation to kayak these waters. The number of bird species is amazing. Common birds are the eagles, geese, herons, crows, ravens(渡鸦), robins, and gulls. The variety of water fowl in the area is tremendous, but they are a bit shyer than the "common" birds already mentioned. The hummingbirds(蜂鸟) are visible hovering around the flowers on the islands, as well.

There are also some other exciting activities such as paddling, camping, cycling and so on. The Gulf Islands National Park is a great place to explore.

Notes

British Columbia 不列颠哥伦比亚省　它位于加拿大西岸,面临太平洋,海岸线长达8850千米,沿岸风景优美,海产丰富,全省面积94.78万平方千米,人口414.13万,仅次于魁北克和安大略省,是全国第三大省,该省南与美国华盛顿州、爱达荷州及蒙大拿州接壤,是加拿大通往亚太地区的门户。该省气候温和,风景秀丽,环境保护良好,依山傍水,省内大部分面积是森林地带,是世界著名的旅游胜地。

Strait of Georgia 乔治亚海峡　位于加拿大不列颠哥伦比亚省西南部大陆与温哥华岛东岸中部之间。海峡的北端是一堆岛屿,将它与更北的约翰斯通海峡和夏洛特皇后海峡隔开。南端以美国华盛顿州的圣胡安群岛为标志。海峡中最大的岛是特克塞达岛和拉斯克蒂岛。大陆沿海有许多小海湾,其中之一是弗雷泽河的河口和温哥华所在地。弗雷泽河入海的水流帮助海峡中的水总体上按逆时针方向运动,当地人称之为海湾。这个海峡在美国华盛顿州西雅图和阿拉斯加州斯卡圭之间形成一段风景优美、有天然屏障的内海航道。1792年乔治·温哥华船长以纪念乔治三世命名此海峡。

porpoise 鼠海豚　一种可以长至1.85米的齿鲸。它背部黑色,腹部白色,生活在北大西洋欧洲、非洲和北美洲东岸以及在黑海和太平洋亚洲和美洲的海岸附近。它以鱼、甲壳动物和乌贼为食。鼠海豚是北海和波罗的海中最常见的齿鲸,但是其数量在不断减少,可能是因为向海中排放毒水以及在渔网中窒息造成的。

Saturna Island 萨图纳岛 地表崎岖不平，是高尔夫群岛中人口最少、自然景观保留最完好的岛屿。岛上近一半的陆地区域属于高尔夫群岛国家公园保护区，来到这里的人们可以一扫都市生活所带来的阴霾和不快。住户与农场都集中在东角公路附近，这条路贯穿着整个小岛，连接着渡轮航线的终点站和东角。萨图纳岛上居民虽然不多，但野生动植物却种类繁多：鹿、野山羊、水獭、猛禽，等等。岛上的地貌也极具特色，例如，奇特的沙岩地层、连绵不绝的古老森林，以及沃伯特派克山。

Mayne Island 梅恩岛 位于温哥华岛和不列颠哥伦比亚省主陆之间的中点。由于其独特的地理位置，梅恩岛在19世纪50年代末至60年代初，自然而然成了矿工们从维多利亚前往卡里布淘金时的休息站。人们大规模移居梅恩岛也正始于此，而该区域许多名胜建筑和历史遗址都可追溯到19世纪末。现在的梅恩岛具有非常鲜明的田园风格。现已开发的区域散布在狭窄的乡间小路两旁，且规模较小。而大部分地区仍然覆盖着茂密的森林。岛上最具知名度的是一系列上世纪初在矿工湾地区新建的古建筑，同时该岛有许多避风港湾和邻近岛屿，提供超一流的划艇体验。

James Bay 詹姆斯湾　加拿大哈得孙湾向大陆延伸的部分，介于魁北克省和安大略省之间。长约440千米，宽约215千米。1610年亨利·哈得孙首先到达詹姆斯湾。湾内岛屿众多。湾岸有许多贸易点，以安大略省的穆索尼最重要。

Channel Islands 海峡群岛　位于英吉利海峡中的群岛。这些岛屿分为两个行政区：根西和泽西。英国对这些地方有宗主权，但是这些地方并不被视为英国本土，而算是英国的属地。群岛风景优美，花卉繁多，草木茂盛，具有温暖宜人的海洋性气候，为游览胜地。第二次世界大战时期，该岛为遭受德国侵占的唯一英国领土，以养牛和输出水果、花卉、番茄、马铃薯而闻名。群岛享受课税主权，出口受英国关税壁垒的保护。

Fraser River 弗雷泽河　加拿大华人称为菲沙河，是加拿大不列颠哥伦比亚省中部大河。加拿大第十长河流。以1808年5月来此探险的西蒙·弗雷泽的姓氏命名。流域面积为23.8万平方千米，其中70%为海拔900米以上的高原。河长1 368千米。源于洛矶山脉西坡，注入乔治亚湾。沿岸景色秀丽。东部支流源出落矶山、奥米尼卡山、卡里布山和莫纳希山，包括麦克雷戈河、鲍伦河、卡斯诺尔河和汤普森河。西部支流源出沿海山脉的东坡，包括尼查科河、西路河、奇尔科廷河、布里奇河和利卢埃特河。弗雷泽流域内的经济活动以林业为主，是不列颠哥伦比亚省第二大木材产地，产品远销海外。农业生产不发达。南部的草原、奇尔科廷高原的绿地及尼古拉(Nicola)峡谷内有少量畜牧业。

SaltSpring Island 盐泉岛 位于温哥华岛东部外海。盐泉岛的牧场让人仿佛置身纽西兰，有矮山、绿草、牧场、绵羊及马匹，让游客感受宁静、恬淡、温馨的农村生活。它是加拿大艺术家与工艺家的天堂，占地面积与台湾澎湖差不多，海岸线长达182千米，人口近1万人，却拥有225座牧场、1万头绵羊，吸引来自世界20多个国家近30位素人艺术家到此移民，也是北美最受欢迎的素人艺术宝库。每年春季与夏季，这些素人艺术家公开展售各种艺术创作品，包括：草篮、蜡烛、时钟、服装、食品、家俱、玻璃、装潢、手饰、乐器、绘画、陶艺、拼布、橡皮图章、纺织、雕塑与木工等，游客可随意参观他们的工作室，与艺术家面对面交谈并选购工艺品。

Translation of Selected Sentences

1. Known as the Southern Gulf Islands, these islands are blessed by Canada's only Mediterranean-type climate—warm and dry in the summer and mild and wet in the winter.

海湾群岛的南部因为特有的地中海气候而闻名，成为加拿大地区唯一拥有这样气候的地方，这里夏天温暖干燥，冬季温和潮湿。

2. Human impact on the islands since the turn of the last century—logging, commercial fishing, residential development, and increased visitor traffic has threatened and continues to threaten, these rich island ecosystems.

人类的活动，如上个世纪末的伐木业、商业捕鱼，以及居民定居和不断增加的游客都对群岛构成威胁，并且还在继续威胁着群岛的生态系统。

3. Facing the silhouettes of skyscrapers by day, and the sparkling lights of a big city by night, the Outer Islands are a serene contrast—so near to and yet so far from the frenzy of modern life.

面对着城市一天天建成的摩天大楼和城市夜晚璀璨的灯光，群岛外岛宁静的氛围和城市形成了鲜明的对比，如此近距离又如此远离这疯狂的城市生活。

4. The sensitive ecosystems on these islands and islets are being afforded the highest level of protection within the national park: authorized access only.

这些岛屿敏感脆弱的生态系统已经被国家公园高度保护起来，只有得到官方的允许，才能够进入这些岛屿。

5. The long-term vision for the site is to restore the buildings, including the repair or replacement of windows, doors, siding and logs, and to furnish them to reflect the way they looked in their respective time periods.

对这一遗址的长远规划就是保存这些建筑，包括修复或替换一些门窗、外壁板和原木，以及去布置这些地方以使它们恢复到各自原来的状态。

Questions

1. What's the climate of Gulf Islands National Park Reserve?

2. How many geographical zones does the Gulf Islands National Park Reserve have? What are they?

3. Which part of the park is thelast bastion against the urban metropolis of mainland British Columbia? What's the condition of this part?

4. Who is the first person to purchase the Roesland? How many years have they lived on the Roseland?

5. What kind of activities does the Gulf Island National Park Reserve have? What are they?

Chapter 11

Gwaii Haanas National Park Reserve and Haida Heritage Site

哈纳斯国家公园和海达文化遗址

【简介】

哈纳斯国家公园和海达文化遗址位于夏洛特女皇群岛的南部,距离不列颠哥伦比亚海岸线 130 千米,温哥华以北 640 千米。该野生保护区位置偏僻,由 138 个小岛构成,南北绵延 90 千米。海达文化遗址是不列颠哥伦比亚最壮观的公园之一。这里有着非凡的岛屿、内海和海湾。这种得天独厚的自然环境为人们提供了原始的雨林、崎岖的山脉、纯净的河流、以及在世界上其他地方都无法找到的一种自然与文化的融合。成千上万的海鸟、鲸鱼、黑熊、海狮、秃鹰和海鹦在这里安家落户。这里原来是海达族世世代代的栖居地,因而这里有数不尽的图腾吸引着游客,它的景色之中也融合了一种世界最伟大的艺术传统。除此之外,公园还为人们提供了各式各样的活动,诸如用爱斯基摩划子航海、远足旅行、划船、野生动物观赏和野营,等等。

Gwaii Haanas National Park Reserve andHaida Heritage Site, often referred to simply as Gwaii Haanas, is located in the southernmost Haida Gwaii (formerly known as Queen Charlotte Islands), 130 kilometers (81 miles) off the mainland of British Columbia, Canada. Gwaii Haanas protects an archipelago(群岛) of 138 islands, the largest being Moresby Island and the southernmost being Kunghit Island. The protected area represents the Pacific Coast Mountains Terrestrial(陆地的) Region and the Hecate Strait and West Queen Charlotte Islands Marine Regions. Gwaii Haanas is located in the Pacific Maritime Ecozone. In June 2010, Gwaii Haanas National Marine Conservation Area Reserve and Haida Heritage Site were officially established. This is a type of marine protected area that conserves and protects so that sustainable uses will continue. Gwaii Haanas is now the only place on earth to be protected from mountain top to deep sea—nearly 5,000 km^2 of land and ocean.

Each national park, national historic site and national marine conservation area in the Parks Canada family tells its own story. Gwaii Haanas contains the essence of the rugged beauty and ecological character of the Pacific coast. It is a celebration of more than ten thousand years of connection between land, sea and Haida culture. Visitors have an opportunity to witness(见证) a living culture, to observe nature and to experience solitude(孤独). For some visitors, this is a spiritual experience and results in a deep connection with the land, sea and people.

Visitors to Gwaii Haanas are drawn to its natural beauty and also to its extraordinary living Haida heritage. The carved poles at SGang Gwaay Llnagaay are breathtaking—but there is an enormous quantity of more subtle archaeological(考古的) evidence from the over 10,000 years of Haida habitation on the Islands. There are over 600 archaeological sites in Gwaii Haanas, alone. Culturally modified trees (cedars with bark removed for weaving(编织) or with test holes, and fish weirs(堰) (posts placed in estuaries to help catch salmon) are examples easily seen by informed visitors.

The climate of Gwaii Haanas is characteristic of the coastal mountainous area—wet and mild. Most of the precipitation comes from storms carried in from the Pacific Ocean. Haida Gwaii has the wettest climate in Canada; the west coast has received 4,200 mm of rain during the year, however, the east coast receives several times less rain. The winds on Haida Gwaii can reach up to 200 km/hr. Gwaii Haanas suffers few extremes in temperature, with an annual average of 7 ℃. This is mainly because of the island's proximity to the ocean which, like all larger bodies of water, tends to moderate the climate of the surrounding mainland. The climate is also influenced by the Alaska and the warm Japanese currents, as well as the coastal mountains to the east, the islands rugged topography(地形), and prevailing westerly winds.

The San Christoval Mountains form the backbone of Gwaii Haanas, rising to 1,123 meters at Mount de la Touche. The higher elevations of this range are dominated by mountain hemlock (铁杉) and alpine(高山的) tundra(苔原) vegetation zones. The lower elevations (Kunghit and Moresby Islands) are mainly cedar, pine and western hemlock. The Hemlock-Sitka spruce(云杉) zone is found on the islands of the eastern part of Gwaii Haanas, including the eastern fringe (边缘) of Moresby Island.

The distinct island flora and fauna have evolved over thousands of years. The species here often differ from those found on the mainland. Many common continental species are not found on the islands at all, or have evolved into unique subspecies such as the black bear and the pine marten (both larger than mainland cousins), the deer mouse, the dusky shrew and the short-tailed weasel(鼬鼠). Other species have been introduced relatively recently—the Sitka black-tailed deer, raccoons(浣熊), squirrels, beaver and three species of rats and now exist in large numbers, much to the detriment of the native plants and animals.

An estimated 1.5 million seabirds nest along some 4,700 km of shoreline on the Islands from May through August. Approximately half of these seabirds can be found in Gwaii Haanas. Many are burrow-nesters, such as the rhinoceros auklet, ancient murrelet, tufted puffin(海鸟), horned puffin, Cassin's auklet, Leach's storm petrel(海燕) and fork-tailed storm petrel. Common murres (厚嘴海鸦) and pelagic cormorants, bald eagles and Peale's peregrine falcons(游隼) also nest along the coastline. Because the islands are situated along the Pacific flyway, dozens of species of migrating birds stop here in spring and fall.

The sea around Gwaii Haanas teems with life. These waters are home to salmon, herring, halibut(大比目鱼), rockfish(岩鱼), mussels(贻贝), crab, starfish(海星), sea urchin(海胆) and octopus(章鱼), along with numerous other species. Twenty species of whales and dolphins have been spotted in these waters. The grey whale spring migration route passes Haida Gwaii on the way to summer feeding grounds in the Bering Sea. Humpback(驼背鲸), orca(虎鲸), and minke (缟鰛鲸) whales are also seen regularly in the waters surrounding Gwaii Haanas, along with dolphins, porpoises, and harbour seals.

In 1974, the dispute over the future of southern Moresby Island began with the advent of logging plans and a public proposal to protect the South Moresby Wilderness Area. In 1985, the Haida Nation designated the area a "Haida Heritage Site". Logging continued in the area in the face of legal and political controversy until July of 1987, when Canada and British Columbia signed the South Moresby Memorandum of Understanding. One year later, the South Moresby Agreement was signed, paving the way to Canada's designation of National Park Reserve.

The Government of Canada and the Council of the Haida Nation signed the Gwaii Haanas Agreement in 1993. The Agreement expresses respect for Canadian and Haida interests and designations, and includes a mutual commitment to the protect Gwaii Haanas.

Based on the Agreement, Gwaii Haanas is cooperatively managed by the Archipelago Management Board (AMB), which is made up of an equal number of representatives from the Council of the Haida Nation and the Government of Canada. On January 16, 2010 the Gwaii Haanas Marine Agreement was signed, expanding the cooperative management partnership to include representation from Fisheries and Oceans Canada. The AMB is responsible for all aspects of planning, operation, and management of Gwaii Haanas.

Gwaii Haanas provides a model of a respectful relationship between human beings and the earth. It is a place where people can learn about a way of life where humans are part of nature, where they can experience Haida culture and respect sacred and spiritual values. Traditional activities continue to take place here as part of the living and evolving culture of the Haida.

Hundreds of generations of Haida have been nourished(养育) by the rich abundance of Haida Gwaii. The islands provide all the essential ingredients for sustenance and growth. Resourcefulness and readily accessible foods allow time for the continuation of a complex society and a unique art form that reflects the people's close interaction with the land and seas.

Monumental cedar poles announce the crests and lineages of the Haida families. Haida house construction has secured its place in the world of architecture, while the ingenuity of Haida technology is demonstrated in the ocean-going canoe. The undisturbed areas of Haida Gwaii continue to inspire Haida artists who command international recognition in today's world.

Contact with "people from away" and their communicable diseases nearly annihilated(消灭) the Haida, the population decreased from many thousands to only a few hundred. The Haida not only survived but continue to live and adapt their culture into the mainstream of 21st century North American society. Today, along with a variety of lifestyles and careers, hunting, gathering and the preparation of foods and feasting are still part of everyday life.

In the past, Haida Watchmen were posted at strategic positions around a village to raise the alarm in advance of an approaching enemy. Watchmen also crowned poles as carved figures protecting the village from evil spirits. Today, from spring until autumn, the Haida Gwaii Watchmen serve as guardians at the following old village sites:

Watchmen live at camps at the village sites ensuring that these sites are respected. The Watchmen, although not tour guides, provide a wealth of knowledge to people who visit each site.

Some historic sites in Gwaii Haanas record the early presence of European and Japanese industrial activities. There are abandoned mining sites, whaling stations and remnants of old trails, logging roads, tramways and shipwrecks throughout the area.

Gwaii Haanas' unique cultural and ecological features are fragile. There are a number of things you can do to help sustain the integrity of these features. Firstly, wherever you find an accessible fringe of shoreline here, it is a safe bet that it has been used in the past by the Haida. There may be grass-grown middens of broken shell, or piled rocks outlining canoe landings; depressions or moss-covered logs may be all that remain of fallen house posts, or totem(图腾) or mortuary(丧葬的) poles. Treat everywhere you land as a potential archaeological site. Do not remove or disturb artifacts, bones or other cultural remains. Do not dig into middens, or strip bark from trees. Secondly, one of the outstanding features of Gwaii Haanas is the number of birds that breed here from mid-April through July. But be aware that peregrine falcons and many seabirds will abandon sites, eggs or chicks if they are disturbed. Please do not approach nests or land on obvious bird colonies. At the same time, avoid using lights—flashlights, campfires, boat lights—near colonies. Some birds come ashore only at night and may be disoriented by any light. Noise such as loud voices, boat engines or music can also disturb breeding birds. Thirdly, the waters around Gwaii Haanas teem with life. They have supported the traditional way of life of the Haida people for generations, but they cannot stand unlimited harvesting. If fishing is to be part of your experience, limit it to catching only what you will eat during your stay. Try not to crush barnacles (藤壶) and other marine life as you explore the seashore. Life is hard enough in sea battered intertidal pools! If you pick up shells, rocks, crab or other creatures, take care to replace them where you found them.

Besides, there are many recreational activities you can enjoy such as the Kayak, camping, hiking and so on. Random camping is the rule in Gwaii Haanas. There are no designated campsites within Gwaii Haanas—you camp where you please, preferably on sand or stone, or wherever else you will have as little impact as possible. Camping is not permitted at T'anuu Village, or on Gandla'kin (Hotspring), Ata Naa (House), Copper, Jeffrey, Rankine, or Skung Gwaii (Anthony) Islands, or Slug Islet. These areas are particularly sensitive cultural sites or important bird-nesting areas. There are no designated or maintained hiking trails in Gwaii Haanas. All hiking is of the "bushwhacking" (丛林旅行) variety. It is easy to get lost due to fog, and density of forest. Hikers should have excellent compass skills and use map and compass at all times.

Notes

Haida 海达　海达人居住在夏洛特皇后群岛、不列颠哥伦比亚省、威尔斯太子岛南部及阿拉斯加的印第安人,操海达语。阿拉斯加海达人又名凯加尼人。据说海达语与阿萨巴斯坎语有关。海达人在文化上与特林吉特人及钦西安人有关。海达人分为两个半偶族,成员以出生定其所属分支,以母系为基础。每一半偶族包括许多地方氏族。这些氏族拥有土地所有权,占有若干单独的村落,村落再由一个或多个家庭组成。氏族有领袖(最高级成员),各家有家长。每一氏族可单独宣战、讲和及举行典礼,并在经济上各自独立。海达人的经济活动以海上捕鱼(尤以鲑鱼、大比目鱼及鳕鱼)及狩猎为主;长于木工技术,以其独木舟制作工艺著称,其艺术也久负盛名,用各种神物装饰实用器皿,风格高度刻板化,其精雕的"图腾柱",柱顶刻有家史上的重要事件。海达人的"散财宴"多在赠与、批准或维护政治地位(为酋长职位)或社会地位时举行。

Queen Charlotte Islands 夏洛特皇后群岛　加拿大不列颠哥伦比亚西部岛群。位于阿拉斯加锅柄地区南面。南北长约 280 千米,约有 150 个岛屿,陆地面积 9 596 平方千米,与阿拉斯加、不列颠哥伦比亚省和温哥华岛之间分别隔着狄克森海峡、赫卡特海峡和夏洛特皇后湾。最大的格雷厄姆和摩尔斯贝两岛,形状不规则,海拔约1 200米。群岛地势崎岖,受暖流影响,冬季气候温和。居民以印第安人为主,主要从事渔业。

Hecate Strait 赫卡特海峡　位于加拿大不列颠哥伦比亚省中部海上,为北太平洋东部通道。北起狄克森海峡,南至夏洛特皇后湾,长达260千米。水面宽64~128千米,将西面的夏洛特群岛与大陆隔开。峡内为捕捞鲑鱼及大比目鱼的渔场。

hemlock 铁杉　一种常绿乔木,通常树高25~30米,胸径40~80厘米。树皮片状剥落,褐灰色,大枝平展,枝梢下垂。树冠塔形,直立高大,树干下部之大枝通常不脱落。侧枝展开,线型的叶在枝上螺旋状排列,基部扭转排成二列,条形,先端纯圆,有凹缺,全缘。叶面绿色有光,叶背淡绿,有两条气孔带。铁杉于每年的4~5月开花,10月间球果成熟。铁杉为耐阴树种,幼树畏惧强烈日照,成年树可在林缘生长。它根系发达,能够抗风、抗雪压,寿命长,但生长较慢。

pine marten 松貂　又名林貂,是分布在北欧的一种鼬科动物。它们约有家猫大小,体长53厘米,尾巴长25厘米。雄貂较雌貂稍大,平均重1.5千克。毛皮呈浅褐至深褐色,在冬天会逐渐变长、变浅。松貂栖息在多草木地区,会在树穴或灌木林中筑巢。它们的爪可以半收缩,让它们可以生活在树上。它们主要在夜间或黄昏活动。

black-tailed deer 黑尾鹿　偶蹄目鹿科动物,是骡鹿在太平洋西北部的一个亚种。

rhinoceros auklet 角嘴海雀　海雀科角嘴海雀属的鸟类。体型略大(32厘米)的暗污色海雀。嘴橘黄或黄色,头部有两道特征性白色条纹,脚黄色。上体有灰色横斑及杂斑。繁殖羽的明显特征为眼后的白色眉纹及眼下的白色髭纹。嘴及脚的橘黄色较浓,上颚基部有一处小的浅色角状物,角嘴海雀因此得名。它繁殖于西伯利亚东部沿海、日本北部、千岛群岛、阿留申群岛、阿拉斯加和美国西北部及相邻海域。

pelagic cormorant 海鸬鹚　分布于北太平洋沿岸和邻近岛屿,最北从俄罗斯东西伯利亚海沿岸的符兰格尔岛和楚科奇海沿岸的楚科奇半岛经白令海峡到萨哈林岛、千岛群岛、日本和中国的辽东半岛一带,以及从北美洲的阿拉斯加、阿留申群岛到加利福尼亚。海鸬鹚体长约73厘米。全身羽毛呈黑色,头、颈部具有紫色光辉,其他部分有绿色光辉。繁殖期内头顶和头后部各具羽冠。嘴基内侧和眼周为红褐色。下体体侧具有较大的白色斑块。眼部为绿色,嘴为黑色。面部和喉部的裸皮呈褐色,并具有橙色小突起,脚为黑色。

Bering Sea 白令海　太平洋沿岸最北的边缘海,介海区呈三角形。北以白令海峡与北冰洋相通,南隔阿留申群岛与太平洋相连。位于太平洋最北端的水域。它将亚洲大陆(西伯利亚东北部)与北美洲大陆阿拉斯加分隔开。1728年丹麦船长白令航行到此海域,因而以他的姓氏命名。

Translation of Selected Sentences

1. Each national park, national historic site and national marine conservation area in the Parks Canada family tells its own story. Gwaii Haanas contains the essence of the rugged beauty and ecological character of the Pacific coast.

在加拿大的每一个国家公园、历史遗迹和国家海洋保护区都有它们自己的历史故事。哈纳斯国家公园和海达文化遗址包含着崎岖的山水美和太平洋海岸的特征。

2. This is mainly because of the island's proximity to the ocean which, like all larger bodies of water, tends to moderate the climate of the surrounding mainland. The climate is also influenced by the Alaska and the warm Japanese currents, as well as the coastal mountains to the east, the islands rugged topography, and prevailing westerly winds.

这主要是因为岛屿距离海洋较近,就像水巨大的身体一样,不断地去缓和内陆的气候。同时,这里和向东的海岸山脉的气候也受到阿拉斯加和日本气流的影响。除此之外,岛屿还受到岛屿崎岖地形的影响,使这一地区盛行西风。

3. Gwaii Haanas provides a model of a respectful relationship between human beings and the earth. It is a place where people can learn about a way of life where humans are part of nature, where they can experience Haida culture and respect sacred and spiritual values.

哈纳斯国家公园和海达文化遗址为我们提供了一个人类与自然界相互尊重和平相处的模型。在这里,人们可以了解到人类是如何成为自然界的一部分,同时还可以体验海达文化,尊重那些神圣的、精神上的文化价值。

4. The islands provide all the essential ingredients for sustenance and growth. Resourcefulness and readily accessible foods allow time for the continuation of a complex society and a unique art form that reflects the people's close interaction with the land and seas.

岛屿提供了生存和发展的所有必要的因素。丰富便捷的食物来源为这个复制的社会繁衍创造了条件,独特的艺术反映了人们和大自然陆地海洋的不断交流。

5. Some historic sites in Gwaii Haanas record the early presence of European and Japanese industrial activities. There are abandoned mining sites, whaling stations and remnants of old trails, logging roads, tramways and shipwrecks throughout the area.

在哈纳斯国家公园和海达文化遗址,一些历史遗迹记录着早期欧洲和日本开发者的工业活动。这里还可以看到被废弃的矿厂、捕鲸站和旧的伐木业的小路、轨道和船只残骸等踪迹。

Questions

1. What kind of special characteristics dose Gwaii Haanas National Park Reserve and Haida Heritage Site have?

2. What is the climate of Gwaii Haanas National Park Reserve and Haida Heritage Site?

3. What is the history of Haidapeople? How do these people live in the park?

4. What suggestions in the text have been put forward to protect the Gwaii Haanas National Park Reserve and Haida Heritage Site?

5. What are the responsibilities of Haida Watchmen from past to now?

Chapter 12

Ivvavik National Park
依华域国家公园

【简介】

依华域国家公园位于北极边，是育空地区各种动物的聚集地，占地10170平方千米。公园植被主要是苔原，它在箭猪、驯鹿的迁徙路线上，也是水禽的主要栖居地。公园有着世界上最伟大的急流河流之一，山谷窄到仅容一支独木舟可过，有许多急流区域。依华域国家公园是第一个经过当地原始居民和加拿大政府协商，为了保护特有的北极环境和历史文化而建立的国家公园。依华域国家公园已经被提名为联合国世界文化遗产。这一地区的主要历史是由最早期居住此地的原住民——印第安人文化。在依华域国家公园可以观赏极光、野生动物、淘金、独木舟泛舟、原住民文化、狗拉雪橇、滑雪等，而极光是公园最著名的自然奇景，最佳的观赏时间为冬季没有月亮的夜晚。

Located in the Yukon Territory, the Ivvavik National Park was the very first national park to be created under the aboriginal(土著的) land claim agreement between the Canadian government and the Aboriginal people. Ivvavik, meaning "a place for giving birth to and raising the young" is situated along the Beaufort Sea adjoining Alaska and covers 10,170 sq km. Ivvavik is a true wilderness area with no facilities or established trails within the park's boundaries.

It encompasses a variety of arctic and subarctic ecosystems, which range from the height of land in the south, through mountains, foothills, and river valleys to the coastal plain and the Beaufort Sea. Park is the summer home of the 120,000 Porcupine caribou herds, which migrates annually from the forested valleys of north-central Yukon to calving grounds on the Beaufort coast. Spring migration peaks at the end of May and early June. Archeologists have discovered the remains of eight distinct cultures that have traveled through the northern Yukon and there is evidence that human habitation could extend back 30,000 years. The park protects an ancient landscape and the memories of ancient cultures. During the last ice age, most of what is now Ivvavik was not glaciated(被冰河覆盖的). People have travelled in this area for at least 8,000 years, a story told by the numerous archaeological sites in the park and by the descendents of the people themselves.

The Arctic is a place of extremes, whether referring to the daylight, the temperatures or the weather. With a month of darkness in the winter and two months of twenty-four hour daylight in the summer, plant and animal life must take advantage of the short growing season, and have the ability to endure the harsh, long winters. The average summer temperature in Ivvavik is 14°C, and the average temperature in winter is -29°C. The annual precipitation(降水量) is small enough that the Arctic is actually classified as "desert". Coastal areas are largely influenced by maritime(海洋的) air masses. More continental subarctic conditions (e. g. less fog, less precipitation, larger fluctuations in temperatures) prevail in the centre and southern areas of the park.

The British Mountains dominate Ivvavik, accounting for about two-thirds of the park's area. These mountains, along with the Barn Range to the east, the Alaska Brooks Range to the west and the Old Crow Flats to the south, were not glaciated but formed a vast ice-free land mass called Beringia. The ice-free area extended into eastern Siberia, crossing the Bering Sea via a wide land bridge. Beringia was a refuge where plants, wildlife and people lived during the last glacial period of 30,000 to 14,000 years ago. As the continental ice sheets melted, colonization of the surrounding lands took place. Many features in the park are typical of non-glaciated landscapes. V-shaped valleys, isolated conical(圆锥形的) hills and side slope rock outcrops(露头) called "tors"(突岩) are the most obvious ones. The principal forces shaping the land are frost shattering, wind and water erosion, and the forces of gravity which slowly move material down slope. The ridges and peaks of the British Mountains range from 860 to 1,680 meters above sea level. These mountains increase in ruggedness from east to west. The Babbage, Firth and Malcolm river valleys carve their way through these mountains in a southwest to northeast direction. For thousands of years these valleys have been travel routes for wildlife and humans alike.

The Firth River provides a natural corridor(走廊) for a rich variety of wildlife, and for park visitors who want to see the diverse arctic landscape. The river offers 130 km of navigable river from Margaret Lake (near the Alaska border), northward to the Beaufort Sea. The flow of the Firth is relatively slow through the initial wide river valley, but increases in force as the valley narrows as it approaches Joe Creek. Travelling past Joe Creek, the river gradient (坡度) increases and numerous boulder(巨石) gardens ranging from Class II and III + are encountered. About halfway to the ocean the Firth River enters a canyon(峡谷) and does not emerge until it reaches the coastal plain. As the coastal plain is approached, the river slows and braids, forming a delta(三角洲) as it empties into the sea at Nunaluk Lagoon.

The coastal plain is a band of flat to rolling topography extending from the Bucklknd Hills north to the Arctic Ocean. The width of the coastal plain varies from 30 km in the east to 10 km in the west. During the last ice age a tongue of the continental ice sheet extended westward from the Richardson Mountains; it advanced along the plain to a point about halfway between the Firth and Malcolm rivers. As the glacier receded it left in its wake piles of sedimentary deposits. Some of these hills of glacial debris scattered along the Buckland Hills harbored chunks of glacial ice which have since melted and collapsed into kettle-like depressions. These depressions, often filled with water, are distinctive features of the landscape, as are the melt-water channels that once contained glacial floodwaters. Today, the coastal plain is influenced by the cooling effects of the Beaufort Sea. Snow and ice arrive by early October and do not melt until late May, and even then the sea is still frozen. Throughout the summer, ice floes(浮冰) and pack ice sit offshore and can blow in anytime. The ever present ice cools the air, delaying the onset and reducing the length of summer. The ice also creates fog banks which can make days cool and wet. As on all arctic coasts, wind is a factor and is cool even in mid-summer when it blows in from the north.

The entire northern Yukon is rimmed(形成边状) by the Beaufort Sea, a section of the Arctic Ocean. This is an ocean that is gripped in ice for at least eight months of the year. Open leads start developing along the coast in April, but the land fast-ice does not break up and drift away until June or even July. The frozen nature of the Arctic Ocean makes it technically difficult and expensive to study and it is the least understood of all ocean basins. The direction of offshore currents affects the shoreline topography as well as the habitats and movement patterns of wildlife. Shingle(小石子) spits and islands, oriented parallel to the shore, offer important nesting habitat to the sea birds. Brackish lagoons, protected from the open ocean by spits, are important feeding areas for both birds and fish. Ocean currents affect the location of open water leads in the land fast-ice each spring. Seals and whales depend on these leads to navigate to areas important for feeding, whelping(下崽) and calving(产仔). The same currents carry driftwood westward, from the mouth of the Mackenzie River, dispersing logs along thousands of kilometers of coastline. The logs provide essential habitat for a variety of small mammals and nesting birds. In the past, humans depended on driftwood for warmth and shelter while on the coast. The many driftwood shelters, sod roofed huts and grave sites are evidence of this.

The arctic climate makes is difficult for many species of plant and animal life to survive in the Ivvavik area. To thrive in Ivvavik, animals and plants must be able to take advantage of a short growing season which includes two months of twenty-four hour sunlight and also endure a long, harsh winter (including a month of total darkness). The three main types of vegetation in Ivvavik National Park are: arctic tundra, alpine tundra and taiga. Taiga is the transition between boreal forest and tundra. It consists of open stands of stunted spruce and balsam poplar. These trees grow to within 30 kilometers of the Beaufort Sea and represent the northernmost extension of their range in Canada. Permafrost(永久冻土) and amount and location of moisture, as well as air, soil temperature and exposure to wind have an impact on how plants are distributed throughout Ivvavik National Park. The Firth River Valley is home to the most delicate plants, whereas on the costal plain permafrost prevents many types of plant life from taking hold. Some of the wildlife that call Ivvavik home are: the porcupine caribou, muskoxen, dall's sheep and arctic ground squirrel. The Porcupine caribou herd migrates annually from the forested valleys of north-central Yukon to calving grounds on the Beaufort coast. Muskoxen prefer the open tundra of the coastal plain, but will sometimes wander south into the mountain valleys. Polar bears spend most of their time north of the park hunting for seals amongst the open leads of the offshore pack ice; however, they do come ashore throughout the year, especially during winter. Grizzly bears range over the entire park. Red and arctic fox are both found in Ivvavik. Dall's sheep is at the northern limit of their range here in the British Mountains. Moose occasionally wander onto the coastal plain but prefer to forage(寻觅食物) in valley bottoms. Wolves and wolverine travel wherever there is enough prey to sustain them. Marten and porcupine are forest dwellers that venture as far north as the tree line. The arctic ground squirrel thrives wherever it can find soft soil for burrowing. You may also see resident or migratory species of birds in the Ivvavik National Park.

There is evidence that human habitation could extend back as far as 30,000 years in Canada'sYukon. We do know for certain that people have lived there for more than 12,000 years. The remains of eight distinct native cultures that have traveled through the Yukon have been found by archaeologists. It is assumed these people crossed from Asia to North America by either ice bridges or the Bering Strait land bridge.

Many signs still remain today that mark the passage of Canada's original people. Most are not visual to the inexperienced eye. A campsite may be defined by a small circle of stones that had once been used as weights to hold down the edges of a teepee(圆锥形帐篷). A pile of rocks may be a cache(隐藏处) where the trophies of the hunt were hidden from scavengers(拾荒者) until the meat was needed. A line of boulders may have been used to drive caribou to waiting hunters. Arrowheads(慈姑) and spear points found on the tundra may have killed a buffalo or caribou.

Harpoon(鱼叉) heads found on beaches indicate the hunting of seal, whale or polar bear.

Two hundred years ago, there were two main Native groups living in the region. These were the Mackenzie Inuit and the Gwilchin(哥威迅人)These people were all but eliminated by disease brought in by the white men who came here to hunt.

In the early 1900's, fur traders set up trading posts. In the 1950's, the DEW line was built by the American army at Komakuk Beach. Since then, two more DEW line sites have been established in northern Yukon. People lived along the northern Yukon coast until the 1960's. Now, most people live in communities along the coast and adjacent to the Mackenzie River.

Artists in the Park are a unique outreach program that Parks Canada delivers. Each year, six artists are given the chance to capture special memories of Ivvavik National Park, whether it is from a photo, piece of art or a film. Accompanying the artists are a few Parks Canada staff along with a local cook. For seven to ten days, each individual has the opportunity to discover the vast beauty that Ivvavik National Park has to offer.

The most popular recreational activity in Ivvavik is whitewater rafting on the Firth River, which includes 130 navigable km from Margaret Lake to the Beaufort Sea. The rapids range from Class I to Class IV. Other activities include backcountry(边远地区) camping, hiking, wildlife viewing, fishing, and kayaking. Also, with twenty-four hours of daylight for essentially the entire summer, the soft light offers endless opportunities for photography.

Notes

Yukon Territory 育空地区　位于加拿大西北边陲,约十分之一位于北极圈内,气候严寒。育空省为加拿大三个行政区之一,是北美唯一有公路可进入北极圈的省份。面积比美国加州大,为483450 平方千米,人口约3.3 万人,以首都怀特霍斯最多,为2.4 万人,第二、三大城市为道森市、瓦森湖镇。坐落在育空西南部圣艾利雅斯山脉的克劳恩国家公园,被联合国教科文组织确立为世界自然文化遗产,园区内的洛根山,为加拿大最高峰。育空特区的主要历史是由最早期居住此地的原住民——印第安人文化,以 1898 客朗戴克拓荒及 1942 阿拉斯加公路兴建工程的开发为主要架构,追溯至上千万年前上一个冰河纪的长毛象时代。目前观光业是育空政府最主要积极致力推广的目标。育空地区的特色是拥有大量野生动物和连绵不绝的大自然风光,是北美唯一有公路可以通向北极地区的省份。育空的旅游项目主要有:极光观赏、淘金、野生动物观赏、独木舟泛舟、原住民文化、狗拉雪橇、滑雪等,而极光是育空地区最著名的自然奇景,最佳的观赏时间为冬季没有月亮的夜晚。

Beaufort Sea 波弗特海　北冰洋边缘海。位于美国阿拉斯加州北部和加拿大西北部沿岸以北至班克斯岛之间,北极群岛以西,楚科奇海以东。海中岛屿稀少,有无岛海之称。从阿拉斯加的巴罗角向东北延伸至帕特里克王子岛的终端地角;东起班克斯岛,西迄楚科奇海。20 世纪 50 年代以来,在近海大陆架发现了丰富的石油、天然气资源。海水中有 70 多种浮游植物和近 80 种浮游动物。底栖动物多达 700 种。沿岸居民以渔猎为生。近海有望发现油气田。1789 年英国探险家亚历山大·马更些沿马更些河到达河口,为第一个发现波弗特海的欧洲人。1806 年英国海军水文地理学家弗兰西斯·波弗特在此估测船只在海上航行时所遇到的不同风速,设计了著名的蒲福风级。此后,各国探险家陆续来此考察。

Alaska 阿拉斯加州　位于美国西北太平洋东岸的州,第 49 个加入美利坚合众国的州,也是美国最大的州、世界最大的飞地地区。阿拉斯加州东接加拿大的育空、不列颠哥伦比亚,南邻阿拉斯加湾、太平洋,西濒白令海、白令海峡、楚科奇海,北临北冰洋。原属俄罗斯,1867 年被俄国人以 720 万美元的价格卖给了美国。

Porcupine caribou 北美皮里驯鹿　身材较小,白色,仅生活在加拿大极北的群岛上。体重在 55 ~ 70 千克。1980 年北美皮里驯鹿的总数估计尚在 3 300 ~ 3 600 头,但现在已经被加拿大濒危野生动物保护委员会列为濒危种类。

Brooks Range 布鲁克斯山脉　美国阿拉斯加州北部,为落矶山脉北端延伸部分。从楚科奇海至育空地区(加拿大),东西延伸 1 000 千米,宽 129 千米。西部山峰平均海拔为 900 米,中东部山峰平均海拔达 2 700 米,最高点伊斯托山为育空河流域和北冰洋流域的分水岭。中部的阿纳克图武克山口是进入育空低地的主要通道。北麓的普拉多湾富石油资源。阿拉斯加石油管道在阿蒂冈山口穿过山脉通至南端终点站瓦尔德斯。

Beringia 白令陆桥　位于白令海,伸延至极限时长达 1 600 千米。白令陆桥连接现今的美国阿拉斯加西岸和俄罗斯西伯利亚东岸、更新世(前 180 万—前 1 万年)时连接的地方数量和变化无法估计。因为来自太平洋的西南风吹至阿拉斯加时已逐渐失去水分,以致白令陆桥降雪量低,所以它未结成冰川。

Siberia 西伯利亚　俄罗斯境内北亚地区的一片广阔地带。西起乌拉尔山脉,东迄太平洋,北临北冰洋,西南抵哈萨克斯坦中北部山地,南与中国、蒙古和朝鲜等国为邻,面积 1276 万平方千米,除西南端外,全在俄罗斯境内。也有人将北冰洋同太平洋水系分水岭作为其东界。

Mackenzie River 马更些河　加拿大第一长河,在北美六大河中仅次于密西西比-密苏里水系。源流阿萨巴斯卡河和皮斯河均源于加拿大洛矶山脉东麓。阿萨巴斯卡河向东北注入阿萨巴斯卡湖,出湖后与皮斯河汇合成奴河,向北入大奴湖;出湖后始称马更些河,向西北流,最后入北冰洋波弗特海。

Taiga 泰加林　指从北极苔原南界树木线开始,向南延伸 1 000 多千米宽的北方塔形针叶林带,为水平地带性植被,是世界上最大的,而且也是独具北极寒区生态环境的森林带类型。又称“寒温带针叶林”或“北方针叶林”。泰加林带主要由耐寒的针叶乔木组成森林植被类型。主要的树种是云杉、冷杉、落叶松等,而且往往是单一树种的纯林。

Bering Strait 白令海峡　位于亚洲最东点的迭日涅夫角和美洲最西点的威尔士王子角之间，西经 169 度零分，北纬 65 度 30 分，约 85 千米宽，深度在 30 - 50 米之间。这个海峡连接了楚科奇海（北冰洋的一部分）和白令海（太平洋的一部分）。它的名字来自丹麦探险家的维他斯·白令，他在 1728 年俄国军队任职时候穿过白令海峡，是第一个穿过北极圈和南极圈的人。

Inuit 因纽特人　也称爱斯基摩人，意为“真正的人”。他们主要居住在北极地区，即包括北冰洋沿岸的亚、欧、北美三洲大陆北部和北冰洋中的许多岛屿上。约 6000 年前，居住在西伯利亚的因纽特人横过白令海峡到达阿拉斯加。他们制造了皮艇、鱼钩、鱼叉等工具捕捉鱼、海豹、海象甚至鲸鱼，使用石制弓箭、刀、刮刀，吃食驯鹿、鹿牛、巨象、鸟类等，用海生动物油点灯，用针在脸上刺花纹。另外一些人则沿阿拉斯加北岸东行至格陵兰，发展了独立的多尔赛特文化，此文化在公元前 1 200—1000 年较繁荣。居住在北极地区的因纽特人将近 20 万，其中约有 6 万来自阿拉斯加、加拿大、格陵兰。居住在加拿大及格陵兰的则多以狗拉雪橇为运输工具。生活在内陆者以狩猎为生，近海者以捕猎海兽、鱼类为主。

Translation of Selected Sentences

1. It encompasses a variety of arctic and subarctic ecosystems, which range from the height of land in the south, through mountains, foothills, and river valleys to the coastal plain and the Beaufort Sea.

它包括北极和亚北极生态系统的多样性，从南部陆地，穿越山脉和山麓小球，再经过河流山谷，最后到海岸平原和波弗特海，各种各样的地形和生态都在公园得以体现。

2. The Arctic is a place of extremes, whether referring to the daylight, the temperatures or the weather. With a month of darkness in the winter and two months of twenty-four hour daylight in the summer, plant and animal life must take advantage of the short growing season, and have the ability to endure the harsh, long winters.

不论是日光、温度还是天气，北极地区都是一个极端气候地区。在冬天有将近一个月的黑暗期，夏天有两个月的 24 小时白昼。所以植物和动物必须利用好着短短的生长季节，才有能力去经受这严酷而漫长的冬季。

3. The frozen nature of the Arctic Ocean makes it technically difficult and expensive to study and it is the least understood of all ocean basins. The direction of offshore currents affects the shoreline topography as well as the habitats and movement patterns of wildlife.

在技术和费用方面，北冰洋冰冻的自然环境使得对它们的研究变得困难而昂贵，人们很难理解所有的海洋流域。近海的水流影响着海岸线地形和野生动物的栖息和迁徙模式。

4. The remains of eight distinct native cultures that have traveled through the Yukon have

been found by archaeologists. It is assumed these people crossed from Asia to North America by either ice bridges or the Bering Strait land bridge.

考古学家已经在育空地区发现了八种不同的本土文化。它证明了这些原住民或是通过冰桥,或是白令陆桥,穿越亚洲来到北美洲的。

5. Each year, six artists are given the chance to capture special memories of Ivvavik National Park, whether it is from a photo, piece of art or a film.

每年,六位艺术家有机会来到依华域公园,以照片、艺术品或是电影的形式捕捉公园独特的风景。

Questions

1. What is the general condition of Ivvavik National Park?
2. Why is it hard to live in the park? What's the climate of the park?
3. How many types of vegetations does the park have? What are they?
4. Who lived in the park two thousand years ago? Who were they?
5. What is the most popular recreational activity in the Ivvavik National Park?

Chapter 13

Jasper National Park
贾斯珀国家公园

【简介】

贾斯珀国家公园位于阿尔伯塔省洛矶山脉最北边，在班芙国家公园的北面及艾德蒙顿的西面。它是加拿大最大的高山国家公园，占地 10 878 平方千米。在贾斯珀国家公园，黑熊是这里的常住民，其他的野生动物，例如，北美红鹿山羊和大角羊等，加起来的数量比居民还多。贾斯珀国家公园内之佩投湖、弓湖、巫药湖、派翠西亚湖、金字塔山、玛琳湖及玛琳峡谷鬼斧神工、美丽如画。最令人惊叹景的是由班芙国家公园的路易斯湖驶往亚伯达贾斯珀的那段称为冰原公路的路，被认为是世界上风景最美丽的车道。1813 年西北公司在现今贾斯珀城镇附近设置补给站。补给站由贾斯珀·赫斯看守，1884 年被废弃。1907 年 9 月加拿大政府设立了贾斯珀森林公园，也就是如今贾斯珀国家公园。1984 年它被联合国科教文组织宣布为世界遗产。

Jasper National Park is the largest national park in the Canadian Rockies, spanning 10,878 km^2. It is located in the province of Alberta, north of Banff National Park and west of the City of Edmonton. Jasper was named after Jasper Hawes, who operated a trading post in the region for the North West Company. Before this it was referred to as Fitzhugh. The park was established on September 14, 1907 as Jasper Forest Park, and was granted national park status in 1930, with the passing of the National Parks Act. This park was declared a UNESCO World Heritage Site in 1984, together with the other national and provincial parks that form the Canadian Rocky Mountain Parks, for the mountain landscapes containing mountain peaks, glaciers(冰川), lakes, waterfalls, canyons(峡谷), and limestone(石灰岩) caves as well as fossils(化石) found here.

Jasper National Park is an area of dramatic mountain peaks and expansive(广阔的) valleys. Many of the mountains rise to elevations above 3000 meters. Mt. Columbia, the highest peak in Alberta, is 3782 meters. The lowest point in the park, 985 meters above sea level, lies near the park's east gate. The mountains in Jasper National Park form ranges running parallel to one another in a northwest to southeast alignment. The westernmost range in this series is the highest. It also forms the continental divide, separating streams that flow east and north to the Arctic Ocean or Hudson Bay from those which drain westward to the Pacific Ocean. In places, gaps or passes form breaks in the steep-walled ranges and allow travel from one watershed to another. Rivers have also eroded passages through this otherwise rugged and precipitous(陡峭的) land.

To remember Rockies geology, take it range by range. The main ranges are the highest ranges (they includeMt. Robson and the group of mammoth (巨大的) peaks surrounding the Columbia Icefield). The main ranges form the continental divide, which splits North America into eastern and western slopes. In Jasper National Park, the main ranges are largely coarse sandstone, limestone, shale (页岩), and a somewhat metamorphic (变质的) rock called quartzite(石英岩), in particular, 'Gog' quartzite.

East of the main ranges are the front ranges of the Rockies. The sea allowed for the profusion (大量) of a great many plants and animals here, probably because of its shallowness. The organic material that resulted contributed lime(石灰) mud, which thickened and hardened into massive amounts of limestone. The front ranges are pale grey and distinct from the reddish/brownish main ranges. Typical front-range peaks in Jasper are the Colin Range and Roche Miette. Farther east, out of the park, we get to the foothills, those gentle mounds that appear just before the flatness of the plains.

You will not see the western ranges in Jasper National Park. They are located south of here, in Kootenay and Yoho National Parks.

The Athabasca River and its tributaries(支流) drain more than four-fifths of the park. From the toe of the Columbia Glacier the Athabasca flows north and east through the park for nearly 150 kilometers. Ten other moderately-sized rivers and countless smaller creeks augment(扩大) its volume over this distance. The Athabasca is one of the more significant rivers which flow from the mountain barrier toward the Arctic Ocean. The portion of the Athabasca which lies within Jasper National Park has been designated as a Canadian Heritage River.

Part of the north end of Jasper National Park lies in a different drainage basin than that of the Athabasca. The Smoky River, which originates in the extreme northwest corner of the park, courses northward to the Peace River in northern Alberta. Water from the most southeasterly corner of the park is discharged via the Southesk and Brazeau rivers, which exit from the mountains and join the North Saskatchewan River.

These watercourses collect the runoff from an area of over 10,000 square kilometers. Peak flow begins in late June, when snowmelt and rainfall combine to fill the river channels to capacity. Melting glaciers keep the water levels high through July. In midsummer, these rivers contribute significantly to the water supply required for municipal and agricultural needs farther downstream.

With the onset of a warmer, drier climate following their last major advance, the retreating glaciers left exposed a barren(贫乏的) and raw landscape. Soon plants and animals began to take advantage of these newly-exposed expanses, infiltrating the mountain valleys from ice-free areas. Some of these organisms invaded from regions to the south that had remained unglaciated. Other organisms arrived from their unglaciated mountain peak refuges, islands of rock that the glaciers had failed to cover.

The climatic conditions of areas adjacent to melting glaciers were far from ideal for survival, but the constant glacial winds kept nearby valley bottoms mostly snow-free and the melting ice provided an abundance of water. Grasses and other hardy, cold-resistant species of vegetation soon took root in glacial silt and colonized the open spaces. As time passed, the climate continued to moderate. The glaciers dwindled(减小) and disappeared from all but the higher areas near the continental divide. As the annual cycle of growth and decay in the grasslands broke down the rock rubble and built up nutrient-rich soils, trees took root and spread through the valleys. During the warmest periods, between about 7000 and 3500 years ago, it is likely that grasslands and drier forests dominated, at least at lower altitudes. This combination of climate and habitat encouraged the growth of bison, elk and mountain-sheep herds.

More recently, during the Little Ice Age (from about A. D. 1200 to 1850), the onset of cooler temperatures and higher snowfall initiated a minor advance of the glaciers from their mountain strongholds. These colder conditions pushed tree lines lower and permitted cold-tolerant spruce and fir forests to dominate the mountain slopes, much as they do today. Caribou, moose and bison, all species adapted to travel in deeper snow, frequented much of the area, while elk and deer were confined to the grasslands still covering the bottom lands of the lower Athabasca Valley and the slopes of the drier front ranges.

During the past 150 years the climate has continued to fluctuate. Vegetation patterns have responded to its changes. Occasional severe winters, drought and periodicfire have played their part in shaping the ecology of the mountain regions. Numbers of animals, not to mention numbers of humans, have all been affected by the prevailing climate.

Although the Pacific Ocean liesin several mountain ranges to the west, ocean winds influence much of Jasper's weather and contribute to the forces of change in the mountains. The winds rush up and over the main ranges. Moisture-laden clouds dump meters of snow and water in highlands such as the Columbia Icefield. Drier winds flow northeast down the Athabasca Valley, lifting sand and silt and sweeping slopes bare in their wake.

When the dominant west wind falters(减弱), Arctic fronts sweep in from the north and east and the temperature plummets. The park's northerly location and distance from the ocean make the climate of Jasper National Park ' continental', with long, cold winters and short, warm summers.

Jasper National Park is divided into threeJasper National Parks—montane (山区的), subalpine(亚高山带的), alpine—which are broad landscape units with characteristic species, communities and physical environments. Climatic differences associated with changes in altitude are the main determinants of these differences in biodiversity. Higher elevations are generally colder and wetter, while lower elevations are warmer and drier. Local differences in slope angle and direction can create local microclimates. Steeper slopes are generally better drained and drier than moderately sloping or flat areas. South-facing slopes are drier and warmer than north-facing ones at the same altitude.

The ranges of plants and animals across this altitudinal range are related largely to their tolerances to cold, heat and drought. Other factors that influence distributions include food species, competition with other species, and soil conditions. The wettest areas are occupied by lakes, ponds, marshes and fens(沼泽). Grasslands occur in the warmest, driest portions of the park and forests in moderate environments. Trees can't grow in the cold conditions of high altitudes above about 2200 meters and so forests are absent, replaced by low shrub and wildflower communities.

For most visitors, a trip to Jasper is about seeing wildlife. Jasper National Park protects a vestige of the wildlife that once blanketed the West. While the last 200 years have seen a dramatic decline in wildlife in most parts of North America, healthy populations of plants and animals have persisted inside the park. Jaspers three ecological zones are home to 1300 species of plants: the dry forested montane in the river valleys that make up 10% of the park, the sub-alpine forests and the rocky alpine tundra. The grassy open slopes and mixed wood forests of the montane zone make ideal grazing. Trembling aspen(颤杨) and Douglas fir grow primarily in the river bottoms where climate is less severe. The montane is also home to arnica (山金车), paintbrush, crocus(番红花), yellow lady's slipper, wild rose and wild strawberry. About half of Jasper is sub-alpine forest characterized by larch(落叶松树), sharp-pointed Engelmann spruce and alpine fir, which can spread down valleys like a black tide. Multi-colored columbine and the orange western wood lily cling to the mountain slopes. Above the tree line, the back country alpine tundra display moss campion(剪秋萝) with small pink flowers growing in clusters, wolf lichen(地衣), old man's beard and dog-ear lichens that may have been growing there for a hundred years.

Open slopes are the terrain of grizzly bears, elk and bighorn sheep; most animals that live in the park inhabit the grasslandsand open forest of the montane zone. Marsh areas in the low-lying parts provide excellent habitat for moose, which can weigh up to 500 kilograms, beaver and waterfowl. Today they can be seen around Pyramid Beach or the Valley of the Five Lakes building lodges and dams and eating the bark of the trembling aspen. Cougars(美洲狮) and the smaller lynx(山猫) have been sighted in the park but both are very rare. Porcupine, wood rat, marmot, squirrel and deer mouse are also montane residents. Canada's grizzly (or brown) bear, the biggest and most easily recognized for the pronounced hump on its back, is neither slow nor limited in its range. Strong enough to kill an elk and drag it uphill or swim in rapids to catch salmon, it is never tame or harmless. People and wolves are their only predators. Sadly the grizzly has been extirpated(消灭) in many of its original habitats and parks such as Jasper may offer its only chance for survival. Black bears are smaller, darker, less endangered, but no less dangerous. Both eat plants, fruit and fish, but will chase down and kill live animals in their Rocky Mountain habitat. In the upper alpine region lives the hairy marmot whose piercing whistle provided the name for the Whistlers just east of the town of Jasper. About 2000 bighorn sheep and considerably fewer mountain goats live in the park and can often be sighted from the roadways. Wolves hunt the elk, deer, caribou and moose. The coyote(丛林狼), smaller than the wolf, has a bushier tail and a yappy(爱叫的) bark; foxes are very rare. Mule deer and elk commonly appear on the open slopes, while the dwindling population of mountain caribou is being carefully monitored and protected by the park. The populations of rodents, golden mantled ground squirrel and rock rabbit attract a number of birds of prey: bald eagle, golden eagle, great horned owl, osprey(鹗), falcon and several types of hawk. The ptarmigan(雷鸟), grouse, jays(松鸦) and hummingbirds all live here. Rainbow trout(鲑), northern pike(狗鱼) and stickleback(棘鱼) are common in the park's lakes and watersheds.

Some of the park's scenic attractions includeMount Edith Cavell, Pyramid Lake with Pyramid Mountain, Maligne Lake, Medicine Lake, and the Tonquin Valley. Other attractions are the Marmot Basin ski area; the Snocoach tours of the Athabasca Glacier, a distributaries of the Columbia Icefield; Athabasca Falls; Whistler Sky-Tram, the Jasper Tramway, and numerous other outdoor related recreational activities (such as hiking, fishing, wildlife viewing, rafting, kayaking and camping). The Miette Hot Springs are located close to the northeast entrance. The

Miette Hot Springs are created by an extremely hot spring cooled by the mountain to temperatures suitable for humans.

Notes

Canadian Rockies 加拿大洛矶山脉　洛矶山脉的一部分，北从加拿大不列颠哥伦比亚起延伸约1600千米，更北沿西北地方和育空地方边界的马更些山脉及塞尔温山脉常划归加拿大落矶山脉。西方的洛矶山沟（地质上的陷落处）分隔了加拿大洛矶山脉前山岭与哥伦比亚山脉，哥伦比亚山脉包含卡里布、塞尔扣克、莫纳希、珀塞尔等山脉，常被视为洛矶山系的一部分。

Banff National Park 班夫国家公园　建于1885年，是加拿大历史最悠久的国家公园。它坐落于洛矶山脉北段，距加拿大阿尔伯塔省卡尔加里以西约110～180千米处。公园共占地6 641平方千米，遍布冰川、冰原、松林和高山。冰原公路从路易斯湖开始，一直连接到北部的贾斯珀国家公园。西面是省级森林和幽鹤国家公园，南面与库特尼国家公园毗邻，卡纳纳斯基斯镇位于其东南方。公园内主要的商业区为弓河山谷的班夫镇。

Edmonton 埃德蒙顿　加拿大阿尔伯塔省省会。位于该省中部，跨北萨斯喀彻温河两岸，距美、加边界560千米，是通向北方的门户。它是阿尔贝塔省人口第二多的城市（排在卡尔加里之后），是加拿大第五大城市。埃德蒙顿是阿尔伯塔省的文化、政府和教育中心，城市拥有轻轨运输，一年有许多节日，并有北美最大的室内购物中心——西部埃德蒙顿购物中心和加拿大最大的历史公园——埃德蒙顿堡垒公园。

Mt. Columbia 哥伦比亚山　位于加拿大不列颠哥伦比亚省东南部。东界落矶山沟，南临哥伦比亚河，西接内陆高原，北滨弗雷泽河。与加拿大境内的落矶山脉平行，一说是它的西部延伸，大致由西北向东南连绵600千米。从自然地理方面观察，哥伦比亚山脉由四条显然分开的山脉——卡里布、莫纳希、塞尔扣克和珀塞尔组成，海拔均超过3 000米。它是包括国立和省立公园的主要游览区。山中有金、银、铜、铅和锌等重要矿物资源。

Mt. Robson 罗布森山 位于加拿大不列颠哥伦比亚省东部，为加拿大境内落矶山脉的最高峰（3 954米）。四周为省立罗布森山公园，加拿大国营铁路的干线在公园中通过。

Columbia Icefield 哥伦比亚冰原　位于加拿大不列颠哥伦比亚省和亚伯达省交界处，是落矶山系中第一大冰原，北美洲最容易进入的冰川群之一。因其主要堆积区位于大陆分水岭，故称河流之母。冰川区介于西部海拔3 747米的哥伦比亚山和东部海拔3 491米的阿萨巴斯卡山之间。南部有贾斯珀国家公园，东部有公路通往亚伯达省的班夫和卡利加利。冰原面积500平方千米，由冰盖和十几处冰川组成。冰盖一般海拔为3 000米，最高段冰层厚度为130米。有两条主要冰舌。北部的阿萨巴斯卡冰川面积约为30平方千米，属于典型的加拿大落矶山型冰川。其融水经阿萨巴斯卡河、阿萨巴斯卡湖、奴河、大奴湖及马更些河注入北冰洋。

Kootenay national park 库特尼国家公园　设置于 1920 年，占地面积 1 406 平方千米。1910 年不列颠哥伦比亚省居民布鲁斯建议省府和加拿大太平铁路公司合作，开辟公路连接班夫和哥伦比亚谷地的温得米湖区，以便他在谷地种植苹果收成后，得以销往阿尔伯塔省。省府接受建议并于翌年开工，但是工程浩大超出想象，修筑了 22 千米后停工。一次大战结束，布鲁斯转向联邦政府提议拨款筑路，不列颠哥伦比亚省府同意将道路两边各让出 8 千米土地，用作国家公园。

Saskatchewan River 萨斯喀彻温河　加拿大阿尔伯塔省和萨斯喀彻温省的最大河系。源出于加拿大落矶山中，有两大河源，即北萨斯喀彻温河和南萨斯喀彻温河。两河在艾伯特王子城以东汇合，向东 550 千米注入温尼伯湖。汇合处有水力发电站。萨斯喀彻温河以波河为源头计算，全长 1 939 千米，流域面积 38.3 万平方千米。长期为重要的毛皮贸易通道。现主要用于发电和灌溉。河上有几座水坝，全流域为加拿大最大的灌溉区。

Maligne Lake　玛琳湖 洛矶山脉中最美丽的湖泊之一，也是加拿大落矶山脉最大的冰河湖。玛琳湖位于杰士伯镇东南方 48 千米处，湖南北长约 22 千米、东西约长 1 千米。玛琳湖有着秀丽的湖光山色，湖面细长，蜿蜒延伸至远方。三面山峰间环抱着一湖碧水，景色宜人。

Medicine Lake 巫药湖 玛琳湖的水流一部分流入地下，一部分屯积形成巫药湖，其水主要来自玛琳湖，夏天冰雪融化往下流时，巫药湖就水满湖碧；但一到冬天结冰，大部分水都由地下流走，巫药湖渐渐干涸，湖底也会裸露出来，所以早年印第安人称之为“消失的小湖”。一些随季节迁徙的动物，如北美红鹿、驯鹿、灰熊和麋等都在湖边及干涸的湖床上觅食。

Athabasca Falls 阿萨巴斯卡瀑布　贾斯珀国家公园内著名的瀑布。虽然阿萨巴斯卡瀑布不高，只有 23 米，但是其特色之处是那又急又有力的水流。它起源于哥伦比亚冰原的阿萨巴斯卡河，是整个贾斯珀国家公园里最大的河流，所以这个瀑布的排水量非同小可。阿萨巴斯卡瀑布是洛矶山中最令人惊心动魄的瀑布。当阿萨巴斯卡河水进入狭窄的河道时，水流湍急，河水两边的石壁被冲刷得光滑如镜。水流携带着沙石，以巨大的力量冲击形成壶穴。

Translation of Selected Sentences

1. The mountains in Jasper National Park form ranges running parallel to one another in a northwest to southeast alignment. The westernmost range in this series is the highest. It also forms the continental divide, separating streams that flow east and north to the Arctic Ocean or Hudson Bay from those which drain westward to the Pacific Ocean.

贾斯珀国家公园的山脉从西北到东南一个接一个地并排成一列，最西边的山脉是这一区域最高的。同时，这一系列的山脉组成了天然的大陆分界线，把向东、北流向北冰洋的溪流或是流向哈德逊湾，一直再向西流向太平洋的河流分开。

2. During the warmest periods, between about 7000 and 3500 years ago, it is likely that grasslands and drier forests dominated, at least at lower altitudes. This combination of climate and habitat encouraged the growth of bison, elk and mountain-sheep herds.

在最为温暖的时期，大约7000至3500年前，至少在低纬度地区遍布着草地和较干燥的森林。这种气候和栖息地的结合给美洲野牛、驼鹿和山羊创造了很好的环境，促进了它们的生长。

3. During the past 150 years the climate has continued to fluctuate. Vegetation patterns have responded to its changes. Occasional severe winters, drought and periodic fire have played their part in shaping the ecology of the mountain regions.

在过去的150年间，气候不断地变化，植物也随之变化。偶尔严酷的冬天、干旱和周期性的火灾塑造着山区的生态系统。

4. Local differences in slope angle and direction can create local microclimates. Steeper slopes are generally better drained and drier than moderately sloping or flat areas. South-facing slopes are drier and warmer than north-facing ones at the same altitude.

各地坡度角度和方向的不同形成了各地微型的气候环境。较陡的坡通常要比较平的或是平地要干涸和干燥。同一海拔，朝南的斜坡要比朝北的要干燥和温暖。

5. For most visitors, a trip to Jasper is about seeing wildlife. Jasper National Park protects a vestige of the wildlife that once blanketed the West. While the last 200 years have seen a dramatic decline in wildlife in most parts of North America, healthy populations of plants and animals have persisted inside the park.

对于参观者来说，游览贾斯珀国家公园就是去看那些野生动植物。公园保护着在西半球曾经快要消失了的野生动植物。在过去的200年间，在北美洲的大部分地方，野生动植物不断地消亡，但是在贾斯珀国家公园，那些健康的野生动植物被很好地保护着。

Questions

1. What's the general condition of Jasper National Park?
2. How did the Jasper National Park get its name?
3. How did the changes of climate impact the growth and development of environment and lives?
4. How many life zones dose the Jasper National Park have? What are they?
5. What kind of special attractions does the Jasper National Park have?

Chapter 14

Kluane National Park and Reserve
克卢恩国家公园和保护区

【简介】

克卢恩国家公园和保护区位于育空地区的西部边缘，占地21 980平方千米，公园的一半覆盖着大面积冰川。圣埃利亚斯山脉横跨自然保护区，整个山脉有极地冰盖以外的世界最大的冰原和世界上移动最快的冰川，南端冰川划入冰川湾国家公园。公园有8座4 500米以上的高山，高山林立，其中包括加拿大最高峰罗根山，海拔有5 959米。它是灰熊和善走山路的北美大野白羊的家园。这项跨国的世界自然遗产坐落在辽阔的北极地区，素有"极地奇观"的美称。这里茫茫野地高低不平，宏伟壮观，是世界上最大的野生保护区之一。此外，克卢恩国家公园也是土著历史文化保护区。

If you've ever imagined standing amidst a sweeping landscape of mountain valleys carved by glaciers and sprinkled with alpine wildflowers, you will get to live your dream in Kluane National Park and Reserve. A gem in the family of Parks Canada's national treasures, Kluane National Park and Reserve of Canada covers an area of 13,657 square miles. It is a land of precipitous(陡峭的), high mountains, immense ice fields and lush(茂密的) valleys that yield a diverse array of plant and wildlife species and provides for a host of outdoor activities. Kluane National Park Reserve is located in the southwest corner of the Yukon Territory, bordering on both British Columbia and Alaska. The reserve was established in 1976, and protects Canada's Northern Coastal Mountains Natural Region. The Alaska and Haines highways skirt the eastern boundary of the park along the frontal Kluane National Park & Reserve ranges. The park includes the highest mountain in Canada, Mount Logan (5,959 m) of the Saint Elias Mountains. Mountains and glaciers dominate the park's landscape, covering 82% of its area. The park's mountains are part of the Wrangell-St. Elias chain. Kluane and its American counterpart, Wrangell-St. Elias National Park, were jointly declared a World Heritage Site by UNESCO in 1979. Kluane National Park and Reserve has for thousands of years been the home of the Southern Tutchone—First Nations people; the name Kluane is derived from a Southern Tutchone word meaning "lake with many fish".

Kluane's climate is a contrasting blend of moisture and temperature conditions due to its size, mountainous terrain, and proximity to the Pacific coast. Much of the park experiences a dry, cold, continental climate, being in the rain shadow of the St. Elias Mountains. The southern part of the park is subjected to a maritime influence, with generally higher temperatures and more precipitation.

Kluane National Park & Reserve is dominated bymountains and ice, which make up 82% of the surface area. The St. Elias Mountains, Canada's highest and most massive mountains, have two ranges separated by a narrow trough, the Duke Depression. The Kluane Ranges, a chain of mountains averaging 2,500m (8,000') in height, are visible to travellers on the Haines Road or the Alaska Highway. Beyond these guardians of the interior, to the west lie the rugged Icefield Ranges, whose peaks soar into the 5,000m (16,000') range. The giants are: Mount Logan (5,959 m, 19,545') the highest mountain in Canada and the second highest peak on the continent; Mount St. Elias (5,488m, 18,005'); Mount Lucania (5,231m, 17,162') and many others. Although many of the highest mountains are not visible from any point along the highway, some of the higher peaks can be spotted in the distance from viewpoints near Kathleen Lake or the Donjek River.

Amid these ranges is a legacy of the lastIce Age—one of the world's largest non-polar icefields. Massive quantities of snow continue to accumulate as moist Pacific air moves over the St. Elias Mountains. Valley glaciers such as Naludi (Lowell Glacier), 65 km. long, radiate from the icefields. Glacial movements are often immense and spectacular. In the past, surges of Naludi (Lowell Glacier) have blocked the Alsek River near Goatherd Mountain with a dam of ice. The resulting glacial lakes extended well back to and over the present site of Haines Junction.

The most recentLake Alsek is said to have drained around 1850 in two days after the ice dam broke, with a flow rate comparable to that of the Amazon River. Huge gravel current ripples from this outflow, as well as wave-cut lake benches, are visible along the Alsek trail, 10 km north of Haines Junction along the Alaska Highway. The flooding associated with the advances and retreat of Naludi (Lowell Glacier) is the subject of many Southern Tutchone native people's legends and stories.

Rising in the mountains of Kluane National Parkand Reserve, its braided upper reaches nestle in a broad valley, providing an oasis(绿洲) for mountain goats, Dall's sheep and other wildlife. Sand dunes and glacial till provide a contrasting environment for the diverse vegetation found here. Downstream, rivers of ice flow into the Alsek's silt-laden waters, calving huge icebergs into the river before it leaves the Park on its journey to the Pacific Ocean on the Alaskan Panhandle.

Theclimatic overlap of the pacific and arctic air masses over Kluane National Park and Reserve has resulted in one of the greatest diversity of plants and wildlife in northern Canada. A montane(山地森林的) forest of white spruce, trembling aspen and balsam poplar covers much of the lower valleys and slopes. Tree line is at 1050 to 1200 m (3,500 to 4,000'), depending upon local conditions. Low-growing or stunted shrubs in the transition zone include: willow, dwarf birch(桦树), and alder, which provide protection for the smaller plants. Summers in alpine tundra (generally above 1400 m or 4600') are a flourish of color, with over 200 varieties of alpine flora.

Kluane National Park & Reserve provides prime habitat for many species of wildlife, particularlyDall sheep, Kluane National Park & Reserve's most abundant large mammal. Visitors often see them foraging on the wind swept slopes of Tachal Dhal in the spring, fall and winter. Mountain goats scramble(攀爬) along rocky cliffs and ledges(岩架) in the south, and a small herd of caribou occasionally wander through the Duke River area. North America's largest subspecies of moose also range through Kluane National Park & Reserve.

Populations ofgrizzly bears(灰熊) move between alpine meadow and valleys with the passing of the seasons. Black bears, which prefer forested areas, are also common in Kluane National Park & Reserve. As well as a transient population of wolves there is also a variety of smaller mammals including: wolverine, muskrat(麝鼠), mink(鼬), marmot(土拨鼠), red fox, lynx, otter(水獭), coyote(丛林狼), beaver, snowshoe hare and arctic ground squirrel.

The south-western Yukon has a great diversity of birds. At least150 species have been observed in Kluane National Park & Reserve, of which 118 nest in the park. On a spring day, sightings could include varied thrushes, yellow-rumped warblers and mountain bluebirds. Kluane National Park & Reserve also provides sanctuary for a variety of birds of prey from the swift flying falcon to the broad soaring wings of bald and golden eagles.

Since time immemorial, the Southern Tutchone (南突修尼族) people, which include Champagne and Aishihik First Nations and Kluane First Nation, have lived throughout the area that is now Kluane National Park and Reserve. Over thousands of years, dän (the people) have developed effective methods for living in this land of extremes. Elders recall a rich history which details events and places spanning many generations.

They once lived a nomadic lifestyle, travelling great distances, in order to take advantage of seasonal abundances of wildlife and plants throughout their traditional territory. This lifestyle required developing a superior skill set of hunting and travelling techniques in an area of harsh climatic and geographic variations.

The accuracy and depth of the Southern Tutchone language was crucial to the survival of the dän. Formal teachings as well as traditional stories told by Elders were and continue to be passed on orally. Place names and the stories associated with them were a source of critical knowledge; they identified where resources could be found, referred to past events, and described the terrain in a way that a traveler would have knowledge of a place without having been there before. The Southern Tutchone name for Kluane Lake is "Łù'àn Män" meaning "big fish lake". The coastal Tlingits, who were trading partners, called the area "ùxh-àni" meaning "whitefish country". The name 'Kluane' (pronounced 'Kloo-wah-nee') was derived from these two names by early settlers.

In 1943, the Kluane Game Sanctuary was established and First Nations were denied access to hunting, fishing and trapping in a significant portion of their traditional territory. This separation caused their special bond with those lands to break resulting in a great cultural and personal loss that has now affected five generations. Today, with the implementation of their land claim and the establishment of Kluane National Park and Reserve, they have reasserted their right to carry out traditional activities in the park and the surrounding Kluane Game Sanctuary lands.

Today, the Southern Tutchone people continue to maintain a rich culture based on strong ties with the natural world. Renewing their relationship with their traditional territory within the park, and providing direction in the management of the natural resources are high priorities for both Champagne and Aishihik First Nations and Kluane First Nation. This cultural reintegration will also help ensure that the First Nations' traditional knowledge of this region is recognized and carried forth into the future.

TheChampagne and Aishihik First Nations Final Agreement and the Kluane First Nation Final Agreement ushered in a new era in the management of Kluane National Park and Reserve. The Champagne and Aishihik First Nations Final Agreement established the creation of the Kluane National Park Management Board and the two final agreements identify specific rights and responsibilities of the two First Nations, the Board and Parks Canada in the management of the park—a cooperative management regime based on shared responsibility.

A day-use area with boat launch, picnic facilities and campground is located at Kathleen Lake, and is operated from mid May to mid September. Hiking is a popular activity, on trails such as St. Elias Lake, Mush Lake Road, Shorty Creek, Cottonwood, Rock Glacier, King's Throne, Kokanee, Auriol, Dezadeash River Trail, Alsek Trail, Sheep Creek Trail, Bullion Plateau Trail, Slims West or Soldiers Summit. Rafting on the Alsek River, mountain biking on old mining roads, horseback riding through the Alsek Pass, boating on Kathleen Lake and Mush Lake as well as fishing for lake trout, arctic grayling, rainbow trout, northern pike and sockeye salmon are also among activities available in the park.

Notes

Mount Logan 洛根山　位于加拿大育空地区的西南角，海拔 5 951 米，是加拿大的最高峰，也是北美第二高峰，仅次于美国阿拉斯加的麦金利山。1925 年首次被攀登峰顶。洛根山是为了纪念加拿大地质勘探队的创建人威廉·洛根而命名的。

Saint Elias Mountains 圣伊莱亚斯山脉 北美洲太平洋海岸山脉的一部分，从弗兰格尔山脉沿美国阿拉斯加-加拿大边界向东南延伸，止于克罗斯海峡。多为 5 200 米以上的山峰，包括圣伊莱亚斯山、洛根山。整个山脉有极地冰盖以外的世界最大冰原。南端划入冰川湾国家保护区。

Wrangell-St. Elias National Park 兰格尔 - 圣伊莱亚斯国家公园 位于美国阿拉斯加州东南部。公园内山峰林立，其中包括加拿大境内的最高峰——海拔 5 951 米的洛根山。从太平洋上吹来的湿润气流，给这里带来大量降雪，并形成大量冰原和冰川。在阿拉斯加，冰川的面积占其总面积的 5%。这里还有世界上移动最快的冰川。目前冰川不断融化，像河流一样向下方缓缓地滑动，速度大约为每天 9 米左右，而且这种速度还有加快的趋势。冰川的移动带来了无数淤泥与石块，不断地改变着这里的地貌。这里的植被从沿海地带一直延伸到山谷森林，主要植物是高山松树。

Alaska Highway 阿拉斯加公路　一条 2 450.5 千米的公路，从不列颠哥伦比亚省的道森克里克往西北延伸至阿拉斯加州的费尔班克斯。最初是美国军队在 1942 年修建的军事供给线，1947 年该公路向非限制车辆开放。

Ice Age 冰河时代　一般所说的冰河时代，主要是指第四纪大冰川的时代。因为它离我们最近，在地貌及沉积物等方面遗留下许多痕迹，使我们对它了解得比较详细。实际上在整个地球发展史中发生过好几次这样的大冰期，有时冰川的范围扩大到目前在赤道附近的北非、印度和澳洲。根据发展的观点来看，地球上今后还有可能发生大冰川的降临。最近一次冰河时代结束于 1 万多年前，在那次冰河时代，冰川从两极一直向赤道地区延伸，在纽约这样纬度的地区，冰层竟也厚达 1 千米，那是个很寒冷的时代，我们人类的祖先就亲眼见证过这个时代。

trembling aspen 颤杨　产自北美洲北部的白杨树，叶片为圆形，风吹来时，因为叶柄为扁平状，叶子可以随风飘扬。

balsam poplar 香脂白杨　又名胶杨，原产自北美洲的东部地区。这种树的叶子长为3英寸(15 厘米)，多长在树的底部。

alder 桤木　又名水冬瓜树、水青风、桤蒿，为桦木科，桤木属植物。北美种植的红桤木为高大乔木，叶下表面有锈色毛；白桤木，即菱叶桤木，是一种早花乔木，细枝和芽橙红色；斑点桤木为小乔木，具有明显的白色，瘤状皮孔；美洲绿桤木，叶芳香；锡特卡桤木较高大；山桤木，即薄叶桤木，是灌木状乔木，叶上有黄色或橙褐色中脉，枝条下垂，树冠穹顶状。欧洲桤木，即黏性桤木，有时被称为黑桤木，因树皮及果序色深而得名，广布欧亚，有几个变种在北美种植。

thrush 鸫鸟　鸫科所属的鸣禽，约300 种。鸫科有几个属是它们的中间等级。喙细长，跗蹠有靴鳞，即下腿前面覆以单片长鳞，而非多片短鳞。体形较鸲类为大；嘴形较窄，嘴长几乎是头长的一半；嘴须发达；翅形尖，一般超过110 毫米；尾较宽且长，变化也大；跗蹠结实而长；翼下不具斑纹。

Translation of Selected Sentences

1. A gem in the family of Parks Canada's national treasures, Kluane National Park and Reserve of Canada covers an area of 13,657 square miles. It is a land of precipitous, high mountains, immense ice fields and lush valleys that yield a diverse array of plant and wildlife species and provides for a host of outdoor activities.

作为加拿大国家自然财富的一颗明珠，克卢恩国家公园和保护区，拥有13 657 平方英里土地。它拥有陡峭高大的山脉、巨大的冰川和茂盛的山谷。公园里有多种多样的动植物品种，同时也可以进行各种好玩的户外活动。

2. Rising in the mountains of Kluane National Park and Reserve, its braided upper reaches nestle in a broad valley, providing an oasis for mountain goats, Dall's sheep and other wildlife. Sand dunes and glacial till provide a contrasting environment for the diverse vegetation found here.

克卢恩国家公园和保护区连绵不断的山脉，在它们相互交织的上端依偎着广阔的山谷，这为山羊等各种野生动物提供了一个绿洲。沙丘和冰碛为各种各样的植物提供了不同的生长环境。

3. They once lived a nomadic lifestyle, travelling great distances, in order to take advantage of seasonal abundances of wildlife and plants throughout their traditional territory. This lifestyle required developing a superior skill set of hunting and travelling techniques in an area of harsh climatic and geographic variations.

他们过着游牧生活，游牧到很远的地区，为了就是得到随着季节变化而不断丰富的野生动植物。这样的生活方式需要发展很高超的捕猎技能和旅行技巧，来度过这里严酷的气候和不同的地形环境。

4. Today, with the implementation of their land claim and the establishment of Kluane National Park and Reserve, they have reasserted their right to carry out traditional activities in the park and the surrounding Kluane Game Sanctuary lands.

今天，随着他们获得这一地区的土地所有权以及克卢恩国家公园和保护区的建立，他们可以再一次宣称他们在公园和克卢恩动物保护区域内拥有进行传统活动的权利。

5. This cultural reintegration will also help ensure that the First Nations' traditional knowledge of this region is recognized and carried forth into the future.

这种文化上的再整合将帮助确保第一国家的居民的传统知识和文化被承认，以及不断地传承到将来。

Questions

1. Why can the Kluane National Park and Reserve be called World Heritage Site? What are the specialties of park?
2. What's the climate of Kluane National Park and Reserve?
3. Are there anyspecies of wildlife in the Kluane National Park and Reserve? What are they?
4. Who lived in the park before? How did they survive in the park?
5. What kind of special activities can be enjoyed in the park?

Chapter 15

Kootenay National Park of Canada
库特尼国家公园

【简介】

库特尼国家公园成立于1920年,位于加拿大不列颠哥伦比亚境内洛矶山脉的西南坡,面积约为1406平方千米。1984年与加拿大洛基山脉东坡亚伯达境内的班夫(Banff)、贾斯珀(Jasper)、沃特顿(Waterton)湖群国家公园一起被联合国统一命名为“洛矶山脉公园”,并列入世界自然遗产。巍峨的山峰、壮丽的冰川、历经近万年所形成的天然彩绘池、神奇的镭温泉,以及完整多样的生态系统构成了库尼特国家公园的独特风貌。

Kootenay National Park (1,406 km^2) is one of 41 national parks in Canada. Together, they celebrate and protect the diversity of the nation's great landscape(景观). Kootenay National Park represents the south-western region of the Canadian Rocky Mountains(洛矶山脉).

From glacier-clad(冰川覆盖的) peaks along the Continental Divide to grasslands of the Rocky Mountain Trench(沟渠), where cactus(仙人掌) grows, Kootenay National Park is noted for its diversity(多样性的) of landscapes, ecology(生态) and climate. Natural features characteristic of the Rocky Mountains include sedimentary rocks(沉积岩) and thrust-faulted(推力扭断的) mountains, landscapes sculptured by glaciers(冰川) and water, and plants and animals typical of alpine(高山), sub alpine(亚高山地带) and montane ecological(生态学的) zones.

For thousands of years, the area which is now Kootenay National Park was part of the traditional lands identified by the Ktunaxa (Kootenay) and Kinbasket (Shuswap) First Nations people. Archaeological evidence(考古发现) suggests the mountains were used primarily as seasonal hunting grounds. Groups also travelled across the mountains periodically to hunt bison on the plains east of the Rockies.

Kootenay National Park was established in 1920 as part of an agreement between the provincial and federal governments to build the Banff-Windermere Highway—the first motor road across the Canadian Rockies. A strip of land five miles (eight kilometres) wide on each side of the highway was set aside as a national park. The completion of the highway in 1922 expanded the new age of motor tourism in the Canadian Rockies and established a commercial link(商业连接)between the Windermere Valley and Calgary.

The 94 km Banff-Windermere Highway (93 South) takes you through the park along the Vermilion and Kootenay rivers and through the narrow gorge(峡谷)of Sinclair Canyon(辛克莱峡谷)to the famous Radium Hot Springs(镭温泉). Many use the highway simply as a scenic route to reach the Windermere Valley, beyond the park's southern boundary (边界). But those who take the time to stop along the way discover some of Kootenay's quiet, colourful secrets—such as the vibrant(充满活力的)new forests growing up in the wake of the forest fires at Vermilion Pass (朱砂隘口)and Mount Shanks, the blue glacial waters and white dolomite(白云石)walls of Marble Canyon, the rust-coloured mineral pools and ochre (赭石)beds of the Paint Pots(彩绘池), the shaggy (毛茸茸的)white mountain goats (the park's wildlife symbol) on the crags (峭壁)and slopes (斜坡) of Mount Wardle, the panoramic (全景的)overlook at Kootenay Valley Viewpoint, the little green gem(宝石)of Olive Lake nestled in Sinclair Pass, and the iron rich cliffs of the Redwall Fault, and so on.

Due to the geographic location of Kootenay National Park, climate and weather vary considerably from one part of the park to another and from valley bottom to mountain top. This variety(多样化)is reflected in the park's interpretive theme: "from cactus(仙人掌)to glacier". At the south end of the park, in the Radium Hot Springs area, the climate is relatively warm and dry, while at the north end of the park, at Vermilion Pass, it is relatively cool and moist(潮湿的).

Located on the western edge of the Rockies, the park's climate is dominated by prevailing moist westerly winds from the Pacific Ocean. Occasionally, drier continental weather patterns extend over the region. Generally, summers are short and cool, with occasional hot spells, while winters are long and snowy. The high mountain ranges of the Continental Divide(大陆分水岭) generally prevent winter arctic(北极)air from spilling over to the western side of the Rockies, though occasionally this does occur, bringing the park's coldest temperatures.

The semi arid climate(半干旱气候)in the Radium Hot Springs area is due to its relatively low elevation (900 m) and the "rain shadow" effect of the Purcell Mountains to the west. Pacific air rising over the Purcell's cools and drops its moisture, so by the time it reaches the Columbia Valley it is relatively dry. Rising over the Rockies, the air cools once more and drops more moisture. Precipitation(降雨量)is heaviest along the Continental Divide, at the north end of the park.

Annual precipitation in Kootenay National Park ranges from less than 380 mm at lower elevations to over 1250 mm at higher elevations. In valley bottoms, average January highs are −7 ℃, while July highs average 22 ℃. Average temperatures decrease(减少)by about 0.5 ℃ for every 100 m of elevation gain(增加). Also, winds are generally stronger at higher elevations.

The character of Kootenay National Park is defined by its ecosystem(生态系统)—the dynamic interaction of its native species and natural processes. Preserving the natural state of the park requires maintaining the integrity(完整性)of its ecosystem. The ecological integrity of Kootenay National Park is threatened by a variety of human impacts such as development, fire suppression, introduction(引进、移植)of non-native plants, habitat fragmentation(分裂、破碎)and human/wildlife conflicts. Parks Canada is taking a number of actions to restore(恢复、还原)and maintain the ecological integrity of national parks.

Kootenay National Park of Canada offers opportunities to experience the Rocky Mountains, learn about their natural and cultural heritage(遗产), and connect with nature. National parks belong to the people of Canada. As such, we all share in the responsibility for their stewardship(职责). An understanding and respect for the land and its people will help ensure the ecological and cultural integrity of these special places are maintained for future generations.

Notes

motor tourism 自驾游　起源于20世纪被称为建立在“轮子上的国家”的美国,并迅速成为发达国家中产阶级喜欢的一种旅游方式。自驾游,顾名思义,是指自己驾驶交通工具外出旅游的一种方式,其中车辆包括汽车(轿车、越野车、旅行房车)、摩托车和自行车,以私有为主,也可以采用租赁、借用及其他方式。在现实中经过长期的发展后,自驾游的内涵发生了一些变化:驾车者可以是自己,也可以是车主或是同行者;车辆主要指的是汽车。本文中提到的1922年班夫德米尔高速公路的竣工,记录了加拿大自驾游的开始时间。

Sinclair Canyon 辛克莱峡谷　数千年前经水流的冲刷侵蚀而形成,两侧峭壁与地面几乎垂直形成呈“V”字形。在洛矶山脉,水流侵蚀而成的峡谷要少于冰川形成的峡谷。

Radium Hot Springs 镭温泉　库特尼国家公园的著名景观,它的名字源于温泉水质中含有的化学元素镭。镭虽然是一种放射性元素,但温泉中的浓度很低,只有23 pCi/L,对人体不但没有危害,还有缓解疲劳的功效。镭温泉水温常年保持在29~44 ℃摄氏度之间,符合国际奥林匹克游泳比赛对水温的要求,是一个天然的游泳池。镭温泉毗邻一个大约有 800 人口的村庄,也用镭温泉命名。

Vermilion Pass 朱砂隘口　海拔 1 680 米,是库特尼国家公园与班夫国家公园的分界线,同时也是著名的大陆分水岭(Continental Divide)的一个主要节点。该景点的主要标志是三面旗帜和一个水泥界碑。三面旗帜为居中的加拿大国旗和两侧的不列颠哥伦比亚省与阿尔伯塔省的省旗,因此朱砂隘口又是不列颠哥伦比亚省与阿尔伯达省的分界线。

Paint Pots 彩绘池　库特尼国家公园内的著名景点。因该景点附近的土壤里富含氧化铁,氢氧化铁等元素,所以土壤与河水都呈现出赭红色。这种赭色土壤可以作为绘画染料的底色,相传最早由活跃于本区的土著印第安人就用它来描画身体。彩绘池是名副其实的矿泉群,由三个相距不远的矿泉池组成。这些矿泉中涌出的冰冷矿泉水富含铁矿物质,每个池口的边缘形成逐渐增高的轮缘,流淌的泉水则将土壤染成红褐黄等各种颜色。

Translation of Selected Sentences

1. Kootenay National Park (1,406 km^2) is one of 41 national parks in Canada. Together, they celebrate and protect the diversity of the nation's great landscapes.

库特尼国家公园(1 406 平方千米)是加拿大 41 座国家公园之一。它和其他国家公园一起颂扬并保护着民族壮丽的景观。

2. For thousands of years, the area which is now Kootenay National Park was part of the traditional lands identified by the Ktunaxa (Kootenay) and Kinbasket (Shuswap) First Nations people.

千百年来,库特尼国家公园所在的这片区域都被视作是加拿大最早两大民族——库尼特族和苏斯瓦普族——生活的传统领地。

3. The character of Kootenay National Park is defined by its ecosystem—the dynamic interaction of its native species and natural processes.

库特尼国家公园的最大特点在于它的生态系统,即本地物种和自然环境的动态互动。

4. The ecological integrity of Kootenay National Park is threatened by a variety of human impacts such as development, fire suppression, introduction of non-native plants, habitat fragmentation and human/wildlife conflicts.

对库特尼国家公园生态系统完整性构成的威胁主要来源于各种人类活动的影响,如开发、火灾、非本土植物的移植、栖息地破碎化和人类与野生动物之间的冲突等。

5. An understanding and respect for the land and its people will help ensure the ecological and cultural integrity of these special places are maintained for future generations.

对一个国家和民族的理解和尊重，将有助于确保这些特殊区域的生态和文化的完整性，应该让子孙后代继承这种品质。

Questions

1. What is the general condition of the Kootenay National Park according to the passage?

2. What is motor tourism and what expanded the new age of motor tourism in the Canadian Rockies?

3. What is the wildlife symbol of Kootenay National Park?

4. Please state three well-known attractions in the Kootenay National park.

5. What should we do for the ecological integrity of Kootenay National Park?

Chapter 16

Mount Revelstoke National Park

雷夫尔斯托克山国家公园

【简介】

雷夫尔斯托克山国家公园成立于1914年，位于不列颠哥伦比亚省的赛尔客克山，公园面积为260平方千米。雷夫尔斯托克国家公园不仅拥有世界上面积最大的内陆温带雨林，还同时拥有亚寒带针叶林和高山冻土带，形成了一山容四季的独特景观。在野生动物保护方面，雷夫尔斯托克国家公园也有杰出的表现，它不仅保护山地驯鹿、野生高山羊，还为北美灰熊、黑熊提供栖息地。

Mount Revelstoke National Park is located adjacent（相邻）to the city of Revelstoke, British Columbia, Canada. The park is relatively small for a national park, covering 260 square kilometres. It is located in the Selkirk Mountains and was founded in 1914. Approximately 600,000 visitors enter Mount Revelstoke and nearby Glacier National Park each year.

Mount Revelstoke National Park is a place of contrasts. Take a drive along the Meadows（草甸）in the Sky parkway through a variety of geographical zones. From dense old-growth rainforest（雨林）of giant cedar（雪松）and pine（松树）, travel up through subalpine（亚高山带）forest, and finally alpine（高山）meadows and tundra（冻土地带）. Marvel（惊异于）at the spectacular（壮观的）view of the ice-clad peaks of the Monashee Mountains（莫纳西山）and, to the east, the Selkirk Mountains（赛尔客克山）. The Giant Cedars hiking trail（徒步旅行栈道）takes you through a stand of ancient Western Red Cedars（红杉树）, and the Skunk Cabbage trail introduces you to jungle-like wetland, a birder's paradise.

Mount Revelstoke National Park contains three distinct life zones created by elevation（海

拔). The higher the elevation, the more often precipitation occurs as snow rather than rain. Deep, lingering snow and colder temperatures at high elevations shorten the growing season, until an elevation is reached beyond which no trees can grow. The three life zones can be thought of as "Rainforest, Snowforest and No Forest".

The park contains part of the world's only temperateinland rain forest. At the lowest elevations in the park, the region's high annual precipitation(降雨量)falls mostly as rain. This is the interior(内在的)rainforest: dense vegetation(植物)and large trees remind many visitors of the coastal Pacific Northwest. Western red cedar, western hemlock, and western white pine(白松树)tower over shrubs(灌木)such as devil's club, Pacific yew(紫衫)and mountain box(黄杨). A rich variety of ferns and mosses carpets the forest floor. Foresters call this zone the Interior Cedar-Hemlock forest. The once-vast forests of British Columbia's interior wetbelt constitute the world's only inland temperate rainforest. You can experience this magnificent(壮丽的)forest type along the Giant Cedars Nature Trail .

As you leave the Trans Canada Highway and begin to travel up Mount Revelstoke, you will notice that by about the Snowforest viewpoint, the forest begins to change in appearance and composition. You are entering the interior subalpine life zone, known to foresters as the Engelmann Spruce(英国针枞)or Subalpine fir zone. The lower subalpine forests are dominated by Englemann spruce, subalpine fir and mountain hemlock. Trees begin to thin in the upper subalpine near treeline, creating park-like, open meadows. The meadows explode into colour in August, with wildflowers including paintbrush, glacier lily, fireweed, lupine and monkeyflower. You can easily visit subalpine meadows in summer by driving up the Meadows-in-the-Sky Parkway to Balsam Lake, near the summit of Mount Revelstoke.

At about 2000m elevation on average, the growing season becomes so short that trees can no longer survive(存活). Fully 50% of the area of Mount Revelstoke National Park is found above treeline. Much of this area is rock. About 12% of the park is permanent(永久的)ice and snow. Tundra(苔原)vegetation includes heather(石楠花)and sedge(莎草)meadows, but in many places only lichen(青苔)survives, encrusting(包裹)the rock. Areas above the treeline provide year-round or seasonal habitat for such wildlife as grizzly bear(灰熊), mountain goat(野生白山羊), white-tailed ptarmigan(白尾雷鸟), hoary marmot(花白旱獭), golden-mantled ground squirrel(金背黄鼠), mountain caribou(北美驯)and pika(鼠兔).

The park also protects a small herd of the threatened mountain caribou and provides habitat for grizzly bear and mountain goat. Specialized habitats found in Mount Revelstoke National Park include avalanche slopes and wetlands. Wetlands are very rare in the steep, narrow valleys of the Selkirk Mountains.

Mountain Caribou make extensive use of both low and high elevation old growth forests due to their winter dependence(依赖) of heavy lichen loads found mainly on 200 year old plus trees. Caribou require decadent cedar and hemlock forests at low elevations in early winter before snow has deepened and firmed up on subalpine winter ranges. They live off lichen-covered blow down trees and branches in the ICH, as well as making use of Mountain Boxwood(黄杨木) in the understory before it is buried by deep snow. Later, in January when the snowpack has consolidated, caribou are able to migrate upslope to their ESSF ranges. There, they are able to reach lichen found higher up trees by standing on the three to four metre snowpack.

Bears symbolize the Canadian wilderness though most Canadians have never seen a bear in its natural habitat. British Columbia is one of the few places in the world that still has significant numbers of bears. However, as our ecological footprint expands, grizzly bears in particular are relegated(游荡) to ever shrinking wilderness areas. Grizzly bears range in colour from dark brown to blonde. Their name in North America comes from the most prevalent(流行的、普遍的) fur colouring which is a dark brown hair with a blond tip. In the Columbia region their weights average 102 kg for adult females to 168 kg for adult males. Larger bears occur on the Pacific coast where they have access to spawning salmon.

Grizzlies have a dished shaped head with ears that are proportionally smaller than black bears. A prominent shoulder hump of muscle can help distinguish grizzly bears at a distance. This muscle mass assists the bear to dig for flower bulbs and ground squirrels with their long , relatively straight claws(爪). Black bears are usually black with a lighter coloured muzzle(鼻口) and sometimes a white chest(胸) patch. In this region about one in ten black bears are cinnamon coloured. Their weights average 63 kg for adult females to 93 kg for adult males. Black bears have short, curved claws well suited to climbing trees. They arrived in North America from Asia about one million years ago and evolved(进化的) to live in mixed forests and floodplains. In spring, they will climb to end branches of poplars to eat catkins(雄花序) leaving tell-tale claw marks on the bark of the tree. With the option of climbing trees, it is thought that black bears are less aggressive(攻击性) than grizzlies.

The region surrounding Yoho, Kootenay, Mount Revelstoke and Glacier national parks is near the front line of the shrinking grizzly range in North America. The pattern of local extinction within the grizzly's Canadian range will undoubtedly continue unless deliberate efforts are made to prevent it. Whether or not we are willing to allow large areas of land to remain in a sufficiently wild condition may be the determining factor in keeping this animal a part of the Canadian landscape.

Notes

inland temperate rainforest 内陆温带雨林　雷夫尔斯托克国家公园中的内陆温带雨林是世界自然基金会(WWF)认定的世界最大的温带雨林。

mountain Caribou 北美驯鹿　产于北极地区,有些生物学家把北美驯鹿分成两种,一种叫北方驯鹿,栖息在北方的冻土地带和针叶树林地带。另一种叫森林驯鹿,栖息在加拿大的森林之中。驯鹿区别于其他鹿的显著特点是,雄性和雌性都有角。雌性的角较短,结构较简单。驯鹿肩高约0.7~1.4米,重量约300千克。和北美林地驯鹿相比,北美荒原驯鹿的体型要较小些,重量在90~135千克,毛色也要略浅些。在一年中的大部分时间里,它们主要生活在从阿拉斯加到巴芬岛的广大冻土地带,共有5个种群,总数大约在120万头。北美荒原驯鹿也具有迁徙性。与北美林地驯鹿不同的是,它们到了春季才聚集在一起,等到下一代出生后,迁往高原冻土地带。

grizzly bear 北美灰熊　北美灰熊为棕熊亚种,分布于北美洲西部,是北美最常见的棕熊。灰熊通常独居生活,但是在沿海地区,灰熊在鲑鱼产卵的季节会聚集在溪流、河流、湖泊和池塘周围。雌性灰熊每隔一年生产1~4头幼崽,通常为2头,每头幼崽重量仅为500克。母熊严密保护其幼崽,在认为自己或幼崽受到威胁时,便会进攻。

ICH 红杉-铁杉内部生态体系　ICH是Interior Cedar Hemlock的缩写。Cedar是红杉树,Hemlock是铁杉树,它们是加拿大内陆温带雨林中最主要的两种树,形成了独特的红杉-铁杉内部生态体系,简称ICH。

Translation of Selected Sentences

1. Mount Revelstoke National Park is a place of contrasts. Take a drive along the Meadows in the Sky parkway through a variety of geographical zones. From dense old-growth rainforest of giant cedar and pine, travel up through subalpine forest, and finally alpine meadows and tundra.

雷夫尔斯托克山国家公园是一个景致差异极其明显的地方。开车沿着地势较低的草甸区向通天的高处驰骋,你会看到丰富多样的地质地貌,从植被浓密的温带雨林,到长满参天红杉和黄杨的亚高山带森林,最终抵达有苔藓生长的高山冻土地带。

2. Mount Revelstoke National Park contains three distinct life zones created by elevation. The three life zones can be thought of as "Rainforest, Snowforest and No Forest".

雷夫尔斯托克山国家公园按照海拔高低分为三个生态区,这三个区分别是雨林区、雪林区和无林区。

3. The park also protects a small herd of thethreatened mountain caribou and provides

habitat for grizzly bear and mountain goat.

公园保护受到威胁的山地驯鹿，并为北美灰熊和野生山羊提供栖息地。

4. Bears symbolize the Canadian wilderness though most Canadians have never seen a bear in its natural habitat.

熊是加拿大荒原的标志，尽管大多数加拿大人从来没有在熊的自然栖息地看见过熊。

5. The region surrounding Yoho, Kootenay, Mount Revelstoke and Glacier national parks is near the front line of the shrinking grizzly range in North America.

包括优鹤、库特尼、雷夫斯托克、冰川等国家公园在内，及其周边地区都是研究北美灰熊的前沿阵地。

Questions

1. What is the general condition of the Mount Revelstoke National Park according to the passage?

2. How many distinct life zonesdoes the Mount Revelstoke National Park contain and what are they?

3. What is the main vegetation in the temperate inland rain forest?

4. What are the main animals in the Mount Revelstoke National Park and what are their living habits?

5. What symbolize the Canadian wilderness?

Chapter 17

Nahanni National Park Reserve
纳汉尼国家公园

【简介】

纳汉尼国家公园成立于1972年,位于加拿大西北区,面积约为5400平方千米(2007)。纳汉尼国家公园是北方荒原河流、峡谷、高山苔原等景点景观的代表。1978年纳汉尼国家公园被联合国教科文组织列入世界遗产名录。1987年南纳汉尼河被认定为加拿大遗产河。雄奇的纳汉尼河吸引了大批探险者和淘金者,1972年帕特森将他在纳汉尼河探险的经历写成了《危险的河流》一书,使纳汉尼河名声大振。

Nahanni National Park Reserve in the Dehcho Region of the Northwest Territories, Canada, approximately 500 km west of Yellowknife. It was set aside by Order-in-Council as a National Park Reserve in 1972, and was gazetted(刊载于报)as a national park reserve in an amendment(修正案)to the National Parks Act in 1976 and was 4,766 km^2 in area. The park was inscribed on UNESCO's World Heritage List(世界遗产名录)in 1978. The South Nahanni River achieved(获得)Canadian Heritage River status(地位)in 1987. In 2003, an agreement between the Dehcho First Nations and Parks Canada gave temporary protection to 23,000 km^2. In August 2007, the federal government added an extra 5,400 km^2

Nahanni National Park Reserve of Canada protects a portion of the Mackenzie Mountains Natural Region offering the adventurous visitor a wilderness experience. A key feature(特色)of the park is the Naha Dehé (South Nahanni River). Four great canyons line this spectacular whitewater river. At Nailicho (Virginia Falls) the river plunges in a thunderous plume. The park's sulphur(硫化)hotsprings, alpine tundra(高山苔原), mountain ranges, and forests of spruce and aspen are home to many species of birds, fish and mammals. A visitor centre in Fort Simpson features displays on the history, culture and geography of the area.

Nahanni National Park Reserve is an outstanding example of northern wilderness rivers, canyons(峡谷), gorges and alpine tundra. The park is centred on the river valleys of the South Nahanni and Flat rivers in the southwest part of the Northwest Territories.

One of the key features of the park is the South Nahanni River. This river originates as a small, clear boulder-strewn stream in the remote Mackenzie Mountains(麦肯齐山). It grows dramatically in size and power as it flows 580 km downstream to the southeast. The silt-laden waters of the South Nahanni meander through scenic mountain valleys and hurtle through a series of deep canyons before calming and joining the Liard River at Nahanni Butte. This dramatic change in character, along with the varied and spectacular(壮观) terrain that it passes through, have given the South Nahanni its international reputation(声誉) as a premier wilderness river. Along the river are many unique features. These include: Rabbitkettle Hotsprings, source of the largest known tufa mounds in Canada; Virginia Falls(佛吉尼亚瀑布), with a vertical drop twice that of Niagara Falls; a series of river canyons up to 1,200 m deep; and caves such as Grotte Valerie with its ancient skeletons of nearly a hundred Dall's sheep. The section of the South Nahanni River protected by the park was proclaimed a Canadian Heritage River in 1987, in recognition of its outstanding wilderness character and recreational value.

The park is home to a variety of wildlife. Dall's sheep, mountain goats, woodland caribou, wolves, black bears, grizzly bears and trumpeter swans all find refuge in Nahanni. The wildlife of Nahanni National Park Reserve is diverse for the relatively high latitude (自由) of the park. The dominant characteristic of the wildlife species(物种) in the park is their adaptation(适应性) to a rugged(崎岖的) and highly seasonal environment. The majority of species adapt by either migrating for the winter to a less severe environment, or spending the winter hibernating. Species which are year-round residents adapt by showing seasonality in their movements. The park's diversity of wildlife is best seen during the relatively short summer season.

Forty-two mammal(哺乳动物) species have been recorded in the park. Although the animals themselves may be elusive, their tracks and scats are often seen. Both black and grizzly bears commonly occur in Nahanni, and the Rabbitkettle Lake area has been identified as seasonally important bear habitat. Wolves are found along the river valleys, and field observations suggest moose and caribou are staple prey items. Several weasel species, including wolverine, occur in the park, as do a variety of rodents and bats. Small and medium-sized mammals found in the park include shrews, voles, Arctic ground squirrels, muskrats, beaver, marmots (土拨鼠), and snowshoe hare. Some mammalian predators(食虫动物) are mink(貂), marten, otter, lynx(猞

猁) and red fox.

A number of ungulate(有蹄类) species occupy a wide range of ecological niches in the park. Moose(麋鹿) are the park's most common and broadly distributed ungulate. Small numbers of mountain goats can be found in a few of the alpine areas, but most of their range is outside the park boundary. Dall's sheep(戴尔绵羊) are commonly observed in the park, but their movements also extend well beyond the park area. A small percentage of woodland caribou are found in the park year-round. Caribou(北美驯鹿) use the park most heavily in the winter. Prime calving summer habitats and autumn ruts for the woodland caribou are located farther to the northwest beyond the park boundary. Mule deer and, more recently, white-tailed deer, have been seen in eastern parts of the park. Mineral licks in Nahanni constitute an important seasonal resource for most species of ungulates.

There are no known reptile(爬行动物) species in the park and very few amphibians(两栖动物) Wood frogs(树蛙) have the widest distribution throughout the north of any amphibian. They are thought to inhabit areas with early, deep snowfalls which provide a protective layer of insulation for hibernating. Studies are needed to determine the extent of amphibian ranges within the park.

Sixteen fish species have been recorded within the SouthNahanni watershed: lake trout(淡水鳟鱼), Dolly Varden(花束鲷), Arctic grayling(北极茴鱼), burbot(江鳕), longnose sucker, northern pike(白斑狗鱼), lake whitefish, round whitefish, mountain whitefish, inconnu, slimy sculpin, lake chub, longnose dace, spoonhead sculpin, trout-perch and spottail shiner. Most of the species are found in the park, although spoonhead sculpin, trout-perch, and spottail shiner are not common. Lake trout occur throughout the park, inhabiting deeper, less turbid standing water. They are vulnerable(脆弱的) to overfishing due to their restricted distribution, slow growth, late maturity and limited reproductive capacity. Dolly Varden is found in the South Nahanni River and many tributary streams above and below Virginia Falls. Young Dolly Varden hatch in the headwater streams of the watershed, and then migrate downstream into the park and beyond during adulthood. Arctic grayling is the most common fish species found, inhabiting most small streams and larger lakes within the watershed. Like Dolly Varden, grayling hatch in the headwaters and then migrate downstream. Because of their hatching location and migration route, Dolly Varden and Arctic grayling are particularly susceptible to overfishing as well as upstream developments that may affect their migration routes and spawning areas. Northern pike are confined primarily to Yohin Lake and the lower South Nahanni and Flat rivers.

Cultural resources in the park include pre-contact sites, as well as remnants of cabins, caches and trails from the era of trapping, prospecting and exploration. Little is known of the origins and significance of these resources. The Dene and their ancestors have lived and hunted in the Nahanni area for thousands of years. In the early 19th century, the first Europeans came to the area, seeking fur and gold, however, it wasn't until the 1950s, with the publishing of Dangerous River by Raymond M. Patterson that the legends about the region were put to rest and the river finally came to prominence(声望). Since that time, the Nahanni has become one of the prime wilderness rivers in Canada, frequented as a whitewater destination by adventure seekers around the world. Albert Faille is perhaps the best-known prospector(探矿者)associated with the park. Faille prospected in the park area for 46 years, from the mid 1920s to the 1950s, and was featured in the National Film Board film Nahanni. His home is still standing in Fort Simpson.

Notes

Dehcho Region 丹科地区 加拿大西北地区区五个行政区之一,它的办事处设在辛普森堡。该地区的主要居民为加拿大土著民族,即第一民族。

Northwest Territories 西北区 加拿大联邦直辖区,位于加拿大北部,与育空地区、努纳武特地区、不列颠哥伦比亚省、阿尔伯塔省、萨斯克切温省交界,首府为耶洛奈夫。

UNESCO 联合国教科文组织 全称为联合国教育、科学、文化组织(United Nations Educational, Scientific and Cultural Organization),是联合国专门机构之一,简称联合国教科文组织(UNESCO)。总部设在法国巴黎。1945 年二战结束后,近 40 个国家的代表在伦敦举行会议,决定成立一个以建设和平文化为宗旨的组织。同年 11 月 16 日,这些国家签署《联合国教育、科学及文化组织组织法》,联合国教科文组织宣告成立。教科文组织的宗旨是促进教育、科学及文化方面的国际合作,以利于各国人民之间的相互了解,维护世界和平。其主要机构有大会、执行局和秘书处。截至 2009 年 10 月,教科文组织已拥有 193 个成员和 7 个准成员,是联合国范围内成员数量最多的组织之一。

Dene 提纳人 居住在加拿大北部的北方和北极地区的土著人群体,是加拿大第一民族的分支。他们有自己的语言阿萨帕斯卡语,也使用自己的文字阿萨帕斯卡字。

Raymond M. Patterson 雷蒙特 · M · 帕特森(1898—1984) 生于英国,毕业于牛津大学,是著名的作家和探险家。帕特森喜欢探险,1927 年他独自划独木舟探索加拿大纳汉尼河,并将他的经历写成了《危险的河流》一书,从此名声大振,吸引了欧洲大批的探险者和淘金者前往此地。

Albert Faille 艾尔伯特 · 罗缎 1887 年出生于美国明尼苏达州,是第一批到达加拿大南纳汉尼河与麦肯齐河的淘金者。在那里他遇到了年轻的帕特森,并为帕特森的《危险的河流》一书提供素材。1962 年罗缎成为一部由加拿大导演唐纳德 · 怀特导演的关于纳汉尼河的纪录片中的主角,镜头跟随着罗缎老人探索纳汉尼河流域的金矿,记录了纳汉尼河的雄奇风光。该片分别在1962年和1963年两次获奖。

Translation of Selected Sentences

1. Nahanni National Park Reserve is an outstanding example of northern wilderness rivers, canyons, gorges and alpine tundra.

纳汉尼国家公园是北方荒原河流、峡谷、高山苔原等景点景观的代表。

2. The section of the South Nahanni River protected by the park was proclaimed a Canadian Heritage River in 1987, in recognition of its outstanding wilderness character and recreational value.

1987年南纳汉尼河国家公园保护段被认定为加拿大遗产河,这是对它的荒原河流特点和娱乐价值的认可。

3. The wildlife of Nahanni National Park Reserve is diverse for the relatively high latitude of the park.

纳汉尼国家公园的野生动物享有高度的自由。

4. The dominant characteristic of the wildlife species in the park is their adaptation to a rugged and highly seasonal environment.

公园里野生动物的最主要特征是对崎岖的、季节性很强的生存环境有较强的适应能力。

5. Cultural resources in the park include pre-contact sites, as well as remnants of cabins, caches and trails from the era of trapping, prospecting and exploration.

公园的文化遗产包括捕猎时代残存的小木屋,诱捕、淘金和探险时代留下的小径等遗址。

Questions

1. What is the general condition of the Nahanni National Park according to the passage?
2. What are the key features of the park?
3. What are the main animals in the Nahanni National Park and what are their appearances and living habits?
4. Who was R. M. Patterson?
5. What are thecultural resources in the Nahanni National park?

Chapter 18

Pacific Rim National Park Reserve
太平洋沿岸国家公园保护区

【简介】

太平洋沿岸国家公园保护区成立于1970年，面积包括511平方千米的陆地和海面。公园背靠温哥华岛，面向太平洋，展示了加拿大西海岸丰富的自然和文化遗产。公园拥有众多的岛屿、绵延的海滩，是皮划艇爱好者和观潮者的天堂。茂盛的温带雨林为众多珍稀的动植物提供了栖息地。

Pacific Rim National Park Reserve is a Canadian national park reserve in British Columbia made up of three separate regions: Long Beach, the Broken Group Islands, and the West Coast Trail. The entire park encompasses 511 km^2 of land and ocean. The park is characterized by rugged(崎岖的)coasts and lush(茂盛的)temperate rainforests. The park encompasses(包含)a thin strip of land located on the south-west coast of Vancouver Island. To the east of the park lies the Vancouver Island Ranges of the Insular Mountains and to the west is the Pacific Ocean. The park is open from mid-March until mid-October.

Backed by the Vancouver Island Range and facing the open Pacific Ocean, Pacific Rim presents the rich natural and cultural heritage of Canada's west coast. Its cool and wet maritime (海洋的) climate produces an abundance of life in the water and on land. Lush coastal temperate rainforest gives way to bountiful and diverse intertidal(潮间带)and subtidal areas(潮下带). These natural wonders are interwoven(混合、交织)with the long and dynamic history of the Nuu-chah-nulth First Nations and European explorers and settlers.

Long Beach is the most visited and most accessible of the three regions. It is made up of the coastal region from Tofino to Ucluelet. The primary feature in this area is Long Beach itself. The area also contains the Green Point campground, which has 94 campsites for both tents and small trailers. There are also numerous trails in this region, running through bogs and temperate rainforest areas.

The Broken Group Islands region is made up of over one hundred small islands and islets in Barkley Sound. The largest forested islands are Effingham, Turret, Turtle, Dodd, Jacques, Nettle and Gibraltar Island. The area is accessible only by boat, and is popular with kayakers(皮划艇运动员). There are eight camping areas scattered on the islands. Campers should be aware that there is no fresh water available in the Broken Group Islands, so any water needed will have to be brought in.

The West Coast Trail is a 75 km trail along the west-coast of Vancouver Island from Port Renfrew to Bamfield. The trail was built to aid in the rescue(营救) of shipwrecked sailors(海难水手). Construction on the trail started in 1907 and by 1910 the "Lifesaving Trail" was complete. The trail was abandoned in the 1950s. By 1970 the trail was transformed into The West Coast Trail, a challenging trail that takes visitors along rocky beaches, through rainforest, and across sometimes rough and muddy terrain. The trail has been improved greatly over the years and can be traversed in 5 – 7 days.

During fall and winter, the area is continually subjected to moist air masses from the Pacific Ocean. The presence of the mountain ranges causes the air masses to rise and deposit large quantities of precipitation(降雨量), a phenomenon known as orographic(山丘的) precipitation. The area averages over 3,000 mm of precipitation per year, a key factor in producing temperate rainforests. Since Canada's rainforest falls between the Arctic Circle(北极圈) and the Tropic of Cancer(北回归线), the forests are temperate. Other countries that have temperate rainforests include Chile, New Zealand and Norway. All temperate rainforests share similar characteristics, but different plants and animals populate each. Many of the world's temperate rainforests of the world are a mix of both deciduous(落叶的) and coniferous(针叶) trees. The dominance of conifers(针叶林) in Pacific Rim National Park Reserve's rainforest is one characteristic that distinguishes the other temperate rainforests. An open forest of trees so twisted and stunted they look like broccoli, reaching up from a mass of thick, wet moss, drenched in rain that collects in pools. A forest of trees so huge, tall and ancient we all look like little children beside them. A world of vivid green moss, arching ferns, and a mass of plants so thick it's impossible to see the soil beneath it, trees that range in age from small seedlings just starting out, to very old and grand trees that have lived many hundreds of years. An abundance of plants are growing on other plants. Large logs are lying on the forest floor and in streams.

Wander along a sandy beach, scramble(攀爬) over headlands, peer into tidepools, marvel at the life beneath a boulder, or watch the fury of the sea against the rocks. The seashore invites visitors all over the world exploration and discovery.

"Intertidal" simply means "between the tides(潮汐)", and refers to all the land or shoreline exposed to the air at low tides, and covered by the ocean at high tides. Pacific Rim National Park Reserve of Canada protects an incredible abundance and diversity of life in the intertidal zone. The sheer number of different species in the intertidal zone is one reason why it's such an exciting place to explore. This national park experiences two high and two low tides in a typical day with roughly six hours between each low and high. The times and heights of the high and low tides change slightly with every succeeding day in response to the relative changes in location of the sun, moon and earth. At low tide the beaches are wider and more of the intertidal creatures(动物) are visible. At high tide launching kayaks(单人皮艇) is usually easier and watching storms is more spectacular.

Pacific Rim National Park Reserve is home to over 60% of the total population of this very rare plant in Canada. The pink sand-verbena was thought to be extinct(灭绝) in Canada due to its shrinking dune habitat. Three plants were discovered in 2001 on the West Coast Trail and thanks to Parks Canada staff, pink sand-verbena is making a comeback. Thus far, thousands of these plants' offspring have been planted and are beginning to repopulate(重新安家) the dunes. Since the repopulation, other pink sand-verbena plants are appearing around the park, making it a true success story.

Establishment of this national park in 1970 stopped some activities within the park. Commercial logging(登记) and the development of private and commercial residences(住宅) are no longer permitted. Still, other activities that stress the rainforest continue. Park visitation, for example, requires roads, trails, buildings, and campsites, compete for valuable space in the crowded rainforest. Parks Canada is committed to maintaining the ecological integrity of Canada's protected areas, as well as repairing damaged ecosystems for the benefit of future generations.

Notes

British Columbia 大不列颠哥伦比亚省　位于加拿大的最西端,其名字是维多利亚女王在 1858 年钦定的,1871 年成为加拿大第六大省。除了首府维多利亚市外,加拿大第三大城市温哥华也在该省。据统计,截止到 2009 年,该省人口达到 4 419 974 人,其中一半人口居住在温哥华。旅游业、伐木业、采矿业是该省主要经济产业。

Nuu-chah-nulth 努查努斯族　加拿大太平洋西北海岸的土著民族,是加拿大第一民族的分支。该族拥有自己的语言。

Tofino 托菲诺　位于西经 125°51' 北纬 49°05',是加拿大著名旅游圣地。

Ucluelet 尤克卢利特　位于西经 125°34' 北纬 48°55',是加拿大著名旅游圣地。

Translation of Selected Sentences

1. Backed by the Vancouver Island Range and facing the open Pacific Ocean, Pacific Rim presents the rich natural and cultural heritage of Canada's west coast.

太平洋沿岸国家公园背靠温哥华岛,面向太平洋,展示了加拿大西海岸丰富的自然和文化遗产。

2. Since Canada's rainforest falls between the Arctic Circle and the Tropic of Cancer, the forests are temperate.

由于加拿大的雨林位于北极圈和北回归线之间,所以它属于温带雨林。

3. Pacific Rim National Park Reserve is home to over 60% of the total population of this very rare plant in Canada.

太平洋沿岸国家公园保护区是超过 60% 的加拿大珍稀植物的家。

4. Establishment of this national park in 1970 stopped some activities within the park. Commercial logging and the development of private and commercial residences are no longer permitted.

自从 1970 年国家公园成立以来,公园内一些活动被禁止:商业采伐和开发,以及私人和商业住宅的建设被禁止。

5. Parks Canada is committed to maintaining the ecological integrity of Canada's protected areas, as well as repairing damaged ecosystems for the benefit of future generations.

加拿大公园管理局致力于维护保护区内生态的完整性,同时修复被破坏的生态系统,造福子孙后代。

Questions

1. What is the general condition of the Pacific Rim National Park Reserve according to the passage?
2. What are the main landscapes of the park?
3. What are the functions of the West Coast Trail?
4. What dose the word "Intertidal" mean?
5. What haven't been permitted in the Pacific Rim National Park Reserve since 1970?

Chapter 19

Prince Albert National Park
艾伯特王子国家公园

【简介】

艾尔伯特王子国家公园成立于1927年，占地面积约3874平方千米，是加拿大第10大国家公园。它位于加拿大萨斯喀彻温省北部，距萨斯卡通城仅200千米之遥，是连接北部寒带针叶林和南部农垦区的重要过渡地带。艾伯特王子国家公园的生态系统被完整的地保护下来。广袤的北方针叶林、独特的羊毛草原、开阔的水域系统为各种野生动物提供了栖息地。加拿大第一位自然主义者、著名作家灰色猫头鹰曾居住于此，并以该地为背景书写了大量歌颂荒野保护的作品，唤醒世人的生态保护意识。他位于艾尔伯特王子国家公园的故居（海狸小屋）每年都会吸引大量游客参观。

Prince Albert National Park was established by order-in-council（议会）on March 24, 1927. It was officially opened on August 10, 1928 by Prime Minister William Lyon Mackenzie King. It covers 3,874 square kilometres in central Saskatchewan（萨斯喀彻温省）, Canada and is located 200 kilometres north of Saskatoon. The park is open all years but the most visited period is from May to September.

Prince Albert National Park of Canada protects a slice of the northernconiferous or 'boreal' forest（北方针叶林）. The park lies in a transition zone（过渡带）between natural regions, with a diversity of wildlife. The park contains many outstanding natural and cultural features, including the only fully protected white pelican（白鹈鹕）nesting colony in Canada, rare fescue grasslands（羊茅草原）, free-ranging（自由放养的）plains bison（野牛）and the isolated, lakeside cabin of conservationist Grey Owl（灰色猫头鹰）.

The noted naturalist (自然主义者) Grey Owl lived in the park during the 1930's and worked for the Dominion Park Service. Through his writing, films and lecture tours he drew the world's attention to the need for conservation(保护). His real name was Archibald Stansfeld Belaney and he was actually of English decent. The Dominion Parks Service became aware of Grey Owl through his articles that spoke passionately of wilderness protection. Grey Owl's writing reflected the philosophy of National Parks and he was hired as the first naturalist. In 1931, a cabin(小屋) was built on a small lake in Riding Mountain National Park for Grey Owl, Anahareo, and their two beavers Rawhide and Jelly Roll. However, due to unsuitable water conditions, Grey Owl and his family were moved to Ajawaan Lake in Prince Albert National Park later that year. Grey Owl wrote three bestselling books while he lived in Prince Albert National Park: Pilgrims of the Wild (1935), Sajo and the Beaver People (1935) and Empty Cabin (1936). These books, plus films produced about Grey Owl and his beavers, not only helped to promote his message of conservation but made him and his cabin, Beaver Lodge(海狸小屋), famous. Hundreds of people visited Grey Owl during the summers.

On a trip through southern Saskatchewan to the park, the landscape gradually exchanges poplar(白杨) bluffs and grain fields for forests of aspen and evergreens. Roadways based upon right angles and parallels give way to routes winding through gentle hills, skirting spruce(云杉) bogs(沼泽) and hugging lake shores.

The unmistakable blending of scents from the forest and across the open water fill your head. You've ventured into the southern edge of the great northern forest and the 3,875 sq. km of Prince Albert National Park. Here, nature's vast web of the living and non-living interact to create what is variously called the 'northern coniferous forest' or 'boreal forest'. In the southern portion of the park, remnant patches of fescue grasslands occur in the dryer locations. Water bodies comprise almost 30% of the area of the park from huge expanses of water like lakes Kingsmere and Crean to the tiny marshes(沼泽), bogs and creeks(小溪) that dot the landscape.

The topography of the park owes its origin to the glaciers. The landscape is a mosaic of uplands and lowlands that range in altitude between 488 meters and 732 metres above sea level. The boreal forest extends northerly into the Canadian Shield area from the agricultural zones of southern Canada. Prince Albert National Park lies south of the Shield in landscapes that were shaped by Pleistocene glaciers(更新世冰川) that deposited glacial till, sand and other materials that were later colonized by trees and shrubs. The park protects a small slice of the boreal ecosystem and within its 'neighbourhood' functions as a benchmark of ecological integrity(生态完整性).

Some of the many animals are elk(麋鹿), moose, black bear, red fox, beaver(海狸), deer, badger(獾), otter(水獭), red squirrels (松鼠), wolf. A herd of plains bison roams in the southern areas of the park where grasslands and woodlands mix. Flycatchers, Tennessee warblers(田纳西州莺), double-crested cormorants(双凤头鸬鹚), red-necked grebes(红颈鸊鷉), brown creepers, nuthatches(五子雀), three-toed woodpeckers, bald eagle, osprey(鹗), great blue herons(鹭鸟), many species of ducks, and the common loon are just a few of the water fowl and birds which make their home in the park. There are 21 species of fish recorded in the park, including Iowa darter(细镖鲈), yellow perch(金鲈), brook stickleback, spottail shiner, cisco, northern pike, walleye, and lake trout.

The park is the tenth largest national park in Canada. It is the largest protected wilderness area in Saskatchewan, approximately 0.6% of the area of the province.

Notes

William Lyon Mackenzie King 威廉·莱昂·麦肯齐·金(1874 年 12 月 14 日—1950 年 7 月 22 日)　20 世纪 20—40 年代加拿大优秀的政治家。他分别在 1921 - 1948 年间三次出任加拿大总理,是加拿大历史上任职时间最久、最优秀的总理。他的主要政绩表现在,成功引领加拿大人民度过第二次世界大战,出台最先进的社会福利政策,为日后加拿大社会福利制度的发展奠定基础。

Saskatchewan 萨斯喀彻温省　简称萨省,被誉为加拿大的"产粮之篮",以牧场和麦田而闻名。萨省位于加拿大中心地带,东西与曼尼托巴省和阿尔伯塔省为邻,南部与美国的蒙大拿州和北达科他州接壤。"萨斯喀彻温"的名称据说是由早期印第安土著民提出的,意思是"遍布境内的湍急河流"。

coniferous or 'boreal' forest 北方针叶林　亦称泰加林(Taiga)。分布于北半球高纬度地区,由松、杉类植物形成的森林,是寒温地带性植被。群落结构简单,树冠整齐、层次分明,乔木以松、云杉、冷杉、铁杉和落叶松等属占优势,多为单优势种森林。分布于欧亚大陆和北美大陆北部,在加拿大分布带相当宽。

Grey Owl 灰色猫头鹰或黑色夜枭　原名阿奇·毕兰尼(Archi Belaney),生于 1888 年 9 月 18 日,卒于 1938 年 4 月 13 日,是著名的作家、演说家和保护动物主义者。阿奇从小就梦想成为印第安人,1906 年他移民加拿大,先是加入了五大湖地区的一个奥吉布瓦族部落,在那里成家,靠做向导和打猎为生,被部落起名为灰色猫头鹰(grey owl)。后来他抛弃家庭,和依落郭安族姑娘安娜哈蕾奥相恋。一次打猎中,他杀死了母海獭,却在哈蕾奥的鼓励下,救下了它的两个孩子,在抚养过程中,与海獭建立起的友好关系使他决定放弃这种残忍的捕猎生涯,从此他努力为海獭等动物寻找庇护所,组建动物保护组织,在全世界范围内展开演讲。1930 年被加拿大政府授予名誉公园管理员，并在阿尔伯特王子国家公园 Ajawaan 湖畔为他和他的海獭建立了家园。他用灰色猫头鹰作笔名，旨在唤起世人对自然保护的意识。他的作品热情赞扬荒野保护，讴歌原生态自然的美好。他通过写作、电影、巡回演讲等方式引导更多人参与自然保护。他的文字反映了国家公园的理念，因此 1931 年他被聘为加拿大第一位自然主义学家。他的自传体作品黑色夜枭被拍成电影，为中国观众所熟知。

Ajawaan Lake Ajawaan 湖 艾尔伯特王子公园的自然湖泊之一。著名自然主义者、作家灰色猫头鹰在 1932—1938 年间曾居住于此，现在他的故居还坐落在 Ajawaan 湖西岸。Ajawaan 湖的生态系统得到了完整的保护，湖中有原生的白斑狗鱼、角膜白斑鱼和海狸。

Translation of Selected Sentences

1. Prince Albert National Park of Canada protects a slice of the northern coniferous or 'boreal' forest.

艾尔伯特国家公园在保护北方针叶林或北方森林方面起到了一定作用。

2. Through his writing, films and lecture tours he drew the world's attention to the need for conservation.

通过他的写作、电影和巡回演讲，他唤起全世界对于生态保护的意识。

3. Grey Owl's writing reflected the philosophy of National Parks and he was hired as the first naturalist.

灰色猫头鹰的作品反映了国家公园的理念，他被聘任为第一位自然主义者。

4. These books, plus films produced about Grey Owl and his beavers, not only helped to promote his message of conservation but made him and his cabin, Beaver Lodge, famous.

这些关于灰色猫头鹰和他的海狸的书籍和电影，不仅仅向世人传递了生态保护方面的信息，还使作者本人和他的故居——海狸小屋，也出了名。

5. Water bodies comprise almost 30% of the area of the park from huge expanses of water like lakes Kingsmere and Crean to the tiny marshes, bogs and creeks that dot the landscape.

公园将近 30% 的面积被水域覆盖，从面积较大的金米尔湖、克林湖到不知名的湿地、沼泽、小溪，它们点缀着公园美景。

Questions

1. What is the general condition of the Prince Albert National Park according to the passage?
2. Who is Grey Owl and why was he hired as the first naturalist?

3. Please state three well-known attractions in the Prince Albert National Park.

4. What are the main animals in the Prince Albert National Park?

5. Which functions does the Prince Albert National park have as a benchmark of ecological integrity?

Chapter 20

Prince Edward Island National Park
爱德华王子岛国家公园

【简介】

爱德华王子岛国家公园位于夏洛特王子岛以北 32 千米。创建于 1937 年,公园沿圣劳伦斯海湾延伸 25 里,占地 18 平方千米,是加拿大最小的一座国家公园。公园拥有众多沙丘、滨外岛屿、海滩、沙岩峭壁、湿地和雨林,是各种野生动物的栖息地。岛上还有悠久灿烂的文化遗产,其中最著名的就是安妮的绿顶小屋,它反映了 19 世纪岛上居民的生活方式,它也是加拿大著名女作家露西·莫德·蒙哥马利的著名小说《小孤女》中女主人公安妮·雪莉的主要活动场所。

Prince Edward Island National Park is a National Park located on Prince Edward Island. Situated along the island's north shore, fronting the Gulf of St. Lawrence, the park measures approximately(大约)60 km in length and ranges from several hundred metres to several kilometres in width. Established in 1937, the park's mandate includes the protection of many broad sand beaches, sand dunes(沙丘)and both freshwater wetlands(淡水湿地)and saltmarshes(盐沼泽). The park has been designated a Canadian Important Bird Area.

Prince Edward Island National Park of Canada is home to sand dunes, barrier islands(障壁岛)and sandspits, beaches, sandstone cliffs, wetlands and forests. These diverse habitats provide a home for a variety of plants and animals, including the threatened Gulf of Saint Lawrence Aster(翠菊)and the endangered(濒临绝种的)Piping Plover(笛鸻). The National Park also features unique(独一无二的)cultural resources, notably Green Gables, part of L. M. Montgomery's Cavendish National Historic Site and Dalvay-by-the-Sea National Historic Site. In 1998, six kilometres of the Greenwich Peninsula(格林威治半岛) were added to the Park to protect unique dune formations, rare(稀有的) plants and animals, as well as archaeological findings dating back 10,000 years.

Prince Edward Island National Park protects a portion of the Maritime(沿海的) Plain(平原) Natural Region, which is characterized by sand dunes, barrier islands and beaches, sandspits, and associated wetlands. The diverse habitats in the Park provide a home for a variety of plants and animals. The woods and shores of the Park are filled with over 300 species of birds and a large variety of plants. The roots and rhizomes(根状茎) of marram grass(滨草) form a living net, which helps hold the dune in place. Research on the effect of trampling(踩踏) on vegetation(草本植物) is a major priority for the Park. Wild rose is one of the first shrubs(灌木) to take root in more sheltered(遮蔽) areas in the dunes. The shiny coating on its leaves protects it from the harsh environment.

Sand dunes are an important natural habitat(栖息地) and act as a natural protective barrier against the effects of storms and waves. Research on dunes and all its associated features—vegetation, wildlife, and wetlands—helps us to understand and protect them better. Even the more stable(稳固的) dunes are fragile(易碎的) and easily damaged. Walking on dunes eliminates(除去) the protective plant cover. Studies have shown that it can take as few as 10 footstTIF through the same area to destroy a marram grass colony. Once the grass is gone, the wind blows away the exposed(暴露的、无遮蔽的) sand and carves small depressions(凹陷) into giant holes called blowouts(风蚀坑). Blowouts turn stable dunes into constantly shifting hills, unable to support vegetation or wildlife.

Prince Edward Island National Park is an example of the Maritime Plain Natural Region of Canada. Over 400 species of plants live in the park, and it is also home to 300 species of birds, including great blue heron(蓝苍鹭) and endangered piping plover. The chance of seeing wildlife in the wild is one of the most exciting things about Prince Edward Island National Park. However, it is important to treat wild animals with the respect they deserve. Approaching them too closely threatens their survival(幸存者). Feeding or approaching animals too closely causes them to lose their natural fear of people. Once habituated(习惯), animals are likely to become increasingly aggressive. Feeding may also: Attract animals to roadside areas where they can be injured or killed by vehicles. Lead to eating garbage. Animals eat almost anything which has the smell of food. No one can predict(预知) when threats(威胁) to human health due to direct contact with wildlife can occur. Known hazards(危险) include rabies(狂犬病) and tick-related(痉挛类的) diseases. It is illegal to feed, entice, or disturb any animal in Prince Edward Island National Park.

Prince Edward Island National Park is steeped with a rich and diverse human history. The park's central theme of "The Sea, People and the Changing Landscape" represents the earliest Aboriginal(土著的)people to live on Prince Edward Island to the French, Acadian(阿卡迪亚人), Scottish, Irish and English settlers who followed, who have all had an inseparable(不可分割的) relationship with the land and the sea. Early Aboriginal people represent the first cultural group to inhabit Prince Edward Island. They relied entirely upon the riches of the land and sea for their survival and prosperity(繁荣). When Europeans first began to visit and later settle on Prince Edward Island, they were greeted by the Mi'kmaq(阿尔冈琴人), who lived in harmony with the land and sea and continue to live on Prince Edward Island to this very day. Archaeological digs(考古挖掘)in the park have found traces left by the major cultures that have existed on Prince Edward Island over the past 10 000 years.

There is no better way to get in touch with Prince Edward Island's rich and diverse cultural heritage(遗产) than to visit any of the historical sites located within Prince Edward Island National Park, or administered by Parks Canada in Prince Edward Island. Two of the more enduring(经久不衰的)and most frequently visited historical sites in the park are Green Gables and Dalvay-by-the-Sea National Historic Site. Green Gables, which once belonged to cousins of L.M. Montgomery's grandfather, is the farm that inspired the setting of Montgomery's Anne of Green Gables. Montgomery was recognized as a person of national historic significance(重要性) by the Historic Sites and Monuments Board of Canada shortly after her death in 1942. Green Gables also portrays(描绘)the lifestyle of many island families in the late 19th century. The Green Gables property was acquired for inclusion in the national park in 1937.

The western tip of Greenwich, a peninsula that separates St. Peters Bay from the Gulf of St. Lawrence, became part of Prince Edward Island National Park in 1998 to protect and preserve the natural and cultural resources found in the area.

The site contains an extensive and fragile coastal(沿海的) dune system, wetlands and various natural habitats in which numerous rare plant species are found. Among the most spectacular natural characteristics to be protected at Greenwich are the unusually large and mobile parabolic(抛物线状的)dunes with their associated counter ridges(脊背) or Gegenwälle(防护堤). This phenomenon is very rare in North America.

Greenwich is also noted for its cultural and historic richness(丰富). Research at the site is ongoing(前进的), and evidence found thus far indicates a landscape that has changed drastically(大幅度的、彻底的)over time. Several archaeological digs conducted between 1983 and 2002 by the Canadian Museum of Civilization and Parks Canada have found traces left by the major cultures that have existed on Prince Edward Island over the past 10 000 years. These include early Aboriginal peoples, the Mi'kmaq, French and Acadian settlers, and Scottish, Irish, and English immigrants.

The Centre features(特征)over 20 exhibits to help people learn and appreciate the amazing natural and cultural history of this special place. The Time Line exhibit depicts(描述)a story of 10,000 years of inhabitation(居住), including artifact(手工产品)which have been found on-site. Visitors can test their own naturalist skills with the Shell Game, Shorebird Challenge, and Dune Plant quiz. The 3D in-floor model displays the Greenwich Peninsula, St. Peters Bay, and surrounding areas. A major component of the exhibit is the 12 - minute multimedia presentation entitled "Wind, Sea and Sand, the story of Greenwich". The Interpretation Centre is fully wheelchair accessible and also includes a gift shop and multipurpose room. Open daily from early June to early September.

Greenwich features three hiking(徒步旅行)trails varying in length from 1.25 km to 4.5 kms. The landscape varies from secluded wooded areas to open abandoned agricultural fields to spectacular vistas over Bowley Pond. All trails are equipped with interpretive signs, which explain the unique ecosystem and rich history of the area. Trails are open from late spring until late fall.

The facility in the park has been constructed using a sustainable(可持续地)design concept. Energy is provided by means of wind and solar power energy. Composting(制肥堆)toilets replace the need for a traditional sewage(污水)system. Services available include washrooms, change rooms, exterior showers, a large picnic shelter, a wooden boardwalk to the beach, and observation tower. Supervised swimming is provided at this location between July 1 and mid August.

Notes

Prince Edward Island 爱德华王子岛　加拿大领土面积最小,人口最少的一个省份。全岛可以分割为三个几乎均等的部分:位于最东边的国王区(Kings)、位置居中的皇后区(Queens)以及西边的王子区(Prince)。爱德华王子岛最初是由探险家雅克奎斯·卡提尔(Jacques Cartier)于1534年沿海岸航行时发现,并命名为圣让(Saint Jean)。之后,阿堪迪亚人(Acadians)于17世纪中叶开始迁移至此——原为迈克麦克原住民(Micmac Amerindian)居住的土地。直到1763年七年战争结束,根据双方签订的《巴黎合约》,该岛归于英国,方才以乔治三世之子——爱德华王子之名重新命名。

Gulf of Saint Lawrence 圣劳伦斯湾　位于加拿大东南部的大西洋海湾,是北美五大湖圣劳伦斯河进入大西洋的入口。它是一个半封闭的海湾,水域覆盖面积约23.6万平方千米。海湾是在地质构造运动中发生沉降作用后形成的,因而湾岸曲折,多暗礁、浅滩,不利于航行。

Green Gables 绿顶小屋　加拿大爱德华王子岛上一座19世纪的农场的名字。随着蒙哥马利作品《小孤女》的问世,绿顶小屋也成为最显著的文学地标式的建筑。随着安妮绿山墙在周围农场的逐渐流行,绿顶小屋也成为爱德华王子公园最重要的历史遗址。

Lucy Maud Montgomery 露西·莫德·蒙哥马利(1874年11月30日—1942年4月24日)　加拿大著名的女作家。她最著名的作品是以自己的故乡爱德华王子岛为故事背景,以《小孤女》为起点,围绕主角安妮·雪莉展开的一系列作品。蒙哥马利在作品中客观地记录了家乡爱德华王子岛的美丽风光,使爱德华王子岛声名远播。爱德华王子岛上与小说主角安妮有关的绿顶小屋和艾利凡小镇也成为爱德华王子岛的著名景观。

Greenwich Peninsula 格林威治半岛　加拿大的格林威治半岛以其独特的沙丘和独特的排列方式著称。为保护这一独特的地址风貌,1988年格林威治半岛长达6千米的一段被划进爱德华王子国家公园加以保护。

Mi'kmaq 阿尔冈琴人　居住在美国东北部新英格兰、加拿大大西洋省份、魁北克加斯佩半岛地区的原土著居民。他们拥有本民族的语言和象形文字。

Translation of Selected Sentences

1. Feeding or approaching animals too closely causes them to lose their natural fear of people.

近距离的喂食或接近动物会使它们消除对人的天生的恐惧感。

2. It is illegal to feed, entice, or disturb any animal in Prince Edward Island National Park.

在爱德华王子公园里喂食、引诱或打扰动物是违法行为。

3. There is no better way to get in touch with Prince Edward Island's rich and diverse cultural heritage than to visit any of the historical sites located within Prince Edward Island National Park.

没有什么比参观爱德华王子国家公园里的历史古迹能够更好接近爱德华王子岛灿烂多样的文化遗产。

4. Green Gables also portrays the lifestyle of many island families in the late 19th century.

绿顶小屋再现了19世纪岛上居民的生活方式。

5. The facility in the park has been constructed using a sustainable design concept.

公园内的设施完全按照可持续性发展的概念设计的。

Questions

1. What is the general condition of the Prince Edward Island National Park according to the

passage?

2. Who is Lucy Maud Montgomery?

3. What is the function of marram grass and wild rose?

4. What kinds of behaviours are illegal inthe Prince Edward Island National Park?

5. What is Greenwich peninsula dune system?

Chapter 21

Pukaskwa National Park
帕卡斯夸国家公园

【简介】

帕卡斯夸国家公园位于加拿大安大略省中部，苏必利尔湖的东北湖岸线。它是安大略省唯一的一座野外公园，同时它也是该省最大的国家级公园。这座公园创建于 1978 年，面积约为 1 878 平方千米，公园里有崎岖不平的加拿大地盾，80 千米长的苏必利尔湖湖岸线，岩石小岛和海湾，壮观的悬崖峭壁，以及很多小型的湖泊和溪流，被誉为“内地海的野生海岸”。公园里保留着一些史前的印第安遗迹，生活着多种多样的野生动物：海狸、麝鼠、北美大灰狼、黑熊、貂鼠、水貂、猞猁、白尾鹿、驼鹿、驯鹿等。广袤的森林生长着白云杉、黑云杉、加拿大短叶松、杨树、桦树等丰富的树种。一直以来，帕卡斯夸国家公园就倡导人们减少对自然界的破坏和改造，包括对五大湖区（Great Lakes）最低限度的修整。此外，帕卡斯夸国家公园还是露营、徒步旅行、划艇的天堂。

Pukaskwa National Park is a national park located south of the town of Marathon, Ontario in the Thunder Bay District of northern Ontario, Canada, which is the only true wilderness park in Canada's National Park system in Ontario. Established in 1978, Pukaskwa is known for its vistas of Lake Superior, boreal forests（北方针叶林）and Canadian Shield.

There are many spellings of the word "Pukaskwa" and many legends about the meaning of the word. Some contend that the word is descriptive terminology concerned with cleaning fish. Others suggest it could mean "eaters of fish", "something evil", or "safe harbour". We may never know the true origin of the name, but today it has become synonymous with the wild shores of Lake Superior known as Pukaskwa National Park.

When discussions were to establish Pukaskwa National Park in the early seventies, the area had few of the concerns of many other national park sites. Pukaskwa had none of the crowd concerns of Banff, nor was the area a small isolated fragment of wilderness like Point Pelee(皮利角) or not so small but isolated fragment of Riding Mountain National Park. In its favor, Pukaskwa was a reasonable size (almost 2000 km^2), it had the support of local people and a co-operative agreement with the local First Nations. Wilderness throughout the park was undisturbed by humans. Most of the 6000 km^2 around the park was also relatively pristine(原始状态的) as very little modern human development had yet occurred even though the Trans-Canada Highway passed within 20 kilometers of the park boundary.

As the park was officially established in 1978, timber adjoining the park was allocated(分配) to a new sawmill in White River. Several years later, one of the largest gold mines in North America appeared in the Hemlo Gold Field, 20 kilometers north of the park. Corporate interests in the region's natural resources had suddenly intensified. Fifteen years later found the mid-nineties and the park, the mill and the mine were operational leaving questions about the future of Pukaskwa National Park. Although most of the natural resources had been allocated, creating the potential for conflict between local park stakeholders(有权益关系者) and large corporations, there were many opportunities for conservation in the light of such concepts as "sustainable development", "ecosystem management"(生态系统管理) and "Greater Park Ecosystems".

Pukaskwa National Park covers an area of 1,878 km^2, of which 95.6 percent is land, and the remaining 4.4 percent is water. Pukaskwa (actually a river name on the park's southern border) was created to protect, for all time, a representative example of the central Boreal Uplands and the Great Lakes Shoreline and to encourage public understanding, appreciating and enjoyment of its heritage so as to leave it unimpaired(未受损害的) for this and future generations. The special preservation areas in Pukaskwa contain woodland caribou, arctic-alpine flora(北极高山植物群), a jack pine(加拿大短叶松) forest, acid-sensitive lakes and Pukaskwa pit structures, remnants of Aboriginal(土著的) use.

Here, the waters are crystal clear and the shoreline is made up of an impressive semi-mountainous hulk of Canadian Shield. Some of the natural highlights we will encounter along the route include fascinating geological formations like basalt(玄武岩) intrusions, rare arctic-alpine (高寒的) plant species, the endangered peregrine falcon(游隼), and woodland caribou in their most southerly habitat.

Pukaskwa is located on the Lake Superior that is a part of the Great Lakes. The Great Lakes region has humid continental climate, and it is characteristic for hot summers and cool winters—up to three months of snow may stay on the ground. However, Pukaskwa climate is continental, with the exception of a maritime(海洋的) effect which Lake Superior exerts up to kilometers inland. The Lake Superior is a freshwater lake but due to its size behaves like a sea, creates sudden storms, and thick fogs. Lake Superior's storms have forced away the shores, leaving an irregular shoreline spread with rock fragment, and exposing the ancient bedrock. Lake Superior influences the weather and vegetation of the lands that surround it. Winters are long, snowy, and cold. The lake's water is cold, with an average year-round temperature of 4°C, and it is a harsh place. Moist, cold air moves inland. The annual temperature is between 1°C and 3°C. The frost free is between 70 to 100 days, and the growing season between 160 to 180 days in a year.

The word boreal comes from the Greek for north wind. It is an appropriate name as one of the biggest challenges is to survive the tough, cold winter. Thick carpet of moss(苔藓) covers the thin soil, generally acidic. Trees that are adapted to those harsh conditions are: spruce(云杉), fir, cedar(雪松), birch, red maple(红花槭), jack pine forest. Boreal forest gorgeousness is represented by small and delicate orchid(兰花) and arctic-alpine flora. Arctic plants left behind by the ice age have been able to survive alongshore in "arctic" microhabitats because of the cooling effects of Lake Superior. One of these, the encrusted saxifrage, produces white clusters of flowers in late June, along the Southern Headlands Trail, easily accessible at Hattie Cove.

Pukaskwa boreal forest is represented by White spruce and balsam fir(香脂冷杉). Other species included jack pine, white and red pine, aspen(山杨), tamarack(美洲落叶松), and birch, but they are shade intolerant and depend on fire. Those trees are in the process of succession and white spruce will be present and eventually dominate the stand.

Boreal forest is a home for many mammals(哺乳动物), birds and amphibians(两栖动物) for it provides variety of food for them. Because 95% of the Pukaskwa park is zoned as "wilderness", we can expect many animals. A series of special preservation areas provides the highest level of protection to woodland caribou. The most common mammals in the park are: moose(驼鹿), wolf, black bear, spruce grouse (枞树鸡), lynx(猞猁) and the usual array of small northern animals—red fox, otter(獭), snowshoe hare(北美野兔), porcupine(豪猪). Sudden booms in the animal population are typically followed by periods of starvation. Migratory birds such as Bay-breasted Warbler(栗胸森莺) arrive in summer, Gray Jay(灰噪鸦), and White-throated Sparrow(白喉带鹀), Ducks and shorebirds may also be seen along the beaches.

Most of the park area is covered with the Aloman and Laurentian(劳伦系岩石) rock such as: granite(花岗岩), quartz(石英石), porphyries(斑状安石岩). Only small area has acid and basic lavas(火山岩), tuff(凝灰岩), volcanic breccia(火山角砾岩), iron formation(含铁建造), and sedimentary rocks(沉积岩). The Pukaskwa terrain is a remaining of old mountains that were shredded by the passing glaciers(冰川), but still until now Pukaskwa has the highest elevation(海拔) in Ontario, which is Tip Top Mountain 639.8 meters and Soldier 609 meters. Generally, the park may be divided into two parks, the coastal and the interior covered with dense forest, and the interior is accessible only on foot or by boat. Away from the coastal shore, the park's terrain is ridged with pockets filled with melt water making hundreds of lakes. The soil is thin due to bedrock control, mainly acidic(酸性的), and often covered with a thick carpet of moss. Pukaskwa National Park was divided into four terrain groups: uplands, depression, flats, and slopes(斜坡).

Visitors to Pukaskwa National Park will have the opportunity to meet over 25 artists from across Ontario between May and October. Every day will provide the opportunity to meet an artist, learn from their talents and see new works inspired by Pukaskwa's breathtaking natural scenery. Works will be available to view and purchase at the Wildshores Gift Shop in the Hattie Cove Visitor Centre.

In the heart of Pukaskwa National Park, experience Anishinaabe culture and values by visiting an authentic encampment featuring a wigwam(棚屋), teepee(印第安人的圆锥形帐篷), cook tent, and meeting area. A First Nations interpreter will be on hand during peak season to demonstrate and explain aspects of Ojibway(奥吉布瓦人) culture. See how Anishinaabe culture is still a vital, spiritual and social part of life.

The heart of the park is located at Hattie Cove campgroundnear the northwestern tip. You'll find 67 front country campsites that are easily accessible, as well as the launch points for most of the parks activities. The parks visitors centre is also located here allowing you to easily get information on the parks backcountry(边缘地区) activities.

The 60kmCoastal Hiking(徒步旅行) Trail winds from Hattie Cove south to the North Swallow River and requires proper planning and equipment (camping areas are every half to a full-day's hike apart). A 15km day hike along this trail can be taken to the White River Suspension Bridge(吊桥). There are also backcountry trails. In winter, cross-country skiers can use 6km of groomed trails or hazard the fast slopes and sharp turns created by the topography(地势). Snowshoers are welcome, too.

In the warmer months you'll find opportunities for hikers, backpackers, wilderness canoeists and intrepid(无畏的) sea kayakers. Hikers have a number of short trails to choose from while backpackers can take advantage of the Coastal Hiking Trail. Experienced canoeists have two wild rivers to run, the White and the Pukaskwa River. Kayakers can paddle the bays around the park or tackle the Superior shoreline if they are experts.

The White and Pukaskwa rivers offer white-water adventure. You can paddle the easily accessed White River any time during the open-water season. Many wilderness adventurers start from nearby White Lake Provincial Park and travel 4 to 6 days to the mouth of the White River and then paddle about an hour north on Lake Superior to Hattie Cove.

The Pukaskwa River is more remote, more difficult (with rugged and long portages and a 260m drop between the headwaters at Gibson Lake and the river mouth at Lake Superior), and navigable(可航行的) only during the spring runoff. The best place to start is where the river crosses Highway 17 near Sagina Lake and paddle to Gibson Lake via Pokei Lake, Pokei Creek, and Soulier Lake. Otherwise, you'll have to fly in from White River or Wawa.

Notes

Ontario 安大略省　位于加拿大的中东部,北至哈得逊湾,东邻魁北克,西接马尼托巴,南部则与美国的明尼苏达、密歇根和纽约州为界。面积1,076,395平方千米,是加拿大第二大省,人口13 210 667人(2010年)。安大略省得名于五大湖中的安大略湖,在易洛魁语中是"美丽的湖"的意思。欧洲人在此定居之前,此地原住民为北美印第安人中的易洛魁和阿尔冈昆部落。1867年7月1日《英属北美法》生效,安大略建省,与魁北克、新斯科舍、新布伦瑞克四省组成加拿大自治领。安大略省中部和西北部为加拿大地盾,耕地贫乏,但矿产丰富,拥有诸多河流与湖泊;东北部是哈得逊湾低地,森林和沼泽遍布;东南部的五大湖——圣劳伦斯河谷地,气候适宜,人口和工业区集中于此。安大略省制造业发达,东南部城市为全国工业中心,以汽车、钢铁、机械、化工和造纸为主。首府多伦多是全国的金融中心,农业和畜牧业是安省的重要产业。尼亚加拉半岛和伊利湖是果蔬生产基地。著名的景点有尼亚加拉瀑布、千岛湖景区、安大略皇家博物馆。

Lake Superior 苏必利尔湖　北美五大湖中面积最大的湖,是世界第二大湖,也是最大的淡水湖。东部和北部与加拿大安大略省相接,西邻美国明尼苏达州,南至美国威斯康辛州和密歇根州。面积为82 414平方千米。在阿尼什纳比语中,苏必利尔湖被称为"Gichigami",意为"巨大的水域"。包括圣路易斯河、尼皮贡河、皮金里弗河、白河、米奇皮科滕河及卡米尼斯蒂奎亚河在内,总共有超过200条河流流入苏必利尔湖。苏必利尔湖与圣玛莉河与休伦湖相连。沿湖多林地,人口稀少,苏必利尔湖水质清澈,湖面多风浪,湖区为局部的海洋性气候。渔猎和旅游是当地主要产业,矿藏资源丰富,盛产毛皮。在苏必利尔湖周围的风景区有阿波斯尔群岛国家湖滨区、卢瓦尔岛国家公园和帕卡斯夸国家公园。

Canadian Shield 加拿大地盾　一种地质盾构,也是最大的地质大陆盾之一。表面土质层稀薄,鉴于历史上长期呈现火山地貌,其主体由火成岩构成,是加拿大中部和东部地区的主要地质结构,并向北延伸至五大湖和北冰洋地区,差不多覆盖半个加拿大以及美国北部海岸(包括阿第伦达克山区和苏必略高地)。在这种地构覆盖的地区,人口稀少,工业落后,但采矿业发达。

Banff 班夫　加拿大亚伯达省西南部一城镇，位于路易斯湖附近的洛矶山脉，是加拿大班夫国家公园中最大的城镇，人口约 1 万人，班夫有一年一度的班夫山地电影节。班夫国家公园建于 1885 年，是加拿大历史最悠久的国家公园，与其他加拿大洛矶山脉的国家和省立公园一起被列入世界遗产名录中。周围山峦叠嶂，冰川、冰原和松林遍布，还有温泉，是著名的冬季休假胜地。它是重要的户外运动区，主要活动有步行、登山、滑雪等。园内名胜有硫磺山、弓河瀑布、惊奇角和奇石观景台等。

Trans-Canada Highway 加拿大横贯公路　又称横加公路，是一条横贯加拿大的公路，也是世界最长的国家级公路。主线西起不列颠哥伦比亚省的维多利亚，东至纽芬兰省的圣约翰斯，全长 8 030 千米。加拿大横贯公路工程于 1950 年开始，1962 年启用，1971 年完工。6.7 ~ 7.3 米宽的路面，宽敞的路肩，低坡度少弯道，沿线尽量减少与铁路交汇，路面承重能力强都是横加公路为人称道之处。贯穿加拿大全部 10 个省，连接了诸多重要城市，通往很多著名的国家公园和省立公园，如班夫国家公园。

Hemlo Gold Field 赫姆洛金矿区　北美最重要的低成本金矿之一，坐落于苏必利尔湖北岸，位于苏圣玛丽和桑德贝之间，临近马拉松镇。20 世纪 80 年代早期被勘探者发现，三个主要矿山：David Bell, Golden Giant 和 Williams 集中于一英里范围内，其金矿石超过 7 500 万吨，价值大于 70 亿美元。20 世纪 90 年代早期，该矿区开采量达到最大，每年生产黄金 50 万盎司，截止 2006 年，已生产超过 600 万盎司黄金。

woodland caribou 丛林驯鹿　驯鹿的亚种，多生活在加拿大的北方针叶林地区和北部加美边境，从纽芬兰与拉布拉多省以西到华盛顿以南的部分地区。丛林驯鹿是野生群居动物，每年都要迁徙 2 000 千米，活动区域约为 1 百万平方千米。上世纪以来，由于人类大量捕猎以及环境变化等原因，到 1950 年，在魁北克和拉布拉多省北部仅存 5 000 只野生丛林驯鹿。20 世纪末本世纪初，由于对丛林驯鹿的保护得当，该族群已恢复到了上百万只的数量。每年因纽特人、克里人和南方的捕猎活动者要杀死大约 3 万头丛林驯鹿，因此对野生丛林驯鹿的保护仍是不容忽视的。

First Nations 第一民族　几个加拿大境内民族的通称，与印地安人同义，指的是在现今加拿大境内的北美洲原住民及其子孙，但是不包括因努伊特人和梅蒂人。原住民是第一民族、因努伊特人和梅蒂人的总称。在加拿大，原住民的国家级代议机构是第一民族议会。印第安人有的有身份，有的无身份。有身份的印第安人是指那些在官方登记处注册过的、在印第安人法令管理下的印第安人，他们自称为“第一民族”，以表明自己是最早居住在美洲大陆的主人。现在，人们通常用“第一民族”泛指所有的印第安人，偶尔也有人用“第一民族”泛指所有的土著民族。

White Lake Provincial Park 白湖省立公园 位于加拿大不列颠哥伦比亚省，巴尔默勒尔东北 10 千米处，坐落于白湖北岸，面积 266 公顷，园内沙滩遍布，湖水温暖可以畅游，特别是以垂钓虹鳟鱼闻名。是省内最适宜垂钓的三个湖泊之一。夏天的拖钓和飞蝇钓，冬天的冰下捕鱼都别有一番趣味。这一地区曾经盛产皮毛和木材,西部锦龟是不列颠哥伦比亚省仅有的池龟,园内的西部锦龟、兰花和一些食虫植物吸引了大批的观光者和科学家。

Translation of Selected Sentences

1. We may never know the true origin of the name, but today it has become synonymous with the wild shores of Lake Superior known as Pukaskwa National Park.

我们也许永远不会知道这个名字的真正起源,但是今天它已经等同于苏必利尔湖的野生湖岸,作为帕卡斯夸国家公园为世人所知。

2. Pukaskwa had none of the crowd concerns of Banff, nor was the area a small isolated fragment of wilderness like Point Pelee or not so small but isolated fragment of Riding Mountain National Park.

帕卡斯夸国家公园不像班夫那样吸引了大量的关注,也不像皮利角或骑马山国家公园那样,只是一个小型的野生独立环境。

3. Pukaskwa (actually a river name on the park's southern border) was created to protect, for all time, a representative example of the central Boreal Uplands and the Great Lakes Shoreline and to encourage public understanding, appreciating and enjoyment of its heritage so as to leave it unimpaired for this and future generations.

帕卡斯夸,实际上也是帕卡斯夸国家公园南部边界上一条河流的名字。建立这座公园,是为了永远地保护这个位于北方高地中部和五大湖沿岸的代表性区域,并鼓励公众理解、重视和享受这份自然遗产,使它无论是现在还是未来都不受到破坏。

4. Some of the natural highlights we will encounter along the route include fascinating geological formations like basalt intrusions, rare arctic-alpine plant species, the endangered peregrine falcon, and woodland caribou in their most southerly habitat.

在这段路途中最精彩的是,我们将会遇到令人啧啧称奇的地质构造,如玄武岩侵入结构、稀有的高寒植物物种、濒危的游隼以及位于南部栖息地的丛林驯鹿。

5. Many wilderness adventurers start from nearby White Lake Provincial Park and travel 4 to 6 days to the mouth of the White River and then paddle about an hour north on Lake Superior to Hattie Cove.

大多数的荒漠冒险是从白湖省立公园附近出发,经过4~6天的行程到达白河河口,再从苏必利尔湖向北划船,行程约1小时,抵达哈蒂湾。

Questions

1. What are the legendsabout the meaning of the word "Pukaskwa"?

2. According to the passage, what is the real purpose of establishing Pukaskwa National Park?

3. In the heart of Pukaskwa National Park, what kind of activities can you participate in to experience Anishinaabe culture and values?

4. Where is the heart of the park? And what will you find there?

5. What is the route of many wilderness adventurers?

Chapter 22

Riding Mountain National Park
骑马山国家公园

【简介】

骑马山国家公园,位于加拿大曼尼托巴省,面积2 976平方千米。园内树木丛生,与周围大草原上的广袤耕地形成鲜明对比。这里是大灰狼、麋鹿、驼鹿、黑熊和北美野牛的栖息地,上百种鸟类和数不清的各类昆虫在这里生活。1929年成为保护区,1933年骑马山国家公园正式建立。二战期间这里沦为战俘营。园内地形结构既有平原,也有高原,还有部分洼地沼泽,形成了独特的生态系统,1986年被联合国教科文组织指定为生物圈保护区。

Riding Mountain National Park is situated in western Manitoba, approximately 99 km north of the city of Brandon. The park boundaries enclose 2976 square kilometres of rolling（延绵起伏的）hills and valleys, stretch eastward from a dramatic rise of land known as Manitoba Escarpment, and mark the transition from the Manitoba Lowlands to the second prairie（大草原）level, the Saskatchewan（萨斯喀彻温省）Plain. In 1929, the Riding Mountain Forest Reserve was set aside as a national park and on July 26, 1933 Riding Mountain National Park was officially opened. The goal is to protect the biodiversity of the biosphere, and to use the resources wisely to promote the long term health of the region plants and animals.

The forested parkland stands in sharp contrast to the surrounding prairie farmland. The park has been described as "an island of forest rising out of a sea of farmland", rising 365 m from the surrounding agricultural plain to a maximum elevation of 600 m above sea level. Bold boreal forest, expanses of meadows（牧场）and lakes, significant tracts of marsh（沼泽）and river-bottom wetland（湿地）, lofty（高耸的）heights and deep gorges（峡谷）of the Manitoba

Escarpment is home to wolves, moose, elk, black bears, hundreds of bird species and even a captive American bison（北美野牛）herd. The Riding Mountain National Park includes the rustic（乡村的）charm of the town of Wasagaming, lakes, trails, and nature. Outside the park but within the biosphere are cultivated areas, farms, and livestock（牲畜）.

Riding Mountain National Park in the southwest corner of Manitoba is the protected core of the UNESCO designated Riding Mountain Biosphere Reserve. It's an area focused on conservation and sustainable resource. Riding Mountain National Park received visitation of almost 425,000 persons in 1994/5 and is a significant provincial tourism destination. Riding Mountain National Park is open year round, with the peak season in July and August.

The escarpment is a tilting（倾斜的）shelf of siliceous shale（硅质页岩）—a form of sedimentary rock（沉积岩）that is harder than the surrounding limestone（石灰岩）—which extends into northern Saskatchewan, where it forms similar upland areas. Although only 756 metres at its highest edge on the northeastern side, relative to the plains that stretch away from it on all sides, it is a mountain. The park includes a diversity of landscapes: evergreen and hardwood forests, rolling hills, valleys, lakes and streams. The park is a crossroads where prairie, boreal and deciduous（落叶的）life colonies intermingle（混合）. Manitoba is famous for thousands of prairie potholes（壶穴）—lakes of all sizes that were left as glaciers retreated—which are now havens for waterfowl（水鸟）, fish, birdlife and insects. Riding Mountain Park is one of the few places in habitable parts of Manitoba where these prairie potholes have not been drained off and plowed under（隐匿不见）.

The soil parent materials of the region are largely clay, giving rise to relatively fertile, alkaline（碱性的）soils and eutrophic（富养分的）limnological（湖泊的）conditions. The annual freeze-thaw（冻融）cycle prevents the cumulative compaction of soils and sustains soil friability. The soil fertility combined with climatological and hydrological factors contributes to the area being able to sustain a relatively high biomass of diverse biota with high population densities of large herbivores（食草动物）.

Three distinctive vegetation zones are recognized, although they do, in fact, intermix（混杂）. These primary recognizable plant communities are the Fescue（羊茅）Prairies, Aspen Woodland, Mixed Boreal and Eastern Hardwood complexes. In the south aspen（山杨）parkland, open, rough fescue grasslands and aspen groves, edge the eastern side of Lake Audy. Bordering the aspen parkland is a zone of mixed-wood forest containing various combinations of coniferous（松类）（white spruce and balsam fir）and deciduous（white birch, aspen, elm, maple and balsam poplar）species. Very few stands of eastern deciduous forest are left in southern Manitoba, because most have been destroyed in clearing land for agriculture. But here you see a rich remnant of this complex ecological zone. In the surrounding forest, there is an understory（森林次冠层叶簇）of vines（藤蔓）, berry bushes（浆果丛）, mushrooms, and flowers that are at the height of their beauty amidst the prairie grasses in June and July. Areas that are mostly boreal forests, in the extreme north of this natural region, appear as seemingly（表面上看）endless stretches of black spruce muskeg（泥岩沼泽）that dominate much of the flat, poorly drained land. This combination of communities in the Riding Mountain portion of the escarpment forms a distinctive Canadian habitat that is unique in the world.

The Biomes that occur and intermix in Riding Mountain National Park are those of the Temperate Zone（温带）reflecting the climate and geographic location of the Park. Variations in the physiological features such as soil type, hydrology, topography and microclimate further dictate the locations suitable for sustaining the specific vegetative types and consequently form the basis for the intermingling of the ecosystems. The resultant complex of interspersed ecosystems makes this a globally unique biological phenomenon, warranting the MAB status of Biosphere Reserve.

To the west, the escarpment's edge gradually merges into a gently rolling landscape of a highland plateau（高原）, where plant and animal communities meet. Large herds of elk gather in sedge（莎草）meadows while wolf packs congregate（聚集）in nearby forests. Lynx and cougar（美洲狮）inhabit the woodlands, bald eagles（秃鹰）and osprey（鱼鹰）nest along the streams and lakes. The aspen parkland is an extremely productive wildlife habitat where white-tailed deer, coyotes, snowshoe hares and ground squirrels（地松鼠）are conspicuous mammals. In the past, vast herds of bison ranged into the aspen zone. Other large mammals include black bears, which can weigh（称重）up to 400 kilograms, moose, elk, and wolves. Efforts continue to monitor the gray wolf population since they are few in number, highly mobile and vulnerable（易受伤的）to human impact. Black bears easily range 100 kilometres a day and many have grown bigger than most grizzlies（北美洲灰熊）. There are now about 3500 beaver dams in the park and over 18 000 beavers; with such an exploding population, problems do exist, such as the dams preventing fish spawning（产卵）in some areas. In 1931 the famous Archie Belaney, aka Grey Owl was hired by Riding Mountain National Park to re-establish the ailing（境况不佳的）beaver population. Grey Owl stayed only six months, before moving to his new post in Prince Albert, claiming the lakes were too dry for the beavers. The beaver population has since been restored and today numbers somewhere around 18,000. Grey Owl's cabin is a popular hiking destination at Beaver Lake Lodge. There are at least 260 bird species including great gray owls（大灰猫头鹰）. The dozens of prairie potholes provide for the highest density of breeding dabbling ducks（钻水鸭）in North America, particularly mallards（野鸭）, shovellers（琵嘴

鸭）and pintails（针尾鸭）.

Large herbivores such as the deer, elk and moose retard（减缓）the encroachment（侵蚀）of aspen and brush, such as beaked hazel（加拿大榛树）into prairie plant communities by heavily browsing on the new shoots of these plants along the margins. The browsing of these Cervids（鹿类）on regenerative aspen saplings（幼苗）has a thinning effect on the forest allowing more sunlight to penetrate to the level where growth of browse species of plants can benefit. The role of the moose in this regard is augmented by its, widespread behavior of breaking and killing aspen and balsam poplar saplings.

The park has a bison compound at Lake Audy where you can observe these wonderful animals close up, in a natural setting. Over 270 species of birds have been observed in the park. Bears, Lynx, Moose are some of the mammals you might observe if you're lucky. Several varieties of amphibians and reptiles can also be seen.

The lakes of Riding Mountain National Park are open for fishing. The cold deep lakes such as Clear Lake, Lake Katherine and Deep Lake are rich in pike（梭鱼）, whitefish, walleye（白斑鱼）and trout（鳟鱼）. Clear Lake, Lake Audy, Moon Lake and Whirlpool Lake offer great northern pike fishing. Other fish species include walleye, white fish and perch（鲈鱼）in Clear Lake, while Lake Katherine and Deep Lake have a limited number of rainbow and brook trout（溪红点鲑）. Winter or ice fishing is allowed on Clear Lake for most of the winter. A national park fishing license is required, and may be picked up in Wasagaming or in surrounding towns.

Recent archeological（考古学的）evidence of habitation, fishing, hunting, tool and pottery（陶器）making and burial activities suggests that aboriginal peoples have inhabited the area for at least 6000 years. More recently the Ojibway migrated from eastern regions to inhabit the Riding Mountain area, previously the home of the Nakota（纳科他人）Nation. The Nakotas, known to travel widely, shifted west and south moving tribal boundaries. The Ojibway, active fur traders, roamed throughout the area trapping, fishing and hunting. By 1800 the highland was surrounded by fur-trading posts. Later, local settlers used the park's resources for timber, cattle grazing, hunting and recreation. Today there are 3 Indian reserves south of the park's boundaries. In 1896, a fishing reserve established on the shores of Clear Lake for the Keeseekoowenin Ojibway First Nations was wrongfully removed in 1930 at the time of the establishment of the park; this land was returned in 1991 after the Band made a land claim. Oral histories of various elders have added not only to our archeological and geographical knowledge, but also to our understanding of the sacred meaning that the mountain and its waters hold.

Visitors can enjoy a diverse range of memorable and engaging（有吸引力的）activities throughout the park and surrounding communities. On the south shore of Clear Lake, lies the townsite of Wasagaming that provides visitors a full range of services including accommodations, restaurants, shopping, activities and cultural attractions. Most amenities are available in the town site, where you'll find a grocery store, a bakery, a gas station, restaurants and a laundromat. The parks visitors centre is also located here.

The 2976 square kilometer park is located approximately 225 kilometers northwest of Winnipeg. There are 21 campsites and 20 trails located in the backcountry. Typical visitors to the backcountry engage in hiking, bicycle riding, horseback riding, and cross-country skiing in the winter. The park has over 400km of trails for hikers, mountain bikers and horse packers. Most of the hiking trails are single use but almost all of the rest of the extensive trail system is designated as multi-use. The trail network in the park is largely interconnected and many trails are accessed from other trails rather than road accessible trail heads. Many of the trails are open to cross country skiers and back country skiers in the winter months. The Manitoba Escarpment provides for exceptional viewpoints over the park and the surrounding region.

The park has a marina at Clear Lake as well as a boat launch. Sail boats are accommodated both by the marina and the prairie winds which sweep across the lake. Canoes and kayaks are also seen plying the park's waters. In the winter cross country skies and snowshoes are the best ways to see the park and observe the wide variety of wildlife.

Backcountry use inRiding Mountain National Park has shown a steady increase over the past few decades, rising from just over 700 visitor nights in 1972 to nearly 3000 in 1999. As a result of this increase, park managers became concerned about potential campsite and trail impacts. Over the course of several years a series of attempts to monitor the backcountry were initiated. Frustration with the lack of progress led the chief backcountry warden (管理员) to contact researchers at the University of Manitoba to develop a monitoring program for the park. The program, described elsewhere, differs from others in that it integrates visitor use data with biophysical data in order to identify how patterns of visitor use may influence backcountry impacts. This allows managers to monitor these patterns and act accordingly to protect the park during times of impact susceptibility (易受影响的状态).

Notes

Manitoba 曼尼托巴省　加拿大中南部的一个重要省份,是草原诸省之一。1870 年 7 月 15 日根据《曼尼托巴法案》加入加拿大联邦,是加拿大的第 5 个省,位于大西洋和太平洋之间,东靠安大略省,南接美国的明尼苏达州和北达科他州,西临温哥华,北至努纳维特地区,面积约 65 万平方千米,全省人口约 121 万(2009 年),官方语言主要为英语,省府为温尼伯市。曼尼托巴省是加拿大主要粮食产区之一，支柱产业为制造业，其次是农业、建筑、水电和采矿业。以“加拿大的阳光之都”而闻名。冬季气候比较寒冷，夏季温暖。省内湖泊众多，大大小小共有38500多个，其中温尼伯湖是世界上最大的内陆淡水湖。全省五分之二的面积为森林覆盖，北部的哈德逊湾是一个游览胜地，海湾南部的野鸭和北部的北极熊皆为有名的景致。主要的旅游景点有曼尼巴托省议会大厦、百年纪念中心、圣博尼费斯大教堂、唐人街、丘吉尔市。

Manitoba Escarpment 曼尼托巴单面山　位于曼尼托巴省东南部，是一处单面陡峭，另一面却坡度很小的山。它形成于冰川期，构成了史前时代阿加西湖的西部冰川湖岸。它朝向东面，比曼省南部的低地地区海拔要高出150～300米，它从美加西部边界上的雷德河向西北延伸560千米到达萨斯喀彻温省的卡罗特里弗地区。自北向南，河流、马鹿群、森林和山谷将它们分割成了成群的峻岭，如帕斯奎亚丘陵、鸭子山、骑马山和彭比纳山，最高峰与萨省的大草原地区海拔同高。现在这片区域内的骑马山国家公园、数个省立公园以及从林地区，已经成为人们户外休闲的胜地。鸭子山省立公园中的鲍尔迪山是曼省的最高峰。

American bison 北美野牛　牛科野牛属食草动物，体格健壮，体重可达1 000千克，生性剽悍，头上有锋利双角，野生天敌主要是灰狼和有毒的响尾蛇。北美野牛曾一度被猎杀，频临灭绝，这种大屠杀不仅严重地危及了北美野牛的物种存活，而且结束了许多美国印第安人的放牧生活。1905年当时的美国总统罗斯福颁布法令，第一次将野牛等一批珍稀动物置于国家的保护之下。庞大的野牛群终于再次遍布北美草原，在许多国家公园都可以看到它们的踪迹。实际上野牛并不是原始的美洲动物，它们的渊源在万里之外的欧亚大陆。如今野牛在欧亚大陆上几乎绝迹了，只有在波兰境内的巴洛维萨森林地带，还可以发现这种珍奇动物生活的踪迹。

Biosphere Reserve 生物圈保护区　教科文组织于1971年发起的人与生物圈计划的核心部分。在这些地区会尝试使用不同的举措对陆地、淡水、海岸和海洋资源及生物多样性进行综合管理。人与生物圈计划是联合国教科文组织科学部门的一项政府间跨学科的大型综合性的研究计划。生物圈保护区具有保护、可持续发展、提供科研教学、培训、监测基地等多种功能。其宗旨是通过自然科学和社会科学的结合，基础理论和应用技术的结合，科学技术人员、生产管理人员、政治决策者和广大人民的结合，对生物圈不同区域的结构和功能进行系统研究，并预测人类活动引起的生物圈及其资源的变化，及这种变化对人类本身的影响。该计划的特点是将人作为生态系统的组分加以研究。

pothole 壶穴　山区河流急流挟带砾石，在构造破碎、岩性软弱处磨蚀、冲刷、旋磨河床所形成的圆形深穴，其上方为岩坎，是急流或瀑布所在。由于水流中携带的砾石对坑穴的侧壁进行不断刮擦，使得坑穴壁光滑如镜，其形似井，地貌学上称之为壶穴；壶穴由于造形特殊，素有“石面桶”之称。壶穴通常发生在河川幼期，往往需经过数百或数千、数万年才可形成。

coyote 郊狼　也叫草原狼、北美小狼，是北美独有的一种犬科犬属的食肉动物，与狼是近亲。郊狼产于北美大陆的广大地区，北起阿拉斯加、南到巴拿马。欧洲探险家最初是在美国西南部发现这种动物的。郊狼一般单独猎食，偶尔也会组成小型的群体。平均寿命为6－10年。尽管受到广泛的捕猎，郊狼是少数在人类迁入后，种群数量扩大的大中型动物之一，起初郊狼只分布在北美的西半部，现在在加利福尼亚、俄勒冈、新英格兰和加拿大东部都十分常见。

Keeseekoowenin Ojibway First Nation 奥吉布瓦第一民族　简称 KOFN，主要人口居住在曼尼托巴省埃尔芬斯通的原始居留地。奥吉布瓦第一民族有三个主要的居留地：最大的是距离布兰登西北约 80 千米的 61 号印第安原始居留地，其次是位于骑马山国家公园内柯利尔湖岸的 61A 号居留地，最小的则是博特尔湖畔的 61B 号居留地，有时候他们也会被叫做骑马山人。如今，这个民族共有注册居民 1200 多人生活在原始居留地中，已经充分融入到了现代社会和文明中去。

cross-country skiing 越野滑雪　以滑雪板和滑雪杖为工具，运用登山、滑降、转弯、滑行等基本技术，在丘陵起伏的山地沿规定的线路进行的雪上竞速项目。起源于北欧，又称北欧滑雪，是世界运动史上最古老的运动项目之一。据记载，1226 年挪威内战时期，两名被称为“桦木腿”的侦察兵，怀藏两岁的国王哈康四世，滑雪翻越高山，摆脱了敌人。现挪威每年举行越野马拉松滑雪赛，距离 35 英里，与当年侦察兵滑行路线的路程相同。1924 年首次列入冬季奥运会比赛项目，也是受人喜爱的户外休闲运动之一。

Translation of Selected Sentences

1. The park boundaries enclose 2,976 square kilometers of rolling hills and valleys, stretch eastward from a dramatic rise of land known as Manitoba Escarpment, and mark the transition from the Manitoba Lowlands to the second prairie level, the Saskatchewan Plain.

公园内遍布延绵起伏的山峦和山谷，面积达 2 976 平方千米，向东延伸，高度大幅度增加形成曼尼托巴单面山，这里成为曼尼托巴低地向草原第二高度萨斯喀彻温平原的过渡地区。

2. The soil fertility combined with climatological and hydrological factors contributes to the area being able to sustain a relatively high biomass of diverse biota with high population densities of large herbivores.

这里的土壤在气候和水文因素影响下，能够支撑一个相当高生物量的多样生物群系，包括高种群密度的大型食草动物。

3. Areas that are mostly boreal forests, in the extreme north of this natural region, appear as seemingly endless stretches of black spruce muskeg that dominate much of the flat, poorly drained

land.

大部分的北方针叶林分布在这一自然区域的极北地区，黑云杉泥岩沼泽看起来无穷无尽地延伸着，占据了大部分平坦的湿地。

4. Variations in the physiological features such as soil type, hydrology, topography and microclimate further dictate the locations suitable for sustaining the specific vegetative types and consequently form the basis for the intermingling of the ecosystems.

生理特点，譬如土壤类型、水文、地形和局部小气候等因素的变化，昭示着这一地理位置适宜这种特有的植物生长类型，并构成了混合型生态系统的基础。

5. The browsing of these Cervids on regenerative aspen saplings has a thinning effect on the forest allowing more sunlight to penetrate to the level where growth of browse species of plants can benefit.

鹿类动物的啃食对山杨幼苗起到了再生作用。它使整个森林得到"稀释"，使得更多的阳光能够穿透森林。这样即使受到啃食，植物的生长也能从中受益。

Questions

1. What do the soil parent materials of the region look like?
2. What are the cold deep lakes rich in?
3. What are thethree distinctive vegetation zones?
4. What kinds ofmemorable and engaging activities do the park and surrounding communities provide?
5. What are the new concerns of the park managers as the result of the increase of the number of the visitors?

Chapter 23

Torngat Mountains National Park
唐加特山国家公园

【简介】

唐加特山国家公园位于拉布拉多省北部，南抵萨格利克峡湾，北至拉布拉多省的奇德利角，西接魁北克省，东临拉布拉多海。2005 年正式建立，目前是加拿大最晚成立的国家公园，也是加拿大第 42 个国家公园。面积覆盖 9 700 平方千米，这一地区植被稀少，气候恶劣，地形复杂，受冰川作用影响，山脉从海岸陡立，形成地势崎岖的峡湾。这里是因纽特人世代生存繁衍的故乡，也是黑熊、北极熊、北极狐、乔治河驯鹿的栖息地。

The Torngat MountainsNational Park Reserve is Canada's 42nd and newest national park, the homeland and settlement area of the Labrador（拉布拉多）Inuit. The spectacular wilderness setting of the Torngat Mountains National Park protects some 9,700 km^2 of the Northern Labrador Mountains natural region, making the park the largest national park in Atlantic Canada; it is also the southernmost national park in the Arctic Cordillera（山系）. The park located on the Labrador Peninsula at the northern tip of Newfoundland（纽芬兰）and Labrador, extends from Saglek Fjord（萨格利克峡湾）in the south, including all islands and islets, to the very northern tip of Labrador, and from the provincial boundary with Quebec in the west, to the iceberg-choked waters of the Labrador Sea in the east.

Interest in establishing a park in the Torngat Mountains first arose in the early 1970s, but was put on hold in 1979 due to outstanding aboriginal land claims. The history of the efforts to establish the National Park Reserve is one that has taken place within the context of an evolving vision of what a national park should be and of changing attitudes as to how we should go about the process of national park establishment. It is a story that includes the hard-learned lessons of the importance of working with Inuit as equal partners, of the need to achieve mutual respect and trust, of lengthy setbacks, and of tough negotiations and compromise among diverse parties that were nevertheless dedicated to the vision of protecting this magnificent area in perpetuity（永久）.

Whereas a true partnership was entered into amongParks Canada, the Labrador Inuit Association and the Government of Newfoundland and Labrador to undertake the national park feasibility study, a very different and strained relationship with Nunavik Inuit, represented by Makivik Corporation, led to a challenge to the park establishment process in the federal court of Canada in 1997. One of the most important details outlined in the Park Impacts and Benefits Agreements that Parks Canada negotiated with the Labrador Inuit Association and the Makivik Corporation is the commitment that the national park will be established, operated and managed through a co-operative management regime that recognizes Inuit as partners and recognizes and honors Inuit knowledge and the special historical and cultural relationship between Inuit and the land.

On December 1, 2005, almost 10 years after the feasibility study was complete, Torngat Mountains National Park Reserve was officially established, making it the first national park to be created in Labrador. It aims to protect wildlife, while offering wilderness-oriented recreational activities.

The park is characterized by towering（高大的）mountains, a rugged coastline, deep fiords, some of the world's oldest rocks and unique wildlife. It is home to one the world's largest caribou herds and a unique population of tundra-dwelling black bears and polar bears. The mountain peaks along the border with Quebec are the highest in mainland Canada east of the Rockies, and are dotted with remnant glaciers.

Torngat Mountains gets its name from Torngarsuak, an Inuktitut term meaning Great Spirit, or controller of the spirits. Inuit mythology tells of the "Torngait", the spirits that a Shaman or spiritual leader looks to for wisdom and power. "Torngat" comes from this Inuit name and the legends which hold that in this region the spirit world overlaps our own. White people have called this area the Ghost Coast and have commented how the sounds of the winds whistling through the rugged mountains bring forth the feeling that one is in another realm. If the earth is home to ancient spirits they would seek out this land where the rocks are among the oldest on the planet and the landforms hold an otherworldly（超自然的）appearance. Perhaps this truly is a place of spirits.

The Torngats have been home to Inuit and their predecessors for thousands of years. There are hundreds of archaeological sites in the park, some dating back almost 7,000 years. There is evidence of occupation by the Maritime Archaic（古代的）Indians, Pre-Dorset and Dorset Paleo-eskimos, and the Thule culture that merged into modern day Inuit. These sites, which include tent rings, stone caribou fences, food caches, and burial sites, tell the story of the people and cultures that have made this special part of Canada their home over the millennia. The Ramah chert（燧石）quarry was for thousands of years a source of glass-like tool material used by indigenous peoples and traded over a wide area, as far south as Maine.

This region contains two distinct, contrasting landscapes: the George Plateau and the spectacular Torngat Mountains. The George Plateau is a level bedrock plain cut by deep river valleys sloping gently to Ungava Bay. The effects of glaciation are everywhere: drumlin（冰丘）fields, kame（冰砾阜）terraces, erratics（漂砾）and eskers（蛇形丘）that snake over the plateau. The Torngat Mountains, among the highest, most rugged mountains in eastern North America and one of the world's most beautiful wild coastlines, provide a spectacular counterpoint to the gentle George Plateau.

The age of the Torngats encompasses eighty percent of the earth's geologic history. The rocks of the park have preserved, in their fabric and minerals, textbook examples of the cycles of mountain building events. Over the past 3.9 billion years there have been at least 5 major events - three in the Archean（太古代）, one in the early Proterozoic（元古代）and one as recently as the late Jurassic（侏罗纪）to Tertiary（第三纪）. These events represent a synopsis of plate tectonics from initial continental rifting, ocean formation and spreading, through ocean closure and destruction, erosion, and deposition before rifting begins again, culminating in the formation of the Labrador Sea and the Atlantic Ocean. This slow repetition of tectonic events is called the Wilson Cycle, and there are few places where it can be seen as clearly in the rock record as in the Torngat Mountains. The highest mountains in Canada east of the Rockies can be found within this beautiful land. Some of the peaks reach to over 1538 meters. The highest is Mt Caubvick/D'Iberville at 1652 m.

These mountains also provide natural resources that humans have relied upon for thousands of years. Starting with the Maritime Archaic people, chert was gathered for tools and exported to others as far south as New England. Soapstone（皂石）was quarried to make oil lamps. Quartzite（石英石）, pyrite（黄铁矿）, slate（板岩）, steatite（绿泥石）, nephrite（软玉）, mica（云母）and graphite（石墨）were used for a variety of purposes, and a mineral called labradorite（闪光拉长石）has been used for adornment. Stone is still quarried for carving, jewellery and other uses that previous generations could not have imagined.

The climate of the park is generally harsh and is influenced by both altitude and latitude and the nearby Labrador Sea. The best weather typically occurs from mid-July to mid-August, with storms bringing new snow to higher peaks by late August or early September. There are four climatic regions in the park: Mountain alpine climate, Coastal fjords and headlands climate, Southern interior valleys climate, and Northern Torngat Mountains climate.

About 330 species ofvascular plants and 220 species of mosses and liverworts are known to grow in the park area. Almost nothing is known yet about lichens (地衣) and fungi (真菌), but there are probably hundreds of species. There is no real forest in the park, except for dwarf spruce close to the Quebec border. At the heads of fjords, raised beaches and terraces are covered in sedge and grass meadows. Coastal species such as lyme grass (披碱草), roseroot (蔷薇根景天), and seabeach sandwort (蚤缀) spread along the upper beach. Low-growing arctic tundra shrubs hug rocky headlands and valleys. Wiry green alder (桤木) and willow (柳树) thickets line streamsides, and the gravelly floodplains of rivers are bright with the flowers of harebell (蓝铃花). Unvegetated rock and tundra each comprise about 50 percent of the upland surfaces. Arctic black spruce with mixed evergreen and deciduous shrubs, underlain by moss, is dominant on wetter sites.

With its rugged barren mountains, rocky tundra valleys, deep cold fjords and stormy maritime climate, the Torngat Mountains National Park provides a serious challenge for wildlife. Despite these challenges, the park has a variety of boreal and arctic species. Wildlife in the park is characterized mostly by small mammals, although the Torngat Mountains do provide seasonal habitat for polar bear and caribou. This region is home to the only global population of tundra-dwelling black bears and completely encompasses the range of the Torngat caribou herd. It also provides habitat for the southernmost denning of polar bears on the North American east coast. When spring comes, black bears emerge from their dens after a six-month hibernation. Torngat Mountain caribou return from their wintering grounds along Ungava Bay to calve in the upper mountain barrens. Bands of George River caribou may wander into the park after calving to the south in June. Wolves follow the caribou. Voles (田鼠) and lemmings (旅鼠) emerge from their burrows under the snow. Red foxes and Arctic foxes, which have eked out a living over winter, now pursue the summer bounty of rodents (啮齿类动物).

In the bays and fjords, seals and whales move northward along the coast, chasing the summer flush of food as it moves north. Ringed seals (环斑海豹) and hooded seals (冠海豹) follow the retreating ice northward. Harp seals (格陵兰海豹) follow a month or so later, but harbour seals (斑海豹) may linger along the coast. Minke whales also tend to linger in bays and fiords, but larger fin and humpback whales may stay offshore. Arctic char (红点鲑) move into the more productive coastal saltwater from their freshwater spawning grounds where they spend the first years of their life. By the time winter arrives, most summer residents are gone or are hibernating, but Arctic hares and rock ptarmigan（岩雷鸟）remain. The plentiful bird life includes the peregrine falcon, golden eagle and short-eared owl lives here. The Harlequin duck, an endangered species, nests along a number of the rivers running into the Labrador Sea giving the Park a key role in the effort to restore numbers of this handsome, curious duck.

The national park offers visitors an exceptional opportunity toexperience the Arctic landscape and environment. It is a landscape that holds tremendous potential for life-altering experiences. Today's visitor may come to the Torngat Mountains for such wilderness-oriented experiences as: Hiking and backpacking, Backcountry skiing, Mountain climbing, Fishing, Sightseeing by aircraft, and exploring cultural and human history.

The Torngat Mountains National Park can be accessed either by chartering a boat or plane from Nain, Labrador, or by chartering a plane from Kuujjuaq, Quebec. Both towns are about 300 km from the park. Because of its remote location, the park differs greatly from its more accessible cousins in the south. The journey requires careful planning, registration with park officials, and the right equipment. There are no roads, no campgrounds and no signs telling you where to go or what to see. Parks Canada recommends that a visitor engage the services of a trained Inuit polar bear guard when hiking in the park. Parks Canada provides interpretative programs, visitor reception, and orientation services, including safety briefings at the Torngat Mountains Base Camp at the southern boundary of the park throughout the summer.

Visitors to Torngat Mountains National Park, Canada's newest National Park, consist of reporters, researchers, tourists, and local Inuit from both Nunatsiavut and Nunavik. Compared to the 21,000 visitors to Labrador, total visitor numbers to the park are low: 272 in 2006, 484 in 2007, and 565 in 2008. The logistical support for these visitors is provided at the Kangidluasuk/St. John's Harbour Base Camp, located in Saglek Bay outside of the park boundaries, but within the Nunatsiavut Inuit Territory. Parks Canada contracted local Inuit from Nain to organize and manage the base camp and used the services of local Inuit owned and operated long liner boats for support.

Notes

National Park Reserve 国家公园保护区　指由于土地所有权悬而未决而被搁置建设成为国家公园的地区。土地所有权磋商期间，这一区域适用《国家公园法案》，禁止土著狩猎、捕鱼和设陷诱捕等一系列传统的活动，但允许当地原住民参与公园保护区的管理。

Newfoundland and Labrador 纽芬兰与拉布拉多省　简称纽芬兰省或纽省，位于加拿大东北角，包括纽芬兰岛和附近岛屿以及拉布拉多的主要陆地和附属岛屿。纽芬兰与拉布拉多省于1949年4月1日以“纽芬兰省”的名称正式成为加拿大的第10个省,被称为加拿大最年轻的省。但从1964年起,省政府开始称呼自己为“纽芬兰与拉布拉多”政府。加拿大政府于2001年12月6日修改宪法,将省份的名字正式由“纽芬兰”改为“纽芬兰与拉布拉多”。面积405,212平方千米,全省51.08万居民,平均每平方千米1.3个人。首府圣约翰斯,渔业及海鲜加工是该省的传统产业,纸浆及造纸工业在经济中占主导地位,石油、天然气开采为新兴支柱产业。该省大部分土地都是人迹罕至的森林、沼泽、湖泊和山岭,以优美的风景,雄伟的冰山,宁静的沿海渔村和繁华的省府城市而闻名。主要景点有:特拉诺瓦国家公园、锡格纳尔山国家公园和格罗莫讷国家公园。

Labrador Sea 拉布拉多海　位于北大西洋西北部,加拿大拉布拉多半岛和格陵兰岛之间,呈倒三角形。北经戴维斯海峡通巴芬湾,西经哈得逊海峡通哈得逊湾,南经贝尔岛海峡通圣劳伦斯湾,是大西洋的陆缘海。面积约84.1万平方千米,最深处约4316米,岸线异常曲折、陡峭,多半岛、岛屿和峡湾,气候寒冷多雾,且有冰山漂至,影响海上航行,著名的泰坦尼克号游轮就沉没在该海域南侧。该海域资源丰富,尤其是南端纽芬兰岛东南浅滩一带是拉布拉多寒流与北大西洋暖流交汇处,是世界大渔场之一。盛产鳕鱼、鲽鱼、鲱鱼、鲑鱼等。

Parks Canada 加拿大公园管理局　加拿大政府部门,授权保护全国重要的自然和文化遗产,保护生态发展,并促进公众对资源保护的理解和拥护,提升满意度。加拿大公园管理局管理着总共27.7万平方千米的陆地及水域,包括42个国家公园(含7个国家公园保护区)、4个国家海上保护区、1个国家标志性建筑和167个国家历史遗迹。加拿大公园管理局成立于1911年,是内务部的一个分支机构,也是世界上第一个国家公园保护机构。加拿大公园管理局主要根据1930年生效的《加拿大国家公园法》来管理。2011年是加拿大公园管理局成立100周年,世界野生动植物基金会向加拿大国家公园管理局颁发了“馈赠地球奖”。

Makivik Corporation 马基维克公司　由加拿大魁北克省努纳维克地区因纽特人创建的一个非盈利社团组织,1978年成立后即与加拿大政府、魁北克水电委员会等签订《詹姆斯湾和魁北克北部协议》,其宗旨是为了捍卫协议的完整性,致力于促进努纳维克地区的政治、社会与经济发展,并代表魁北克的因纽特人同政府进行一系列的谈判。总部设在魁北克的库巨尔可,5人组成的执行委员会和16人组成的董事会来负责马基维克公司的日常管理和运行。

Shaman 萨满　源自通古斯语 saman 与北美印第安语 shamman,原词含有:智者、晓彻、探究等意,即最能通达、晓彻神意的人,后逐渐演变为萨满教巫师的专称,也被理解为萨满之神的代理人和化身。主要通行于欧亚大陆与北美大陆北方各民族中。他们与其他宗教神职人员最大的不同是能够以个人的躯体作为人与鬼神之间实现信息勾通的媒介。萨满是北方民族的原始信仰,起源甚早,是在人类原始社会阶段自发产生的,在母系制度的社会里已经非常发达与成熟了。萨满文化是世界性的文化现象,其流行区域集中在亚洲北部和中部,乃至欧洲北部、北美、南美和非洲。

Dorset Paleo-eskimos 多赛特旧石器时期爱斯基摩人　公元前500年左右,极地气候经历了一次巨变,比以往更寒冷了。气候的转冷导致猎物和淡水资源的分布变迁,古爱斯基摩人传统的住房和捕猎技术已经不能抵御外界的寒冷和获取足够的猎物,这要求他们必须快速适应环境,否则就会被饿死或冻死。坚强的古爱斯基摩人把捕获的主要对象由陆生动物转向了海洋哺乳动物,并学会了用这些动物的油脂点灯,以得到热量和光明。有人认为,圆顶冰屋可能也是这时候开始使用的。在一些地区,开始出现了用石头和草皮建造的半地下式的大房子,这可以住更多的人,而且也便于更加长久地居住。由此还发展了更丰富的文化艺术和技术,制造出许多新式武器和工具。人类学家把这些古爱斯基摩人称作多塞特人,与他们相应的文化则称为多塞特文化。多赛特人的身躯由于自然规律的演化,比现代人大得多。可惜的是,他们只能适应相当寒冷的环境,后来,当天气渐渐转暖时,他们逐渐消亡,并最终从地球上消失了,被现代的爱斯基摩人所代替。

Thule culture 图勒文化　4500年前,因纽特人的祖先图勒人跨越白令海峡来到北美大陆,多塞特人后来被图勒人所取代。图勒文化在两个方面与以前的文化不同——图勒人养狗并捕鲸,事实上鲸是这些人经济生活的主要依靠。图勒人是因纽特人的直接祖先。图勒文化向因纽特人文化的转变约发生在公元1750年前后。

Wilson Cycle 威尔逊旋回　1973年加拿大地球物理学家和地质学家约翰·图佐·威尔逊提出威尔逊旋回的假说,将大洋盆地的演化归纳为六个发展阶段:大洋胚胎期、大洋幼年期、大洋成年期、大洋老年期、大洋消亡期和大洋遗痕期,相对应的实例为东非裂谷、红海亚丁湾、大西洋、太平洋、地中海和喜马拉雅山。大洋的演化呈现为张开和关闭的旋回形式。由于大洋盆地是全球最大的构造地貌单元,因此大洋的旋回主宰了地球表层活动和演化的全局。鉴于威尔逊首先认识到大洋开闭的深远意义,1974年英国构造地质学家约翰·弗雷德里克·杜威和K. C. A.伯克建议将这种在大陆岩石圈中复杂交替的大洋开闭旋回命名为“威尔逊旋回”。

vascular plants 维管植物　指具有木质化维管系统的陆地光合自养生物。根据化石记录,原始的维管植物大约在距今3.6 – 3.7亿年以前的中泥盆纪,就已分成多条路线演化。维管植物是地球上最奇特的生物类群之一,就现今的生物圈而言,它占总生物量的97%,现存的维管植物大约有25 – 30万种,包括极少部分苔藓植物、蕨类植物、裸子植物和被子植物。维管系统的发生是植物从水生到陆生长期适应环境的结果。

Translation of Selected Sentences

1. It is a story that includes the hard-learned lessons of the importance of working with Inuit as equal partners, of the need to achieve mutual respect and trust, of lengthy setbacks, and of

tough negotiations and compromise among diverse parties that were nevertheless dedicated to the vision of protecting this magnificent area in perpetuity.

这是一个坎坷的学习过程,学习和了解将因纽特人作为平等伙伴共同工作的重要性,实现彼此尊重和信任的必要性,以及与形形色色的党派进行谈判和达成妥协的艰苦。尽管如此,仍以此致力于永久保护这一地区的壮观景色。

2. One of the most important details outlined in the Park Impacts and Benefits Agreements that Parks Canada negotiated with the Labrador Inuit Association and the Makivik Corporation is the commitment that the national park will be established, operated and managed through a co-operative management regime that recognizes Inuit as partners and recognizes and honors Inuit knowledge and the special historical and cultural relationship between Inuit and the land.

公园的《影响和收益协议》彰显出的最重要的细节是:加拿大公园管理局与拉布拉多因纽特人协会、马基维克公司的谈判,承诺通过合作管理体制来建立、运行和管理国家公园,认同因纽特人作为伙伴,承认和尊重因纽特人的知识及其和这片土地特殊的历史和文化关系。

3. These events represent a synopsis of plate tectonicsfrom initial continental rifting, ocean formation and spreading, through ocean closure and destruction, erosion, and deposition before rifting begins again, culminating in the formation of the Labrador Sea and the Atlantic Ocean.

这些事件体现了板块构造的梗概:从最初的大陆断裂,海洋的形成和伸展,经历海洋的闭合和消亡,侵蚀和沉积,再到大陆断裂重新开始,周而复始,最终形成了拉布拉多海和大西洋。

4. The Harlequin duck, an endangered species, nests along a number of the rivers running into the Labrador Sea giving the Park a key role in the effort to restore numbers of this handsome, curious duck.

濒危物种丑鸭,在一些汇入拉布拉多海的河流沿岸筑巢,使得公园在恢复这一"美丽的"、"奇妙的"鸭子数量的努力中起到了关键作用。

5. Parks Canada provides interpretative programs, visitor reception, and orientation services, including safety briefings at the Torngat Mountains Base Camp at the southern boundary of the park throughout the summer.

加拿大公园管理局提供解释程序、游客接待和定向辅导服务,包括位于公园南部边界的唐可特山营地在夏天举办的"安全简介会"。

Questions

1. What were the difficulties in the process of the establishment of Torngat Mountains National Park?

2. What isone of the most important details outlined in the Park Impacts and Benefits Agreements that Parks Canada negotiated with the Labrador Inuit Association and the Makivik Corporation?

3. What are the two distinct, contrasting landscapes in the region?

4. How is the climate in this park?

5. How can visitors get to the Torngat Mountains National Park?

Chapter 24

Wood Buffalo National Park
伍德布法罗国家公园

【简介】

伍德布法罗国家公园是加拿大最大的一个国家公园，主体位于北勃里尔平原，它把加拿大北部的广袤无垠体现得淋漓尽致。公园占地面积有44 807平方千米，包括北部辽阔的森林及大草原，还有一些北美洲幸存的未受破坏的巨大草场和莎草牧场。公园建于1922年，它最终成为19世纪末那些躲避了大屠杀的幸存野牛的家园，公园也是濒危的美洲鹤在世界上最后的一块保留地。如今，伍德布法罗国家公园保护了世界上最大的自由繁衍的牛群。公园里另一个引人入胜的景点是皮斯河和阿萨巴斯卡河之间世界上最大的内陆三角洲。这里随处可见广阔的大草原、森林、湿地、盐沼，以及不多见的内陆大三角洲。1983年联合国教科文组织将伍德布法罗国家公园作为自然遗产，列入《世界遗产名录》。

Wood Buffalo National Park is Canada's largest national park and one of the largest in the world. The park was established in 1922 to protect the free-roaming bison(野牛) herds of the area. Today, the park supports and protects many unique natural and cultural resources, from diverse ecosystems(生态系统) and rare species to the traditional activities of Aboriginal residents. As a remote wilderness park and World Heritage Site, Wood Buffalo Naticnal Park attracts Canadian and international visitors who wish to experience and learn about the unique cultures, landscapes and wildlife of the boreal north.

The park was established in 1922, largely as a home for the remnant(剩余部分) population of bison that escaped the slaughter in the late 19th century that reduced the monarch of the plains from 60 million to a relative handful. The plains in the park are on the historic northern limit of the bison's range, and several thousand plains bison were shipped north to join the native wood bison herd already there. The park is also home to the world's last remaining wild flock of endangered whooping cranes.

Wood Buffalo is Canada's largest national park, and the very embodiment of the space and wilderness that symbolize Northern Canada. Its 44,807 square kilometers include huge tracts of boreal forest and plains, and some of the largest undisturbed grass and sedge meadows left in North America. Those meadows sustain the largest free-roaming herd of bison, commonly known as buffalo, in the world.

The park ranges in elevation from 183 m (600 ft) at theLittle Buffalo River to 945 m (3,100 ft) in the Caribou Mountains. The park headquarters is located in Fort Smith, with a smaller satellite office in Fort Chipewyan, Alberta. The park contains one of the world's largest fresh water deltas, the Peace-Athabasca Delta, formed by the Peace, Athabasca and Birch Rivers. It is also known for its karst sinkholes(喀斯特天坑) in the north-eastern section of the park. Wood Buffalo is located directly north of the Athabasca Oil Sands.

Great rivers grace the park as the mighty Peace flows through the heart of the park and the Slave and Athabasca form the eastern border. Where the Peace and Athabasca issue into Lake Athabasca is the world's largest inland freshwater delta. The endless streams and wetlands of the delta accommodate(提供栖息地) countless waterfowl(水鸟), including ducks, geese, swans, loons(潜鸟) and grebes.

Within Canada's system of national parks, Wood Buffalo National Park protects representative examples of the Northern Boreal Plains, Southern Boreal Plains, and Northwestern Boreal Uplands. The majority of the park is within the Northern Boreal Plains natural region.

Wood Buffalo National Park was created to protect the last remaining herds of wood bison in northern Canada. Plains bison were shipped to the park from Wainwright, Alberta, between 1925 and 1928. The imported bison promptly moved south of the Peace River into the Peace-Athabasca Delta area. In 1926 the park boundaries were expanded to include this new bison range. Today Wood Buffalo National Park protects one of the largest free-roaming, self-regulating bison herds in the world.

Archeological(考古的) evidence shows that Aboriginal people have inhabited the Wood Buffalo region for more than 8000 years, long before fur traders arrived in the early 1700s. The Europeans called the people they met in this region Beaver, Slavey and Chipewyan. The Beaver and Slavey left the area as the fur trade moved west. Today, the communities around the park are mostly made up of Cree, Chipewyan, Metis and non-aboriginal people.

Subsistence hunting, fishing and trapping still occur in Wood Buffalo National Park, as they have for centuries, and commercial trapping continues as a legacy of the fur trade. Traditional use of certain park resources by local Aboriginal groups is an important part of the park's cultural history.

A thorough understanding of traditional and scientific information is critical to protecting the ecological and cultural integrity of the Wood Buffalo area. Park staff and local traditional users are working together co-operatively towards this goal.

The park has four main landscape features: a glacially eroded plateau; glaciated plains; a major freshwater delta formed by three major rivers; and alluvial(冲积的)river lowlands. The lowlands and floodplains of Peace, Athabasca and Slave rivers and the delta in Lake Athabasca exhibit classic fluvial landforms, with a complex series of meander(蜿蜒而行)scars, oxbow(U字形弯曲)lakes and former river terraces, and good examples of birds foot delta development. During dry periods, the mudflats of one plain are dominated by mineral salts. These salt plains are unique in Canada.

Vegetation is typical of the boreal forest zone with white spruce(云杉), black spruce, jack pine and tamarack(美洲落叶松)predominant. Many watercourses have stands of balsam poplar and some upland has almost pure stands of aspen(大齿杨). Extensive stands of white spruce forests cover the banks of Peace, Athabasca and Birch rivers. The upper surface of the plateau is about 1,500 m above the rest of the park and supports a spruce-willow-birch upland tundra community. Some areas of prairie occur.

Shrub(灌木)lands of willow and alder(赤杨)occur where wet marsh soils meet drier forest soils. There is also extensive muskeg(青苔沼泽地)in the west and north of the park, an association of black spruce, sphagnum moss(泥炭藓)and northern heath plants.

The park contains the largest undisturbed grass and sedge meadows in North America. The park was created specifically to protect North American bison, one of the largest free-roaming, self-regulating herds in existence, and consisting of a cross between "wood" bison and "plains" bison.

This is one of a few sites where the predator-prey relationship between wolves and bison still exists. A total of 46 other mammal species have been recorded including black bear, woodland caribou(驯鹿), Arctic fox, moose(北美麋), grey wolf, lynx(山猫), snowshoe hare, muskrat(麝鼠), beaver(海狸)and mink(鼬). Occasionally animals more common to southern Canada are seen, such as red fox, porcupine(豪猪)and white-tailed deer. The caves of karst lands provide essential hibernation sites for bats.

A total of 227 bird species have been recorded which include great grey owl and snowy owl, willow ptarmigan(柳雷鸟), redpoll crossbill(交喙鸟)and boreal chickadee(山雀). This is the only breeding site of whooping crane; peregrine falcon and bald eagle also breed within the park. The Peace-Athabasca Delta is an important area for migrant waterfowl including snow geese, white-fronted geese and Canada geese, whistling swan, diver, all seven species of North American grebe and species of duck.

Reptiles(爬行动物) and amphibians(两栖动物) are severely limited in numbers, but Canadian toad, leopard frog and red-sided garter snake reach their northern limits here. Boreal chorus frog and wood frog are also found in aquatic habitats(水生栖息地). The fish fauna has been poorly studied, although there are a wide variety of aquatic habitats. 36 species have been recorded to date, four of them introduced.

This area was designated aUNESCO World Heritage Site in 1983 for the biological diversity of the Peace-Athabasca Delta, one of the world's largest freshwater deltas, as well as the population of wild bison.

In 1983, Wood Buffalo National Park became the eighth site in Canada to be granted World Heritage status by UNESCO (United Nations Educational, Scientific and Cultural Organization). Inscription on this list confirms the exceptional universal value of a cultural or natural site which deserves protection for the benefit of all humanity.

The outstanding characteristics of Wood Buffalo that led to its designation include:

- one of the largest free-roaming and self-regulating bison herds in the world
- the last remaining natural nesting area for the endangered whooping crane
- the Peace-Athabasca Delta, one of the largest inland freshwater deltas in the world
- some of the finest examples of gypsum karst landforms in North America
- unique salt plains
- vast undisturbed expanses of boreal wilderness.

In 1982, theInternational Union for the Conservation of Nature (IUCN) recognized that Wood Buffalo protected two wetland areas of international significance—the Peace-Athabasca Delta and the whooping crane nesting area. These areas were designated as Ramsar sites—a designation by the Ramsar Convention which focuses on identification and protection of critical habitat for migratory birds.

The Peace-Athabasca Delta was named a Ramsar site because it is one of the largest inland freshwater deltas in the world and a major nesting and staging area for migratory waterfowl in North America. Migratory waterfowl from all four North American flyways pass through the delta in the spring and fall.

The last remaining nesting area for the endangered whooping crane is the other designated Ramsar site, located in the remote north-central corner of the park. It is a fragile complex of marshes, shallow ponds and lakes, streams and bogs.

Notes

Fort Smith 史密斯堡　加拿大城镇，位于西北地区南部奴河畔。史密斯堡是加拿大皇家骑警队的总部，是西北地区各地矿藏以及九个居民点医疗站的汇总中心，是通往更边远地区航线的加油站。

Fort Chipewyan 奇普怀恩堡　坐落于西北边疆以南几百英里的地方。它是阿尔伯塔省最古老的社区之一，也是进入世界第二大国家公园——伍德布法罗国家公园的跳板之地。这里有极其丰富的野生动物，有特殊的原住民的民俗习性，还有闻名世界的阿萨巴斯卡梭鱼。

Alberta 阿尔伯塔省　加拿大的一个省，是加拿大草原三省之一。1905 年 9 月 1 日脱离西北地区，建立了阿尔伯塔省，加入加拿大联邦政府，是加拿大第八个加入联邦政府的省份。它北靠西北地区，南靠美国的蒙大拿州，东临萨斯喀彻温省，西临不列颠哥伦比亚省，西部为落矶山脉，东部为平原，是加拿大仅有的两个不靠海的省份之一。

Peace River 皮斯河　处在不列颠哥伦比亚省和阿尔伯塔省北部，是马更些河主要源流之一，源自加拿大洛矶山东麓，源河名叫芬利河，河流先向东南方向流，然后转向东流，继而转向东北方向流，在克莱尔湖-阿萨巴斯卡湖附近注入奴河。

Athabasca River 阿萨巴斯卡河　位于阿尔伯塔省的北部，是加拿大马更些河水系最南部的河流。源出加拿大落矶山区的哥伦比亚冰原，流经贾斯珀国家公园，继而向东北穿过阿尔伯塔省，注入阿萨巴斯卡湖。沿河多湍流，仅麦克默里堡以上河段可通航。

Athabasca Oil Sands 阿萨巴斯卡油砂　位于加拿大阿尔伯塔省内，是世界上最大的油砂矿藏。

Slave River 奴河　位于加拿大阿尔伯塔省北部和西北地区南部，是马更些河水路的一部分，1771—1772 年赫恩对其进行过勘察。奴河从皮斯河和自亚大巴斯卡湖流出的几条小河的汇合处开始，北流 415 千米，在雷索卢申堡附近注入大奴湖。除 26 千米一段急流可利用陆运连接外，全部可以通航。

Lake Athabasca 阿萨巴斯卡湖　位于加拿大境内，是加拿大第四大湖，跨阿尔伯塔省和萨斯喀彻温两省边界。长 335 千米，宽 51 千米，面积 7936 平方千米，最深 124 米。有皮斯河和阿萨巴斯卡河从西南注入，湖水经奴河从西北导出。沿河重要居民点有珀怀恩堡和铀城，两地之间，6 月到 10 月通汽船。

Chipewyan 奇佩维安人　加拿大北部说阿萨巴斯卡语的最大的印第安部落。他们生活在一个大三角形地带，其主要部分沿着邱吉尔河。奇佩维安部落分成许多独立的宗族，随着驯鹿的季节性活动而进行游牧生活。驯鹿是他们的主要食物来源。鹿皮可制成衣服、帐篷及绳索等。另外还猎取野牛、麝牛、驼鹿及各种小动物、禽类及鱼类。

Cree 克里族　克里族人在加拿大生活了几千年。起初,他们主要吃北美驯鹿、鱼和其他动物。他们用弓、箭、矛狩猎。夏天,他们坐着独木舟在河流和湖泊上旅行;冬天,他们用平底雪橇和雪鞋在冰天雪地里穿行。

Metis 梅蒂斯人　加拿大的梅蒂斯人是指印第安人和白人通婚的后代。最早的一批梅蒂斯人是印第安人女子与位于今天马尼托巴省雷德河地区的经营皮毛生意的欧洲商人所生的子女。1869 年当加拿大人准备接管西北地区时,梅蒂斯人在他们的领袖路易·里尔的领导下对加拿大人进行了武装反抗,并建立了临时政府。1870 年该临时政府与加拿大政府谈判,并最终建立了马尼托巴省,成为加拿大联邦的第五个省。绝大多数的梅蒂斯人散居在全国各地,只有在阿尔伯塔省,有少量的梅蒂斯人居住在为他们设立的定居点内。梅蒂斯人很长时间内一直不具有法律地位,直到 1982 年才被加拿大新宪法所承认。梅蒂斯人在历史、司法、政治、语言和文化等诸多方面都构成一个特殊群体。

International Union for the Conservation of Nature（IUCN）国际自然保护联盟　目前世界上最大的、最重要的世界性保护联盟,是政府及非政府机构都能参与合作的少数几个国际组织之一。该组织成立于 1948 年 10 月,致力于寻找解决当前迫切环境与发展问题的实用解决方式。该组织发布 IUCN 红色名录,根据严格准则评估数以千计物种及亚种的绝种风险。准则是根据物种及地区厘定,旨在向公众及决策者反映保育工作的迫切性,并协助国际社会避免物种灭绝。它是全球动植物物种保护现状最全面的名录。世界自然保护联盟支持科学研究,并协调管理全球范围内政府、非政府组织、联合国机构、公司以及地方社群间各项合作计划,共同推行政策、法规和最佳的实际行动。IUCN 是世界上历史最悠久、规模最大的全球环境保护系统。它拥有超过 1 000 个政府和 NGO 组织会员,以及来自 160 多个国家的超过 1.1 万名志愿科学家团队。联盟总部位于瑞士日内瓦附近的格兰德。IUCN 的愿景是展望"一个重视和保护自然的正义世界"。联盟的任务是"影响、鼓励和支持社会在世界范围内保持自然生物多样性的完整,保证自然资源利用方式的公正和生态上的可持续性"。

Ramsar 拉姆萨尔　伊朗马赞德兰省的城镇,位于里海沿岸,当地人说吉拉基语。这里部分地区有辐射性的温泉,是全球最高自然背景辐射水平的地区之一,最高剂量为每年每人超过 260 毫西佛特。

Ramsar Convention《拉姆萨尔公约》　1971 年 2 月在伊朗的拉姆萨尔召开了"湿地及水禽保护国际会议",会上通过了《国际重要湿地特别是水禽栖息地公约》,简称《拉姆萨尔公约》。《拉姆萨尔公约》主张以湿地保护和"明智利用"（wise use）为原则,在不损害湿地生态系统的范围内以期持续利用湿地。其内容主要包括:缔约国有义务将境内至少几个以上有国际重要意义的湿地列入湿地名单,并加以保护;缔约国应根据本国的制度对所登记的湿地进行保护和管理,并在其生态学特征发生变化时向秘书处报告。

Translation of Selected Sentences

1. Today, the park supports and protects many unique natural and cultural resources, from diverse ecosystems and rare species to the traditional activities of Aboriginal residents.

现在,该公园成为独特的自然和文化资源的保护地,这里有多样的生态环境、稀有的物种和土著居民的传统活动。

2. Within Canada's system of national parks, Wood Buffalo National Park protects representative examples of the Northern Boreal Plains, Southern Boreal Plains, and Northwestern Boreal Uplands.

在加拿大的国家公园系统中,伍德布法罗国家公园主要保护的是北方平原、南方平原和西北高地的代表性物种。

3. Traditional use of certain park resources by local Aboriginal groups is an important part of the park's cultural history.

当地土著居民对某些公园资源的传统使用也是公园文化史的重要组成部分。

4. A thorough understanding of traditional and scientific information is critical to protecting the ecological and cultural integrity of the Wood Buffalo area.

对传统和科学信息的全面理解,对于保护伍德布法罗地区的生态和文化的完整性是至关重要的。

5. This area was designated a UNESCO World Heritage Site in 1983 for the biological diversity of the Peace-Athabasca Delta, one of the world's largest freshwater deltas, as well as the population of wild bison.

由于皮斯-阿萨巴斯卡三角洲——世界上最大的淡水三角洲——的生物多样性和数量众多的野牛,1983 年联合国教科文组织将该地区列为自然遗产。

Questions

1. What are the outstanding characteristics of Wood Buffalo National Park?
2. Why was Peace-Athabasca Delta named a Ramsar site?
3. What are the communities around the Wood Buffalo National Park made up of?
4. What was the Wood Buffalo National Park created for?
5. What are the four landscape features of Wood Buffalo National Park?

Chapter 25

Yoho National Park
优鹤国家公园

【简介】

优鹤国家公园建立于1886年,占地面积达1 310平方千米。YOHO是印地安语,是印第安人表示惊叹时发出的声音。优鹤国家公园内拔地而起的岩壁和奔腾飞溅的瀑布都令人惊叹不已,公园因而得名。

优鹤国家公园位于洛矶山脉的右翼,公园内拥有加拿大最高的瀑布奇景、世界级的化石遗迹地以及鬼斧神工的天然地形景观。它与班夫、杰士伯等国家公园齐名,共同被联合国教科文组织列为世界遗产。优鹤国家公园名气或许小了一点,但是它所提供的露营、滑雪、登山、钓鱼等野趣活动,却是城市人逃离水泥丛林、奔向自然怀抱的最好选择。

Yoho National Park is located in theCanadian Rocky Mountains along the western slope(山坡)of the Continental Divide(大陆分水岭)in southeastern British Columbia. Yoho national parksis bordered(比邻、交界)by Kootenay National Park on the southern side and Banff National Park on the eastern side. The name Yoho comes from a Cree word expressing amazement.

Yoho covers 1,313 km^2 and it is the smallest of the four contiguous national parks(NP). Yoho, together with Jasper NP, Kootenay NP and Banff NP, along with three British Columbia provincial parks—Hamber Provincial Park, Mount Assiniboine Provincial Park, and Mount Robson Provincial Park—form the Canadian Rocky Mountain Parks World·Heritage Site. The park's administrative and visitor centre are located in the town of Field, British Columbia, beside the Trans-Canada Highway.

Yoho's craggy(陡峭的)peaks(山峰)and steep rock faces posed(构成)an enormous(巨

大的)challenge for Canada's early explorers. The mountains that were the curse(诅咒)of railway builders are responsible for the park's many waterfalls(瀑布)including Laughing Falls(欢笑瀑布), Twin Falls(双子瀑布), Wapta Falls and one of Canada's highest at 254 m, Takakkaw Falls(塔卡考瀑布). Silt carried by streams from melting glaciers high on the mountains is responsible for the deep, rich turquoise colour of Emerald Lake and Lake O'Hara.

Water is responsible for creating a natural rock bridge over the Kicking Horse River(马踢河). Torrents have worn through a solid rockbed leaving a flat-rock bridge. Water erosion has also formed another Yoho marvel: balanced boulders on tall pillars of glacial till, called Hoodoos, found in the western end of the park.

Many of British Columbia's plants and animals reach their eastern extension in Yoho. The high peaks of the Continental Divide wring out the precipitation remaining in clouds moving eastward from the Pacific Ocean. This creates pockets of wet belt forest where coastal species(沿海物种)such as devil's club, western red cedar(红雪松)and western hemlock(铁杉)thrive(茁壮成长).

One of the world's most important fossil(化石)finds, the Burgess Shale(布尔吉斯页岩), is located in Yoho. Designated a World Heritage Site in 1981, the Burgess Shale Formation contains the fossilized remains of more than 120 marine animal species(物种)dating back 515 million years. The bizarre animals preserved in the shales represent a complete ecosystem that existed for only a very short time after the first explosion of multicellular life on earth. Since their discovery, the fossils of the Burgess Shale have provided scientists with a wealth of information about the variety of Cambrian life forms, some of which are difficult to classify into familiar categories of modern or fossil taxonomy. These life forms have given scientists a valuable glimpse into the nature of evolution itself. The fossils provide evidence that suggests the evolution of life on Earth has been characterized by rapid diversification and random extinction rather than by an orderly progression of "survival of the fittest". The Burgess Shale World Heritage Site is now incorporated into the larger Canadian Rocky Mountain Parks World Heritage Site with Yoho, Kootenay, Banff and Jasper National Parks and Mt. Assiniboine, Mt. Robson and Hamber Provincial Parks.

The park is home to most abundant large mammals. Lynx (山猫)live in Banff, Yoho and Kootenay National Parks, but not just anywhere in those parks. They establish home ranges in high evergreen(常绿)forests occupied by snowshoe hare(雪兔). Further north, in the boreal forest(寒带森林), hare populations rise and fall dramatically every 10 years, shadowed closely by lynx populations.

The mountain pine beetle(山松甲虫)is a naturally occurring insect of the Rocky Mountain ecosystem. These small cylindrical insects attack and kill mature trees by boring through the bark and mining the phloem—the layer between the bark and wood of the tree. When conditions are favourable, populations of the mountain pine beetle can increase and large numbers of trees can be infected, killing large areas of mature lodgepole pine. Mountain pine beetle activity in west Yoho National Park is expected to increase significantly in the next few years.

As protected areas, the mountain national parks make a significant contribution to grizzly bear(灰熊)conservation in the Rocky Mountain ecosystem. But the mountain parks alone cannot sustain grizzly bears. Instead they help anchor conservation efforts in the broader landscape by providing a core area where ecological integrity has priority for land use planning. Because the Bow Valley watershed is one of the most developed la ndscapes in North America where grizzly bears still persist, this area could be a tipping point for the regional population. Populations that once ranged well across the plains and down to Mexico are long gone. The species is vulnerable to further loss for a number of reasons.

Points of interest inYoho National Park.

Wapta Falls

In 1858, near Wapta Falls, a pack horse kicked explorer James Hector in the chest, and the Kicking Horse River got its name. Wapta Falls is the full width of the river and drops 30m. A short drive off the TransCanada Highway takes you to the start of an easy 2.4 km trail to the falls.

Hoodoos

It's worth the steep hike to see these capped pillars(柱子)of glacial debris(碎片). The 1.6 km trail starts in Hoodoo Creek Campground.

Natural Bridge

The Kicking Horse River has carved a natural bridge through solid rock, 1.6 km from the Trans-Canada highway on the Emerald Lake Road.

Emerald Lake

Emerald Lake is popular for sightseeing, canoeing(皮划艇)and hiking(远足).

Yoho Valley Road

Travel 13 km on this narrow, steep road beside the Yoho River to see Takakkaw Falls. Leave trailers at the parking lot across from Monarch Campground. Facilities on the Yoho Valley Road include Monarch and Kicking Horse Campgrounds, seasonal accommodation and viewpoints. The road is snowbound from mid-October to mid-June; it is used by cross country skiers in the winter.

Takakkaw Falls

With a free-fall of 254 metres, Takakkaw Falls(塔卡考瀑布) is one of the highest waterfalls in Canada.

Lower Spiral Tunnel Viewpoint

A fascinating display explains the history and operation of the spiral tunnels. From the platform, visitors have a spectacular view of the Yoho Valley, Yoho Glacier and the lower spiral tunnel portals in Mt. Ogden.

Abbot Pass Refuge Cabin National Historic Site of Canada

Built in 1922, this sturdy shelter is a unique and enduring monument to the Swiss guides, who first came to the Rocky Mountains in 1899 under the auspices of the Canadian Pacific Railway. It was patterned on examples in the Swiss Alps. All materials, apart from the stone, were carried by pack horses past the Lower Victoria Glacier, then by the guides to the pass summit. Never again was such an arduous feat undertaken in the national parks. The shelter has served as a high-altitude base for generations of climbers, here in the cradle of Canadian mountaineering.

The cabin is located on the Continental Divide between Banff and Yoho National Parks in a col known as Abbot Pass. It is located at an altitude of 2 925 metres and about 1 372 metres above Lake Louise. Abbot Pass was named in honour of Philip Stanley Abbot, who became the first mountaineering fatality in the Canadian Rockies when he fell to his death in 1895 while attempting to climb Mount Lefroy.

Howse Pass National Historic Site of Canada

This pass, which links the North Saskatchewan and Columbia river systems, was known to the Indians of the area long before the arrival of the Europeans. It was probably used from the 18th century onwards by Kootenay Indians to gain access to the buffalo herds on the plains east of the mountains. In 1807, David Thompson and a North West Company party traversed the pass which was used by the Canadian fur traders(皮毛商人) until 1810 to explore and establish posts west of the Rockies. Joseph Howse, the Hudson's Bay Company employee after whom the pass is named, first crossed in 1809.

Howse Pass is located on the border between British Columbia and Alberta. The pass can be reached by hiking 26 km west from the Icefields Parkway (Highway 93 North).

Kicking Horse Pass National Historic Site of Canada

First recorded in the report of the Palliser expedition of 1857 – 60, this pass takes its name from an incident in which Dr. James Hector, surgeon to the expedition, was kicked(踢、踩踏)by his horse while exploring in this vicinity(周边地区). The pass was virtually unused until after 1881 when the Canadian Pacific Railway decided to adopt it as their new route through the Rockies, foregoing the earlier preference for the more northerly Yellowhead Pass. This decision altered the location of the line across western Canada and dramatically affected the development of the West.

The Trans-Canada Highway travels through the summit of Kicking Horse Pass at the boundary of Yoho and Banff National Parks. Visit Spiral Tunnels viewpoint in Yoho National Park to learn more about the history of Kicking Horse Pass (马踩踏出来的小径).

Skoki Ski Lodge National Historic Site of Canada

A rare and little-changed link with the early days of ski tourism(滑雪旅游), Skoki Ski Lodge is a fine example of the log building traditions long associated with the Banff region. When it opened for business in 1931, skiing was just emerging as a major form of winter recreation across Canada. Skoki Ski Lodge was built as a destination for backcountry holidays by a group of ski enthusiasts, using timbers cut in the vicinity. It was the first such facility to operate on a commercial basis in Canada. It assumed its current appearance with additions carried out in 1935 – 1936. The lodge is located in the backcountry of YOHO and Banff National Park.

Twin Falls Tea House National Historic Site of Canada

This charming tea house is a well-preserved example of rustic(乡村的)design in Canada's national parks. The Canadian Pacific Railway constructed the building in stages, beginning around 1908 with a one-storey cabin for patrons(赞助商)taking backcountry tours(游客). In 1923 – 1924, the company added a two-storey structure(两层木结构)to create a larger and more attractive chalet(度假木屋). Skilfully crafted, the enormous logs create the building's rustic character. The chalet remains a vivid reminder of the early days of hiking, mountaineering and trail riding in the national parks. The tea house is located in the Yoho River Valley in Yoho National Park and can be reached by hiking 8.5 km from the Takakkaw Falls parking lot.

Edouard Gaston Deville 1849 – 1924

Born in France, Deville served as a hydrographer(水道测量学家)with the French Navy before coming to Canada in 1874. He took part in the Dominion survey of the newly acquired NorthWest Territories, becoming Inspector(检查员)of Surveys(测量)for Canada in 1881 and Surveyor General in 1885. A pioneer in the new science of photogrammetry, he introduced the technique to the mapping(绘图)of the Canadian Rockies between 1888 and 1896. He gained an international reputation as author, designer of equipment for photogrammetric mapping, and improver of map reproduction techniques. Mount Deville in Yoho National Park bears his name. He died in Ottawa. The E. G. Deville commemorative plaque(纪念碑)is located in Yoho National Park near the entrance to Monarch Campground.

Notes

theContinental Divide 大陆分水岭　美洲大陆分水岭或大分水岭(Great Divide)是指将美洲水系按其入海口划分的一系列山脊或鸿沟。美洲水系入海口一般为(1)沿鸿沟流入大西洋(包括墨西哥湾和加勒比海)的水系;(2)沿鸿沟流入北冰洋的水系。

Burgess Shale 布尔吉斯页岩　布尔吉斯页岩生物群是美国著名科学家、古生物学家维尔·卡特于1909年在加拿大洛矶山脉的布尔吉斯山的一次旅行中偶然发现的。经过发掘和研究发现,布尔吉斯页岩生物群的生活时代大约在中寒武纪(距今5.05亿年),也就是我们常说的寒武纪大爆发时期。在此期间,大量动物种群在较短时期内(也许仅几百万年内)突然出现,几乎现在所有的生物种类都可以在这里找到自己的祖先,而在此之前,动物世界还只是由多细胞后生动物及小壳动物等简单生物所组成。布尔吉斯页岩生物群成为寒武纪大爆发的重要证据。布尔吉斯生物群的发现给当时的科学界带来极大震撼,它使科学家们第一次清楚地认识到,在寒武纪海洋中具有骨骼的动物仅占少数,绝大多数是不易保存的软躯体动物门类,从而纠正了人们对寒武纪仅存有三叶虫等少数硬体动物的错误认识。

James Hector 詹姆斯·赫克托(1834年3月16日—1907年11月16日)　苏格兰地质学家、博物馆学家和外科医生。在1858年陪同帕利斯尔(Palliser)在加拿大洛基山脉的一次考察中,他的马不慎掉进旁边的河水中,他费尽全力将马拉上岸。由于马受到了严重惊吓,上岸后踢中詹姆斯胸部将其踢昏。考察队的同伴以为詹姆斯已经去世,就在河岸边为他修筑坟墓,不想詹姆斯竟奇迹般复活。正是由于这个传奇故事,旁边的河被命名为马踢河。

Translation of Selected Sentences

1. Yoho National Park is located in theCanadian Rocky Mountains along the western slope of the Continental Divide in southeastern British Columbia.

优鹤国家公园位于加拿大不列颠哥伦比亚省西南部大陆分水岭西坡的洛矶山脉。

2. Yoho's craggy peaks and steep rock faces posed an enormous challenge for Canada's early explorers.

优鹤国家公园高耸的山峰和陡峭的岩壁对早期到加拿大的探险家们来说是巨大的挑战。

3. Designated a World Heritage Site in 1981, the Burgess Shale Formation containsthe fossilized remains of more than 120 marine animal species dating back 515 million years.

1981 年吉尔斯页岩地层被认定为世界自然遗产,页岩地层中包含了超过 120 个海洋动物物种化石,其年代可以追溯到 5.15 亿年前。

4. In 1858, near Wapta Falls, a pack horse kicked explorer James Hector in the chest, and the Kicking Horse River got its name.

1858 年在瓦普达瀑布附近,一匹驮行李的马踢中了探险家詹姆斯·赫克托的胸部,旁边的马踢河也因此得名。

5. Emerald Lake is popular for sightseeing, canoeing and hiking.

翡翠湖是受欢迎的观光、皮划艇和远足圣地。

Questions

1. What is the general condition of the Yoho National Park according to the passage?
2. Which water fall is the highest water fall in Canada?
3. What was designated a World Heritage Site in 1981?
4. What are the main animals in the Yoho National Park and what are their and living habits?
5. What is Kicking Horse Pass?